STUDENT SOLUTIONS MANUAL

Richard N. Aufmann
Palomar College

Vernon C. Barker
Palomar College

Joanne S. Lockwood
Plymouth State College

Christine S. Verity

PREALGEBRA

FOURTH EDITION

Aufmann/Barker/Lockwood

HOUGHTON MIFFLIN COMPANY   BOSTON   NEW YORK

Senior Sponsoring Editor: Lynn Cox
Senior Development Editor: Dawn Nuttall
Assistant Editor: Melissa Parkin
Editorial Assistant: Noel Kamm
Senior Manufacturing Coordinator: Florence Cadran
Senior Marketing Manager: Ben Rivera

Copyright © 2005 by Houghton Mifflin Company. All rights reserved.

No part of this work may be reproduced or transmitted in any form or by any means, electronic or mechanical, including photocopying and recording, or by any information storage or retrieval system without the prior written permission of Houghton Mifflin Company unless such copying is expressly permitted by federal copyright law. Address inquiries to College Permissions, Houghton Mifflin Company, 222 Berkeley Street, Boston, MA 02116-3764.

Printed in the U.S.A.

ISBN: 0-618-37265-2

7 8 9-CRS-08 07 06 05

Table of Contents

Chapter 1: Whole Numbers — 1

Chapter 2: Integers — 22

Chapter 3: Fractions — 42

Chapter 4: Decimals and Real Numbers — 83

Chapter 5: Variable Expressions — 108

Chapter 6: First-Degree Equations — 126

Chapter 7: Measurement and Proportion — 158

Chapter 8: Percent — 181

Chapter 9: Geometry — 204

Chapter 10: Statistics and Probability — 241

STUDENT SOLUTIONS MANUAL

Chapter 1: Whole Numbers

Prep Test

1. 8
2. 1 2 3 4 5 6 7 8 9 10
3. a and D; b and E; c and A; d and B; e and F; f and C
4. 0
5. fifty

Go Figure

On the first trip, the two children row over. The second trip, one child returns with the boat. The third trip, one adult rows to the other side. The fourth trip, one child returns with the boat. At this point, one adult has crossed the river. Repeat the first four trips an additional four times, one time for each adult. After completing the fifth time, there have been twenty trips taken, and the only people waiting to cross the river are the two children. On the twenty-first trip, the two children cross the river. So the minimum number of trips is twenty-one.

SECTION 1.1

Objective A Exercises

1. Students should note that the whole numbers include 0, the natural numbers do not.

3. [number line with point at 2]

5. [number line with point at 10]

7. [number line with point at 5]

9. [number line showing arrow of length 4 ending at 5]

    5 is 4 units to the left of 9.

11. [number line showing arrow of length 3 ending at 5]

    5 is 3 units to the right of 2.

13. [number line showing arrow of length 7 ending at 0]

    0 is 7 units to the left of 7.

15. $27 < 39$
17. $0 < 52$
19. $273 > 194$
21. $2,761 < 3,857$
23. $4,610 > 4,061$
25. $8,005 < 8,050$
27. 11, 14, 16, 21, 32
29. 13, 48, 72, 84, 93
31. 26, 49, 77, 90, 106
33. 204, 399, 662, 736, 981
35. 307, 370, 377, 3,077, 3,700

Objective B Exercises

37. five hundred eight
39. six hundred thirty-five
41. four thousand seven hundred ninety
43. fifty-three thousand six hundred fourteen
45. two hundred forty-six thousand fifty-three
47. three million eight hundred forty-two thousand nine hundred five
49. 496
51. 53,340
53. 502,140
55. 9,706
57. 5,012,907
59. 8,005,010
61. $7,000 + 200 + 40 + 5$
63. $500,000 + 30,000 + 2,000 + 700 + 90 + 1$
65. $5,000 + 60 + 4$
67. $20,000 + 300 + 90 + 7$
69. $400,000 + 2,000 + 700 + 8$
71. $8,000,000 + 300 + 10 + 6$

Objective C Exercises

73. 7,108 ⌐Given place value
         └8 > 5

7,108 rounded to the nearest ten is 7,110.

75. 4,962 ⌐Given place value
         └6 > 5

4,962 rounded to the nearest hundred is 5,000.

77. 28,551 ⌐Given place value
         └5 = 5

28,551 rounded to the nearest hundred is 28,600.

79. 6,808 ⌐Given place value
         └8 > 5

6,809 rounded to the nearest thousand is 7,000.

81. 93,825 ⌐Given place value
         └8 > 5

93,825 rounded to the nearest thousand is 94,000.

83. 629,513 ⌐Given place value
         └5 = 5

629,513 rounded to the nearest thousand is 630,000.

85. 352,876 ⌐Given place value
         └2 < 5

352,876 rounded to the nearest ten-thousand is 350,000.

87. 71,834,250 ⌐Given place value
         └8 > 5

71,834,250 rounded to the nearest million is 72,000,000.

Objective D Exercises

89. Strategy To find the person with the greater number of stolen bases, compare the numbers 892 and 937.

Solution 892 < 937
Billy Hamilton had the greater number of stolen bases.

91. Strategy To find the greater number of performances, compare the numbers 2,844 and 3,242.

Solution 2,844 < 3,242
"Fiddler on the Roof" was performed the greater number of times.

93. Strategy To find the food which contains more calories, compare the numbers 190 and 114.

Solution 190 > 114
Two tablespoons of peanut butter contain more calories.

95. Strategy To find the shorter distance, compare the two numbers 1,892 and 1,833.

Solution 1,892 > 1,833
The shorter distance is from St. Louis to San Diego.

97. Strategy To find the smaller planet, compare the two numbers 32,200 and 30,800.

Solution 32,200 > 30,800
Neptune is the smallest planet.

99. Strategy
   a. To determine the length of the State of the Union Address in 2001, read the number in the bar graph above the 2001 bar.

   b. To determine during which year the length of the State of the Union Address was longest, use the bar graph to find the year with the largest number above its corresponding bar.

Solution
   a. The length of the State of the Union Address in 2001 was 67 minutes.

   b. The largest number appears above the bar corresponding to 2000.
   The year in which the State of the Union Address was longest in 2000.

101. **Strategy** To find the land area to the nearest ten-thousand acres, round 161,546 to the nearest ten-thousand.

    Solution 161,546 rounded to the nearest ten-thousand is 160,000.
    To the nearest ten-thousand acres, the land area of the Appalachian Trail is 160,000 acres.

103. **Strategy**
    a. To determine during which school year enrollment was lowest, use the line graph to find the school year corresponding to the lowest point on the line.

    b. To determine if enrollment increased or decreased between 1975 and 1980, read the graph at 1975 and at 1980.

    Solution
    a. Student enrollment was lowest during the 1985 school year.

    b. Enrollment was less in 1980 than in 1975. Enrollment decreased between 1975 and 1980.

105. **Strategy** To find the estimate of the speed of light, round the speed (299,800) to the nearest thousand.

    Solution 299,800 rounded to the nearest thousand is 300,000.
    To the nearest thousand, the speed of light is 300,000 km/s.

Critical Thinking 1.1

107. The largest three-digit number is 999. (The next largest whole number is 1,000, which has four digits.) The smallest five-digit number is 10,000. (The next smallest whole number if 9,999, which has 4 digits.)

109. The key to the visual illusion shown on page 18 is that if only the left side of the picture is observed, it looks like three pipes, while if only the right side is observed, it looks like there are two rectangular extensions.
Another visual illusion involves the two lines shown at the below. Both lines are the same length, but the arrowheads at the ends of the lines make line *a* appear longer and line *b* appear shorter.

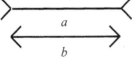

Another visual illusion involves the circles shown at the below Although the bottoms of all the circles lie on a straight line, the difference in the diameters of the circles makes it appear that the larger circles are positioned higher up than the smaller circles.

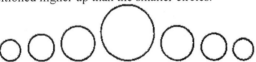

The T shown below appears to be taller than it is wide, although this is not true. Because the vertical line is drawn from the center of the horizontal line, it appears to stretch farther.

111. The Egyptian numeration system, like our decimal system, was based on ten. The symbols used varied over time, but common hieroglyphics represented 1 as ∕, 10 as ∩, 100 as ❓ and 1,000 as ⸸. The Egyptians used the basic symbols to form numbers. For example, the number 1,326 would be written

⸸ ❓❓❓ ∩∩ ∕∕∕

The Egyptians did not have a symbol for zero.

SECTION 1.2

Objective A Exercises

1. Answers will vary.

3. $\begin{array}{r} 732,453 \\ +\ 651,206 \\ \hline 1,383,659 \end{array}$

5. $\begin{array}{r} \scriptstyle 1\ 1\ 1 \\ 2,879 \\ +\ 3,164 \\ \hline 6,043 \end{array}$

7. $\begin{array}{r} \scriptstyle 1\ 1\ \ 1 \\ 45,825 \\ +\ 66,327 \\ \hline 112,152 \end{array}$

9. $\begin{array}{r} \scriptstyle 1\ 1 \\ 4,037 \\ 3,342 \\ +\ 5,169 \\ \hline 12,548 \end{array}$

11. $\begin{array}{r} \scriptstyle 1\ 1\ 1 \\ 67,390 \\ 42,761 \\ +\ 89,405 \\ \hline 199,556 \end{array}$

13. $\begin{array}{r} \scriptstyle 2\ 1\ 1\ 2 \\ 54,097 \\ 33,432 \\ 97,126 \\ 64,508 \\ 78,310 \\ \hline 327,473 \end{array}$

15. $\begin{array}{r} 88,123 \\ +\ 80,451 \\ \hline 168,574 \end{array}$

17. $\begin{array}{r} \scriptstyle 1 \\ 7,293 \\ +\ 654 \\ \hline 7,947 \end{array}$

19. $\begin{array}{r} \scriptstyle 1\ \ 1 \\ 216 \\ 8,707 \\ +\ 90,714 \\ \hline 99,637 \end{array}$

21. $\begin{array}{r} 585 \\ 497 \\ 412 \\ +\ 378 \\ \hline 1,872 \end{array}$

The total number of undergraduates enrolled at the college in 2001 was 1,872.

23. $\begin{array}{r} 6,742 \\ +\ 8,298 \\ \hline 15,040 \end{array} \to \begin{array}{r} 7,000 \\ +\ 8,000 \\ \hline 15,000 \end{array}$

25. $\begin{array}{r} 972,085 \\ +\ 416,832 \\ \hline 1,388,917 \end{array} \to \begin{array}{r} 1,000,000 \\ +\ 400,000 \\ \hline 1,400,000 \end{array}$

27. $\begin{array}{r} 387 \\ 295 \\ 614 \\ +\ 702 \\ \hline 1,998 \end{array} \to \begin{array}{r} 400 \\ 300 \\ 600 \\ +\ 700 \\ \hline 2,000 \end{array}$

29. $\begin{array}{r} 224,196 \\ 7,074 \\ +\ 98,531 \\ \hline 329,801 \end{array} \to \begin{array}{r} 200,000 \\ 7,000 \\ +\ 100,000 \\ \hline 307,000 \end{array}$

31. $x + y$
    $574 + 698$
    $\begin{array}{r} \scriptstyle 1\ 1 \\ 574 \\ +\ 698 \\ \hline 1,272 \end{array}$

32. $x + y$
    $359 + 884$
    $\begin{array}{r} \scriptstyle 1\ 1 \\ 359 \\ +\ 884 \\ \hline 1,243 \end{array}$

33. $x + y$
    $4,752 + 7,398$
    $\begin{array}{r} \scriptstyle 1\ 1\ 1 \\ 4,752 \\ +\ 7,398 \\ \hline 12,150 \end{array}$

35. $x + y$
    $38,229 + 51,671$
    $\begin{array}{r} \scriptstyle 1\ 1 \\ 38,229 \\ +\ 51,671 \\ \hline 89,900 \end{array}$

37. $a + b + c$
    $693 + 508 + 371$
    $\begin{array}{r} \scriptstyle 1\ 1 \\ 693 \\ 508 \\ +\ 371 \\ \hline 1,572 \end{array}$

Section 1.2

39. $a + b + c$
    $4{,}938 + 2{,}615 + 7{,}038$
    $$\begin{array}{r} 12 \\ 4{,}938 \\ 2{,}615 \\ + 7{,}038 \\ \hline 14{,}591 \end{array}$$

41. $a + b + c$
    $12{,}897 + 36{,}075 + 7{,}038$
    $$\begin{array}{r} 1122 \\ 12{,}897 \\ 36{,}075 \\ + 7{,}038 \\ \hline 56{,}010 \end{array}$$

43. The Commutative Property of Addition

45. The Associative Property of Addition

47. The Addition Property of Zero

49. $28 + 0 = 28$

51. $9 + (4 + 17) = (9 + 4) + 17$

53. $15 + 34 = 34 + 15$

55. $42 = n + 4$
    $42 \mid 38 + 4$
    $42 = 42$
    Yes, 38 is a solution of the equation $42 = n + 4$.

57. $2 + h = 16$
    $2 + 13 \mid 16$
    $15 \neq 16$
    No, 13 is not a solution of the equation $2 + h = 16$.

59. $32 = x + 2$
    $32 \mid 30 + 2$
    $32 = 32$
    Yes, 32 is a solution of the equation $32 = x + 2$.

Objective B Exercises

61. Answers will vary.

63. $$\begin{array}{r} 7\,13 \\ 8\overline{8}3 \\ - 467 \\ \hline 416 \end{array}$$

65. $$\begin{array}{r} 2\,15\,10 \\ \overline{3}\,\overline{6}0 \\ - 172 \\ \hline 188 \end{array}$$

67. $$\begin{array}{r} 5\,15 \\ 6\overline{5}7 \\ - 193 \\ \hline 464 \end{array}$$

69. $$\begin{array}{r} 3\,9\,17 \\ 4\overline{0}7 \\ - 199 \\ \hline 208 \end{array}$$

71. $$\begin{array}{r} 7\,10\,14 \\ 6{,}8\overline{1}\overline{4} \\ - 3{,}257 \\ \hline 3{,}557 \end{array}$$

73. $$\begin{array}{r} 4\,9\,9\,10 \\ \overline{5}{,}\overline{0}\overline{0}\overline{0} \\ - 2{,}164 \\ \hline 2{,}836 \end{array}$$

75. $$\begin{array}{r} 2\,\,13\,9\,10 \\ \overline{3}{,}4\overline{0}\overline{0} \\ - 1{,}963 \\ \hline 1{,}437 \end{array}$$

77. $$\begin{array}{r} 2\,9\,\,9\,9\,14 \\ \overline{3}\overline{0}{,}0\overline{0}\overline{4} \\ - 9{,}856 \\ \hline 20{,}148 \end{array}$$

79. $$\begin{array}{r} 1\,\,15\,2\,16 \\ 2{,}\overline{8}3\overline{6} \\ - 918 \\ \hline 1{,}618 \end{array}$$

81. $$\begin{array}{r} 0\,12\,7\,9\,14 \\ \overline{1}\overline{2}{,}8\overline{0}\overline{4} \\ - 5{,}426 \\ \hline 7{,}378 \end{array}$$

83. $$\begin{array}{r} 7\,14\,13\,11\,13 \\ \overline{8}\overline{5}{,}\overline{4}\overline{2}\overline{3} \\ - 67{,}875 \\ \hline 17{,}548 \end{array}$$

85. $$\begin{array}{r} 8\,10 \\ 9\overline{0} \\ - 75 \\ \hline 15 \end{array}$$

    The difference between the maximum heights is 15 ft.

87. $$\begin{array}{rcr} 7{,}355 & \rightarrow & 7{,}000 \\ - 5{,}219 & \rightarrow & 5{,}000 \\ \hline 2{,}136 & & 2{,}000 \end{array}$$

89. $$\begin{array}{rcr} 59{,}126 & \rightarrow & 60{,}000 \\ - 20{,}843 & \rightarrow & - 20{,}000 \\ \hline 38{,}283 & & 40{,}000 \end{array}$$

91. $\begin{array}{r} 36,287 \\ -5,092 \\ \hline 31,195 \end{array} \rightarrow \begin{array}{r} 40,000 \\ -5,000 \\ \hline 35,000 \end{array}$

93. $\begin{array}{r} 224,196 \\ -98,531 \\ \hline 125,665 \end{array} \rightarrow \begin{array}{r} 200,000 \\ -100,000 \\ \hline 100,000 \end{array}$

95. $x - y$
    $50 - 37$
    $\begin{array}{r} {}^{4}\!\!\not{5}^{10} \\ -37 \\ \hline 13 \end{array}$

97. $x - y$
    $914 - 271$
    $\begin{array}{r} {}^{8}\!\!\not{9}{}^{1}1 \\ 914 \\ -271 \\ \hline 643 \end{array}$

99. $x - y$
    $740 - 385$
    $\begin{array}{r} {}^{6}\!\!\not{7}{}^{13}\!\!\not{4}{}^{10} \\ 740 \\ -385 \\ \hline 355 \end{array}$

101. $x - y$
    $8,672 - 3,461$
    $\begin{array}{r} 8,672 \\ -3,461 \\ \hline 5,211 \end{array}$

103. $x - y$
    $1,605 - 839$
    $\begin{array}{r} {}^{0}\!\!\not{1},{}^{15}\!\!\not{6}{}^{9}\!\!\not{0}{}^{15}\!\!\not{5} \\ -839 \\ \hline 766 \end{array}$

105. $x - y$
    $23,409 - 5,178$
    $\begin{array}{r} {}^{1}\!\!\not{2}3,{}^{13}\!\!\not{4}{}^{3}\!\!\not{0}{}^{10}\!\!\not{9} \\ -5,178 \\ \hline 18,231 \end{array}$

107. $29 = 53 - y$
    $29 \mid 53 - 24$
    $29 = 29$
    Yes, 24 is a solution of the equation
    $29 = 53 - y$.

109. $t - 16 = 60$
    $\overline{44 - 16 \mid 60}$
    $28 \neq 60$
    No, 44 is not a solution of the equation
    $t - 16 = 60$.

111. $82 - z = 55$
    $\overline{82 - 27 \mid 55}$
    $55 = 55$
    Yes, 27 is a solution of the equation
    $82 - z = 55$.

Objective C Exercises

113. Strategy   To find the sum of the whole numbers less than 21, add the numbers 0 to 20.

    Solution $0 + 1 + 2 + 3 + 4 + 5 + 6 + 7 + 8 +$
    $9 + 10 + 11 + 12 + 13 + 14 + 15 +$
    $16 + 17 + 18$
    $+ 19 + 20 = 210$
    The sum of the numbers is 210.

115. Strategy   To find the difference, subtract 99 (the largest two-digit number) from 1,000 (the smallest four-digit number).

    Solution $\begin{array}{r} 1,000 \\ -99 \\ \hline 901 \end{array}$
    The difference is 901.

117. Strategy   To find the number of calories, add the number of calories in one apple (80), one cup of cornflakes (95), one tablespoon of sugar (45), and one cup of milk (150).

    Solution $80 + 95 + 45 + 150 = 370$
    The breakfast contained 370 calories.

119. Strategy   To find the perimeter of a rectangle, replace L with 24 and W with 15 in the given formula and solve for P.

    Solution $P = L + W + L + W$
    $P = 24 + 15 + 24 + 15$
    $P = 78$
    The perimeter is 78 m.

121. **Strategy** To find the perimeter of a triangle, replace a, b, and c with 16, 12, and 15 in the given formula and solve for P.

   Solution P = a + b + c
   P = 16 + 12 + 15
   P = 43
   The perimeter is 43 in.

123. **Strategy** To determine the length of the hedge, by finding the perimeter of the playground, replace L with 160 and W with 120 in the given formula and solve for P.

   Solution P = L + W + L + W
   P = 160 + 120 + 160 + 120
   P = 560
   The length of the hedge is 560 ft.

125. **Strategy** To find the difference in the number of orbits, subtract the number of orbits made by Apollo-Saturn 7 (163) from the number of flights made by Gemini-Titan 7 (206).

   Solution
      206
   − 163
       43
   Gemini-Titan 7 made 43 more orbits than Apollo-Saturn 7.

127. **Strategy** To find the difference in seating capacity, subtract the seating capacity of Fenway Park (33,871) from the seating capacity of SAFECO Field (47,600).

   Solution
     47,600
   − 33,871
     13,729
   The difference in seating capacity is 13,729 seats.

129. **Strategy** To find the total cost of the computer system, add the cost of the operating system (830), the monitor (245), the keyboard (175), and the printer (395).

   Solution
     830
     245
     175
   + 395
   1,645
   The total cost of the computer system is $1,645.

131. **Strategy** To find the estimate of the number of miles driven:
   → Round the two readings of the odometer.
   → Subtract the rounded reading from the beginning of the year from the rounded reading at the end of the year.

   Solution 77,912 →   80,000
   58,376 → − 60,000
                   20,000
   The car was driven approximately 20,000 mi during the year.

133. **Strategy** To determine between which two months car sales increased the most, find the difference, if there was an increase, between sales for January and February, February and March, and March and April for 2005.

   Solution Between January and February, 2005: 132 − 108 = 24
   Between February and March, 2005: 152 − 132 = 20
   Between March and April, 2005: not an increase
   Car sales increased the most between January and February of 2005. The increase was 24 cars.

135. **Strategy** To find the value of the investment, replace P by 12,500 and I by 775 in the given formula and solve for A.

   Solution A = P + I
   A = 12,500 + 775
   A = 13,275
   The value of the investment is $13,275.

137. **Strategy** To find the mortgage loan, replace S by 290,000 and D by 29,000 in the given formula and solve for M.

    Solution $M = S - D$
    $M = 290,000 - 29,000$
    $M = 261,000$
    The mortgage loan on the home is $261,000.

139. **Strategy** To find the ground speed, replace a by 375 and h by 25 in the given formula and solve for g.

    Solution $g = a - h$
    $g = 375 - 25$
    $g = 350$
    The ground speed of the airplane is 350 mph.

141. **Strategy**
    a. To determine how many drivers were traveling at 70 mph or less, add the numbers traveling 66–70 mph (3,717), 61–65 mph (2,984), and less than 61 mph (2,870).

    b. To determine how many drivers were traveling at 76 mph or more, add the numbers traveling 76–80 mph (2,503) and more than 80 mph (1,708).

    Solution
    a. $3,717 + 2,984 + 2,870 = 9,571$
    9,571 drivers were traveling at 70 mph or less.

    b. $2,503 + 1,708 = 4,211$
    4,211 drivers were traveling at 76 mph or more.

143. It is not possible to tell how many motorists were driving at less than 70 mph. The data shows that 3,717 drivers were traveling at 66–70 mph. The number 3,717 includes motorists driving at 70 mph as well as those driving less than 70 mph (at speeds of 66–69 mph). We cannot separate those driving at less than 70 mph from those driving at 70 mph.

Critical Thinking 1.2

145. If you roll two ordinary dice, the possible sums are the numbers 2 (rolling two 1's) through 12 (rolling two 6's). Therefore there are 11 different sums possible.
(The numbers 2, 3, 4, 5, 6, 7, 8, 9, 10, 11, and 12).

147. a. Substituting any whole number for a in $a - 0 = a$ will result in a true equation. The statement is always true.

    b. Substituting any whole number for a in $a - a = 0$ will result in a true equation. The statement is always true.

149. The U.S. Bureau of the Census publishes projections of the population of the United States and of the separate states in *Current Population Reports*. Estimates of U.S. population growth vary according to the projections used for the birth rate, the average life span of the population, and immigration figures. Population growth of states is further affected by internal migration.

SECTION 1.3

Objective A Exercises

1. Students should note that there are five 6's. Therefore, the addition $6 + 6 + 6 + 6 + 6$ can be written as 5 times 6: 5×6.

3. $\begin{array}{r} \overset{2\ 6}{127} \\ \times\ \ \ 9 \\ \hline 1,143 \end{array}$

5. $\begin{array}{r} \overset{4\ 6}{6,709} \\ \times\ \ \ \ \ 7 \\ \hline 46,963 \end{array}$

7. $\begin{array}{r} \overset{7\ 6\ 5\ 7}{58,769} \\ \times\ \ \ \ \ \ \ \ 8 \\ \hline 470,152 \end{array}$

9. $\begin{array}{r} 683 \\ \times\ 71 \\ \hline 683 \\ 47\ 81\ \ \\ \hline 48,493 \end{array}$

11. $\begin{array}{r} 7,053 \\ \times 46 \\ \hline 42\,318 \\ 282\,12 \\ \hline 324,438 \end{array}$

13. $\begin{array}{r} 3,285 \\ \times 976 \\ \hline 19\,710 \\ 229\,95 \\ 2\,956\,5 \\ \hline 3,206,160 \end{array}$

15. $500 \cdot 3 = 1,500$

17. $40 \cdot 50 = 2,000$

19. $400 \cdot 3 \cdot 20 \cdot 0 = 1,200 \cdot 20 \cdot 0$
     $= 24,000 \cdot 0$
     $= 0$

21. $q \cdot r \cdot s = qr \cdot s$
     $= qrs$

23. $3,467 \rightarrow 3,000$
     $359 \rightarrow 400$
     $3,000 \cdot 400 = 1,200,000$
     $3,467 \cdot 359 = 1,244,653$

25. $39,246 \rightarrow 40,000$
     $29 \rightarrow 30$
     $40,000 \cdot 30 = 1,200,000$
     $39,246 \cdot 29 = 1,138,134$

27. $745 \rightarrow 700$
     $63 \rightarrow 60$
     $700 \cdot 60 = 42,000$
     $745 \cdot 63 = 46,935$

29. $8,941 \rightarrow 9,000$
     $726 \rightarrow 700$
     $9,000 \cdot 700 = 6,300,000$
     $8,941 \cdot 726 = 6,491,166$

31. ab
     $465 \cdot 32 = 14,880$

33. $7a$
     $7 \cdot 465 = 3,255$

35. xyz
     $5 \cdot 12 \cdot 30 = 60 \cdot 30$
     $= 1,800$

37. $2xy$
     $2 \cdot 67 \cdot 23 = 134 \cdot 23$
     $= 3,082$

39. The Multiplication Property of One

41. The Commutative Property of Multiplication

43. $19 \cdot 30 = 30 \cdot 19$

45. $45 \cdot 0 = 0$

47. $\begin{array}{r} 4x = 24 \\ \hline 4(6) \mid 24 \\ 24 = 24 \end{array}$
     Yes, 6 is a solution of the equation $4x = 24$.

49. $\begin{array}{r} 96 = 3z \\ \hline 96 \mid 3(23) \end{array}$
     $96 \neq 69$
     No, 23 is not a solution of the equation $96 = 3z$.

51. $\begin{array}{r} 2y = 38 \\ \hline 2(19) \mid 38 \\ 38 = 38 \end{array}$
     Yes, 19 is a solution of the equation $2y = 38$.

Objective B Exercises

53. $2 \cdot 2 \cdot 2 \cdot 7 \cdot 7 \cdot 7 \cdot 7 \cdot 7 = 2^3 \cdot 7^5$

55. $2 \cdot 2 \cdot 3 \cdot 3 \cdot 3 \cdot 5 \cdot 5 \cdot 5 \cdot 5 = 2^2 \cdot 3^3 \cdot 5^4$

57. $c \cdot c = c^2$

59. $x \cdot x \cdot x \cdot y \cdot y \cdot y = x^3 y^3$

61. $2^5 = 2 \cdot 2 \cdot 2 \cdot 2 \cdot 2 = 32$

63. $10^6 = 1,000,000$

65. $2^3 \cdot 5^2 = (2 \cdot 2 \cdot 2) \cdot (5 \cdot 5) = 8 \cdot 25 = 200$

67. $3^2 \cdot 10^3 = (3 \cdot 3) \cdot (10 \cdot 10 \cdot 10)$
     $= 9 \cdot 1,000 = 9,000$

69. $0^2 \cdot 6^2 = (0 \cdot 0) \cdot (6 \cdot 6) = 0 \cdot 36 = 0$

71. $2^2 \cdot 5 \cdot 3^3 = (2 \cdot 2) \cdot 5 \cdot (3 \cdot 3 \cdot 3)$
     $= 4 \cdot 5 \cdot 27 = 20 \cdot 27 = 540$

73. $12^2 = 12 \cdot 12 = 144$

75. $8^3 = 8 \cdot 8 \cdot 8 = 64 \cdot 8 = 512$

77. a^4

79. $x^3 y$
     $2^3 \cdot 3 = (2 \cdot 2 \cdot 2) \cdot 3$
     $= 8 \cdot 3$
     $= 24$

Chapter 1: Whole Numbers

81. ab^6
$5 \cdot 2^6 = 5 \cdot (2 \cdot 2 \cdot 2 \cdot 2 \cdot 2 \cdot 2)$
$= 5 \cdot 64$
$= 320$

83. $c^2 d^2$
$3^2 \cdot 5^2 = (3 \cdot 3) \cdot (5 \cdot 5)$
$= 9 \cdot 25$
$= 225$

Objective C Exercises

85. Answers will vary.

87. $\begin{array}{r} 307 \\ 9 \overline{)2{,}763} \\ \underline{-27} \\ 6 \\ \underline{-0} \\ 63 \\ \underline{-63} \\ 0 \end{array}$

89. $\begin{array}{r} 309 \text{ r } 4 \\ 5 \overline{)1{,}549} \\ \underline{-15} \\ 4 \\ \underline{-0} \\ 49 \\ \underline{-45} \\ 4 \end{array}$

91. $\begin{array}{r} 2{,}550 \\ 6 \overline{)15{,}300} \\ \underline{-12} \\ 33 \\ \underline{-30} \\ 30 \\ \underline{-30} \\ 0 \\ \underline{-0} \\ 0 \end{array}$

93. $\begin{array}{r} 21 \text{ r } 9 \\ 32 \overline{)681} \\ \underline{-64} \\ 41 \\ \underline{-32} \\ 9 \end{array}$

95. $\begin{array}{r} 147 \text{ r } 38 \\ 62 \overline{)9{,}152} \\ \underline{-62} \\ 295 \\ \underline{-248} \\ 472 \\ \underline{-434} \\ 38 \end{array}$

97. $\begin{array}{r} 200 \text{ r } 8 \\ 37 \overline{)7{,}408} \\ \underline{-74} \\ 0 \\ \underline{-0} \\ 8 \\ \underline{-0} \\ 8 \end{array}$

99. $\begin{array}{r} 404 \text{ r } 34 \\ 78 \overline{)31{,}546} \\ \underline{-312} \\ 34 \\ \underline{-0} \\ 346 \\ \underline{-312} \\ 34 \end{array}$

101. $\begin{array}{r} 16 \text{ r } 97 \\ 476 \overline{)7{,}713} \\ \underline{-476} \\ 2953 \\ \underline{-2856} \\ 97 \end{array}$

103. $\begin{array}{r} 907 \\ 8 \overline{)7{,}256} \\ \underline{-7\ 2} \\ 5 \\ \underline{-0} \\ 56 \\ \underline{-56} \\ 0 \end{array}$

105. $\begin{array}{r} 881 \text{ r } 1 \\ 7 \overline{)6{,}168} \\ \underline{-56} \\ 56 \\ \underline{-56} \\ 8 \\ \underline{-7} \\ 1 \end{array}$

107. $\dfrac{c}{d}$

109. $36{,}472 \rightarrow 40{,}000$
$\quad\quad 47 \rightarrow \quad\quad 50$
$40{,}000 \div 50 = 800$
$36{,}472 \div 47 = 776$

111. $389{,}804 \rightarrow 400{,}000$
$\quad\quad 76 \rightarrow \quad\quad 80$
$400{,}000 \div 80 = 5{,}000$
$389{,}804 \div 76 = 5{,}129$

113. $38{,}984 \rightarrow 40{,}000$
     $79 \rightarrow 80$
     $40{,}000 \div 80 = 500$
     $38{,}984 \div 79 = 493 \text{ r}37$

115. $332{,}004 \rightarrow 300{,}000$
     $219 \rightarrow 200$
     $300{,}000 \div 200 = 1{,}500$
     $332{,}004 \div 219 = 1{,}516$

117. $\dfrac{x}{y}$
     $\dfrac{48}{1} = 48$

119. $\dfrac{x}{y}$
     $\dfrac{79}{0}$ is undefined, because division by zero is undefined.

121. $\dfrac{x}{y}$
     $\dfrac{39{,}200}{4}$

     $9{,}800$
     $4)\overline{39{,}200}$
     $\underline{-36}$
     32
     $\underline{-32}$
     0
     $\underline{-0}$
     0
     $\underline{-0}$
     0

123. $\dfrac{36}{z} = 4$
     $\dfrac{36}{9}\ \Big|\ 4$
     $4 = 4$
     Yes, 9 is a solution of the equation.

125. $56 = \dfrac{x}{7}$
     $56\ \Big|\ \dfrac{49}{7}$
     $56 \neq 7$
     No, 49 is not a solution of the equation.

Objective D Exercises

127. $10 \div 1 = 10$
     $10 \div 2 = 5$
     $10 \div 5 = 2$
     The factors of 10 are 1, 2, 5, and 10.

129. $12 \div 1 = 12$
     $12 \div 2 = 6$
     $12 \div 3 = 4$
     $12 \div 4 = 3$
     The factors of 12 are 1 2, 3, 4, 6, and 12.

131. $8 \div 1 = 8$
     $8 \div 2 = 4$
     $8 \div 4 = 2$
     The factors of 8 are 1, 2, 4, and 8.

133. 13 is prime number.
     The factors of 13 are 1 and 13.

135. $18 \div 1 = 18$
     $18 \div 2 = 9$
     $18 \div 3 = 6$
     $18 \div 6 = 3$
     The factors of 18 are 1, 2, 3, 6, 9, and 18.

137. $25 \div 1 = 25$
     $25 \div 5 = 5$
     The factors of 25 are 1, 5, and 25.

139. $56 \div 1 = 56$
     $56 \div 2 = 28$
     $56 \div 4 = 14$
     $56 \div 7 = 8$
     $56 \div 8 = 7$
     The factors of 56 are 1, 2, 4, 7, 8, 14, 28, and 56.

141. $28 \div 1 = 28$
     $28 \div 2 = 14$
     $24 \div 4 = 7$
     $28 \div 7 = 4$
     The factors of 28 are 1, 2, 4, 7, 14, and 28.

143. $48 \div 1 = 48$
     $48 \div 2 = 24$
     $48 \div 3 = 16$
     $48 \div 4 = 12$
     $48 \div 6 = 8$
     $48 \div 8 = 6$
     The factors of 48 are 1, 2, 3, 4, 6, 8, 12, 16, 24, and 48.

145. $54 \div 1 = 54$
     $54 \div 2 = 27$
     $54 \div 3 = 18$
     $54 \div 6 = 9$
     $54 \div 9 = 6$
     The factors of 54 are 1, 2, 3, 6, 9, 18, 27, and 54.

147. $2\overline{)4}$ gives 2
 $2\overline{)8}$
 $2\overline{)16}$
 $16 = 2 \cdot 2 \cdot 2 \cdot 2 = 2^4$

149. $2\overline{)6}$ gives 3
 $2\overline{)12}$
 $12 = 2 \cdot 2 \cdot 3 = 2^2 \cdot 3$

151. $3\overline{)15}$ gives 5
 $15 = 3 \cdot 5$

153. $2\overline{)10}$ gives 5
 $2\overline{)20}$
 $2\overline{)40}$
 $40 = 2 \cdot 2 \cdot 2 \cdot 5 = 2^3 \cdot 5$

155. 37 is a prime number.

157. $5\overline{)65}$ gives 13
 $65 = 5 \cdot 13$

159. $2\overline{)14}$ gives 7
 $2\overline{)28}$
 $28 = 2 \cdot 2 \cdot 7 = 2^2 \cdot 7$

161. $3\overline{)21}$ gives 7
 $2\overline{)42}$
 $42 = 2 \cdot 3 \cdot 7$

163. $3\overline{)51}$ gives 17
 $51 = 3 \cdot 17$

165. $2\overline{)46}$ gives 23
 $46 = 2 \cdot 23$

Objective E Exercises

167. **Strategy** To find the number of calories, multiply the number of calories in one ounce of cheese (115) by the number of ounces (4).

 Solution $4 \cdot 115 = 460$
 Four ounces of cheese contain 460 calories.

169. **Strategy** To find the amount of jet fuel used, multiply the amount of fuel used each hour (865) by the time of flight (5).

 Solution $865 \cdot 5 = 4,325$
 The flight used 4,325 gal of jet fuel.

171. **Strategy**

 a. To determine the perimeter of a rectangle, substitute 24 for L and 15 for W in the formula below.

 b. To determine the area of a rectangle, substitute 24 for L and 15 for W in the formula below.

 Solution

 a. $P = 2L + 2W$
 $P = 2 \cdot 24 + 2 \cdot 15$
 $P = 78$
 The perimeter of the rectangle is 78 m.

 b. $A = LW$
 $A = 24 \cdot 15$
 $A = 360$
 The area of the rectangle is 360 m^2.

173. **Strategy** To find the length of fencing needed to surround a square, substitute 24 for s in the perimeter formula below.

 Solution $P = 4 \cdot s = 4 \cdot 24 = 96$
 The length of fencing needed to surround a square is 96 ft.

175. **Strategy** To find the area of the two-car garage, substitute 24 for s in the area formula below.

 Solution $A = s^2 = 24^2 = 576$
 The area of the two-car garage is 576 ft^2.

177. Strategy To find the amount of fabric needed for the flag, substitute 192 for L and 308 for W in the area formula below.

Solution $A = LW = 192 \cdot 308 = 59,136$
The amount of fabric needed for the flag is 59,136 cm².

179. Strategy To find the estimate of the total cost:
→Round the number of suits to hundreds and the cost of each suit to tens.
→Multiply the rounded numbers.

Solution $215 \rightarrow 200$
$83 \rightarrow 80$
$200 \cdot 80 = 16,000$
The total cost of the suits is approximately $16,000.

181. Strategy To find the total amount paid, replace M by 285 and N by 24 in the given formula and solve for A.

Solution $A = MN$
$A = 285 \cdot 24$
$A = 6,840$
The total amount paid on the loan is $6,840.

183. Strategy To find the time to drive the distance, replace d by 513 and r by 57 in the given formula and solve for t.

Solution $t = \dfrac{d}{r}$
$t = \dfrac{513}{57}$
$t = 9$
It takes 9 h to drive 513 mi.

185. Strategy To find the value per share, replace C by 10,500,000 and S by 500,000 in the given formula and solve for V.

Solution $V = \dfrac{C}{S}$
$V = \dfrac{10,500,000}{500,000}$
$V = 21$
The stock has a value of $21 per share.

Critical Thinking 1.3

187. There are 7 days in one week. There are 52 weeks in one year.
$7 \cdot 52 = 364$ days.
There are 364 days in 52 weeks.
There are 365 days in one year (366 days in a leap year).
"52 weeks in one year" is an approximation.

189. We need to find the smallest three-digit multiple of 6 that is a palindromic number. The smallest three digit palindromic number is 101; the next is 121; the next is 131.
Any three-digit palindromic number that has a 1 in the hundreds' place must have a 1 in the ones' place.
No number that has a 1 in the one's place is divisible by 6.
The smallest three-digit palindromic number with a 2 in the hundreds' place is 202, which is not divisible by 6.
The next smallest is 212, which is not divisible by 6.
222 is the smallest three-digit multiple of 6 that is a palindromic number.

191. To determine the number of accidental deaths each hour, divide the number of minutes in one hour (60) by the frequency in minutes of each accidental death (5):
$60 \div 5 = 12$
At the rate of one accidental death every 5 min, there will be 12 accidental deaths each hour.

To determine the number of accidental deaths each day, multiply the number of accidental deaths each hour (12) by the number of hours in one day (24):
$24(12) = 288$
At the rate of one accidental death every 5 min, there would be 288 accidental deaths each day.

To determine the number of accidental deaths each year, multiply the number of accidental deaths each day (288) by the number of days in one year (365):
$365(288) = 105,120$
At the rate of one accidental death every 5 min, there would be 105,120 accidental deaths each year.

SECTION 1.4

Objective A Exercises

1. $x + 9 = 23$
   $x + 9 - 9 = 23 - 9$
   $x + 0 = 14$
   $x = 14$
   The solution is 14.

3. $8 + b = 33$
   $8 - 8 + b = 33 - 8$
   $0 + b = 25$
   $b = 25$
   The solution is 25.

5. $3m = 15$
   $\frac{3m}{3} = \frac{15}{3}$
   $1m = 5$
   $m = 5$
   The solution is 5.

7. $52 = 4c$
   $\frac{52}{4} = \frac{4c}{4}$
   $13 = 1c$
   $13 = c$
   The solution is 13.

9. $16 = w + 9$
   $16 - 9 = w + 9 - 9$
   $7 = w + 0$
   $7 = w$
   The solution is 7.

11. $28 = 19 + p$
    $28 - 19 = 19 - 19 + p$
    $9 = 0 + p$
    $9 = p$
    The solution is 9.

13. $10y = 80$
    $\frac{10y}{10} = \frac{80}{10}$
    $1y = 8$
    $y = 8$
    The solution is 8.

15. $41 = 41d$
    $\frac{41}{41} = \frac{41d}{41}$
    $1 = 1d$
    $1 = d$
    The solution is 1.

17. $b + 7 = 7$
    $b + 7 - 7 = 7 - 7$
    $b + 0 = 0$
    $b = 0$
    The solution is 0.

19. $15 + t = 91$
    $15 - 15 + t = 91 - 15$
    $0 + t = 76$
    $t = 76$
    The solution is 76.

21. $4 + a = 25$
    $4 - 4 + a = 25 - 4$
    $0 + a = 21$
    $a = 21$
    The solution is 21.

23. $c + 17 = 50$
    $c + 17 - 17 = 50 - 17$
    $c + 0 = 33$
    $c = 33$
    The solution is 33.

Objective B Exercises

25. The unknown number: n

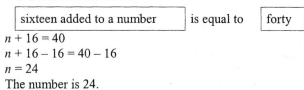 is equal to forty

$n + 16 = 40$
$n + 16 - 16 = 40 - 16$
$n = 24$
The number is 24.

27. The unknown number: n

five times a number is thirty

$5n = 30$
$\dfrac{5n}{5} = \dfrac{30}{5}$
$n = 6$
The number is 6.

29. The unknown number: n

fifteen is three more than a number

$15 = n + 3$
$15 - 3 = n + 3 - 3$
$12 = n$
The number is 12.

31. The unknown number: n

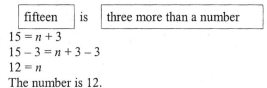 equals seventy-two

$n + 14 = 72$
$n + 14 - 14 = 72 - 14$
$n = 58$
The number is 58.

33. **Strategy** To find the width of the rectangle, write and solve an equation using w to represent the width.

Solution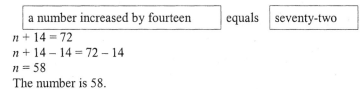

$17 = w + 5$
$17 - 5 = w + 5 - 5$
$12 = w$
The width of the rectangle is 12 in.

35. **Strategy** To find the distance from Forth Worth to Austin, write and solve an equation using x to represent the distance from Forth Worth to Austin.

Solution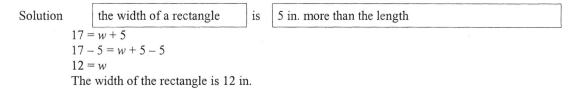

$212 = x + 22$
$212 - 22 = x + 22 - 22$
$190 = x$
The distance from Forth Worth to Austin is 190 mi.

37. **Strategy** To find the number of payments, replace A by 13,968 and M by 582 in the given formula and solve for N.

    Solution $A = MN$
    $13{,}968 = 582N$
    $\dfrac{13{,}968}{582} = \dfrac{582N}{582}$
    $24 = N$
    The number of payments is 24.

39. **Strategy** To find the time, replace d by 1,120 and r by 140 in the given formula and solve for t.

    Solution $d = rt$
    $1{,}120 = 140t$
    $\dfrac{1{,}120}{140} = \dfrac{140t}{140}$
    $8 = t$
    It would take 8 h to travel 1,120 mi at a speed of 140 mph.

Critical Thinking 1.4

41. Answers will vary. For example,
    (a) $5x = 0$, (b) $8x = 8$.

SECTION 1.5

Objective A Exercises

1. We need an Order of Operations Agreement to ensure that there is only one correct answer to a problem involving simplifying an arithmetic expression.

3. $8 \div 4 + 2 = 2 + 2$
   $= 4$

5. $6 \cdot 4 + 5 = 24 + 5$
   $= 29$

7. $4^2 - 3 = 16 - 3$
   $= 13$

9. $5 \cdot (6 - 3) + 4 = 5 \cdot 3 + 4$
   $= 15 + 4$
   $= 19$

11. $9 + (7 + 5) \div 6 = 9 + 12 \div 6$
    $= 9 + 2$
    $= 11$

13. $13 \cdot (1 + 5) \div 13 = 13 \cdot 6 \div 13$
    $= 78 \div 13$
    $= 6$

15. $6 \cdot 3^2 + 7 = 6 \cdot 9 + 7$
    $= 54 + 7$
    $= 61$

17. $14 + 5 \cdot 2^3 = 14 + 5 \cdot 8$
    $= 14 + 40$
    $= 54$

19. $10 + (8 - 5) \cdot 3 = 10 + 3 \cdot 3$
    $= 10 + 9$
    $= 19$

21. $2^3 + 4(10 - 6) = 2^3 + 4 \cdot 4$
    $= 8 + 4 \cdot 4$
    $= 8 + 16$
    $= 24$

23. $6(7) + 4^2 \cdot 3^2 = 6(7) + 16 \cdot 9$
    $= 42 + 16 \cdot 9$
    $= 42 + 144$
    $= 186$

25. $18 + 3(7) = 18 + 21$
    $= 39$

27. $6(8 - 3) - 12 = 6(5) - 12$
    $= 30 - 12$
    $= 18$

29. $16 - (13 - 5) \div 4 = 16 - 8 \div 4$
    $= 16 - 2$
    $= 14$

30. $11 + 2 - 3 \cdot 4 \div 3 = 11 + 2 - 12 \div 3$
    $= 11 + 2 - 4$
    $= 13 - 4$
    $= 9$

31. $17 + 1 - 8 \cdot 2 \div 4 = 17 + 1 - 16 \div 4$
    $= 17 + 1 - 4$
    $= 18 - 4$
    $= 14$

33. $x - 2y$
    $8 - 2 \cdot 3 = 8 - 6$
    $= 2$

35. $x^2 + 3y$
    $6^2 + 3 \cdot 7 = 36 + 3 \cdot 7$
    $= 36 + 21$
    $= 57$

37. $x^2 + y \div x$
    $2^2 + 8 \div 2 = 4 + 8 \div 2$
    $= 4 + 4$
    $= 8$

39. $4x+(x-y)^2$
$4\cdot 8+(8-2)^2 = 4\cdot 8+6^2$
$= 4\cdot 8+36$
$= 32+36$
$= 68$

41. $x^2+3(x-y)+z^2$
$2^2+3(2-1)+3^2 = 2^2+3\cdot 1+3^2$
$= 4+3\cdot 1+9$
$= 4+3+9$
$= 7+9$
$= 16$

43. $11+(8+4)\div 6 = 11+12\div 6$
$\qquad = 11+2$
$\qquad = 13$
$12+(9-5)\cdot 3 = 12+4\cdot 3$
$\qquad = 12+12$
$\qquad = 24$
Since $24 > 13$,
$12+(9-5)\cdot 3 > 11+(8+4)\div 6$

Critical Thinking 1.5

45. $27\div 9+8 = 3+8 = 11$
$81-8^2 = 81-64 = 17$
$5(10-2)\div 4 = 5(8)\div 4 = 40\div 4 = 10$
$4+3\cdot 12 = 4+36 = 40$
$50-6(8) = 50-48 = 2$
$2(1+4)^2 \div 10 = 2(5)^2 \div 10$
$\qquad = 2(25)\div 10$
$\qquad = 50\div 10$
$\qquad = 5$
Since $40 > 17 > 11 > 10 > 5 > 2$,
$4+3\cdot 12 > 81-8^2 > 27\div 9+8$
$> 5(10-2)\div 4 > 2(1+4)^2 \div 10 > 50-6(8)$

47. The addends within the parentheses can be paired so that each pair equals 100 $(47+53 = 100, 48+52 = 100, 49+51 = 100)$. The sum of the addends is $3\cdot 100 = 300$. Therefore $(47+48+49+51+52+53)\div 100 = 300\div 100 = 3$.

CHAPTER REVIEW

1. [number line with point at 7, marks 0–12]

2. $10^4 = 10,000$

3. $\quad 4,207$
$\underline{-1,624}$
$\quad 2,583$

4. $3\cdot 3\cdot 5\cdot 5\cdot 5\cdot 5 = 3^2 \cdot 5^4$

5. $\quad 319$
$\quad 358$
$\underline{+712}$
$1,389$

6. $38,\!729$ — Given place value; $2 < 5$
$38,729$ rounded to the nearest hundred is $38,700$

7. $247 > 163$

8. $32,509$

9. $2xy$
$2\cdot 50\cdot 7 = 100\cdot 7$
$\qquad = 700$

10. $\quad\quad 2,607$
$6\overline{)15,642}$
$\underline{-12}$
$\quad 36$
$\underline{-36}$
$\quad\quad 4$
$\underline{\;\;-0}$
$\quad\quad 42$
$\underline{\;-42}$
$\quad\quad\quad 0$

11. $\quad 6,407$
$\underline{-2,359}$
$\quad 4,048$

12. $482 \to \quad 500$
$319 \to \quad 300$
$570 \to \quad 600$
$146 \to \underline{+100}$
$\qquad\quad 1,500$

13. $50\div 1 = 50$
$50\div 2 = 25$
$50\div 5 = 10$
$50\div 10 = 5$
The factors of 50 are 1, 2, 5, 10, 25, and 50.

14. $24-y = 17$
$24-7 \mid 17$
$17 = 17$
Yes, 7 is a solution of the equation.

15. $16 + 4(7-5)^2 \div 8 = 16 + 4 \cdot 2^2 \div 8$
    $= 16 + 4 \cdot 4 \div 8$
    $= 16 + 16 \div 8$
    $= 16 + 2$
    $= 18$

16. The Commutative Property of Addition

17. Four million nine hundred twenty-seven thousand thirty-six

18. $x^3 y^2$
    $3^3 \cdot 5^2 = (3 \cdot 3 \cdot 3) \cdot (5 \cdot 5)$
    $= 27 \cdot 25$
    $= 675$

19. Strategy
    a. To determine by many times more PG-13 films were released than NC-17 films:
       →Find the number of PG-13 films (112) and the number of NC-17 films (7) from the pie chart.
       →Divide the number of PG-13 films by the number of NC-17 films.

    b. To determine by many times more R rated films were released than NC-17 films:
       →Find the number of R rated films (427) and the number of NC-17 films (7) from the pie chart.
       →Divide the number of R rated films by the number of NC-17 films.

    Solution
    a. $\dfrac{112}{7} = 16$
       There were 16 times more PG-13 films released than NC-17 films.

    b. $\dfrac{427}{7} = 61$
       There were 61 times more R rated films released than NC-17 films.

20. $\,67 \text{ r } 70$
    $92\overline{)6,234}$
    $\underline{-552}$
    714
    $\underline{-644}$
    70

21. 659
    $\underline{\times 4}$
    $2,636$

22. $x - y$
    $270 - 133 = 137$

23. $3\overline{)15}^{\,5}$
    $3\overline{)45}$
    $2\overline{)90}$
    $90 = 2 \cdot 3 \cdot 3 \cdot 5 = 2 \cdot 3^2 \cdot 5$

24. $\dfrac{x}{y}$
    $\dfrac{480}{6} = 80$

25. $1 \cdot 82 = 82$

26. $36 = 4x$
    $\dfrac{36}{4} = \dfrac{4x}{4}$
    $9 = x$
    The solution is 9.

27. $x + y$
    $683 + 249 = 932$

28. 18
    $\underline{\times\, 24}$
    72
    $\underline{36}$
    432

29. $(a+b)^2 - 2c$
    $(5+3)^2 - 2 \cdot 4 = 8^2 - 2 \cdot 4$
    $= 64 - 2 \cdot 4$
    $= 64 - 8$
    $= 5$

30. Strategy To find the person with the greater number of rebounds, compare the numbers 17,440 and 16,279.

    Solution $17{,}440 > 16{,}279$
    Kareem Abdul-Jabbar had more rebounds.

31. Strategy To find the total cost, multiply the number of square feet of floor space (2,800) by the cost per square feet (65).

    Solution $2{,}800 \cdot 65 = 182{,}000$
    The total cost of the contractor's work will be $182,000.

32. Strategy
    a. To find the perimeter of the rectangle, substitute 25 for L and 12 for W in the formula below.
    b. To find the area of the rectangle, substitute 25 for L and 12 for W in the formula below.

    Solution
    a. P = 2L + 2W
       P = 2 · 25 + 2 · 12
       P = 50 + 24
       P = 74
       The perimeter is 74 m.

    b. A = LW
       A = 25 · 12
       A = 300
       The area is 300 m².

33. Strategy To determine between which decade the number of students enrolled in college increased the most, and the amount of increase, find the amount of increase for each decade. Then compare the numbers.

    Solution 1960 to 1970: 8,581,000 − 3,789,000 = 4,792,000
    1970 to 1980: 12,097,000 − 8,581,000 = 3,516,000
    1980 to 1990: 13,819,000 − 12,097,000 = 1,722,000
    1990 to 2000: 14,889,000 − 13,819,000 = 1,070,000
    4,792,000 > 3,516,000 > 1,722,000 > 1,070,000

    a. The number of students enrolled in college increased the most from 1960 to 1970.
    b. The amount of increase was 4,792,000 people.

34. Strategy To find the distance traveled, substitute 3 for t and 14 for r in the given formula and solve for d.

    Solution $d = rt$
    $d = 14 \cdot 3$
    $d = 42$
    The cyclist traveled 42 mi.

35. Strategy To find the markup, substitute 2,224 for S and 1,775 for C in the given formula and solve for M.

    Solution $M = S - C$
    $M = 2,224 - 1,775$
    $M = 449$
    The markup on the copy machine is $449.

CHAPTER TEST

1. 3,297 · 100 = 329,700

2. 2 · 2 · 2 · 2 · 10 · 10 · 10 = 16,000

3. 4,902
    −873
    ─────
    4,029

4. $x \cdot x \cdot x \cdot x \cdot y \cdot y \cdot y = x^4 y^3$

5. 23 = p + 16
   23 − 16 = p + 16 − 16
   7 = p
   Yes

6. 2,961 ← Given place value
   6 > 5
   2,961 rounded to the nearest hundred is 3,000.

7. 7,177 < 7,717

8. 8,490

9. three hundred eighty-two thousand nine hundred four

10. 392 → 400
    477 → 500
    519 → 500
    +648 → 600
    ───── ─────
    2,036 2,000

11. 1,376
    ×   8
    ─────
    11,008

12. 36,479 → 40,000
    50 → 60
    40,000 · 60 = 2,400,000

13. $92 \div 1 = 92$
    $92 \div 2 = 46$
    $92 \div 4 = 23$
    $92 \div 23 = 4$
    The factors of 92 are 1, 2, 4, 23, 46, 92.

14. $3\overline{)15}^{\,5}$

    $2\overline{)30}$

    $2\overline{)60}$

    $2\overline{)120}$

    $2\overline{)240}$

    $240 = 2 \cdot 2 \cdot 2 \cdot 2 \cdot 3 \cdot 5 = 2^4 \cdot 3 \cdot 5$

15. $x - y$
    $39,241 - 8,375$
    $= 30,866$

16. The Commutative Property of Addition

17. $\dfrac{x}{y}$

    $\dfrac{3,588}{4}$
    $= 897$

18. $27 - (12 - 3) \div 9$
    $= 27 - 9 \div 9$
    $= 27 - 1$
    $= 26$

19. $46,300 - 32,400$
    $= \$13,900$

20. $60 = 17 + d$
    $60 - 17 = 17 + d - 17$
    $51 = d$
    The solution is 51.

21. $176 = 4t$
    $\dfrac{176}{4} = \dfrac{4t}{4}$
    $t = 44$
    The solution is 44.

22. $5x + (x - y)^2$
    $5 \cdot 8 + (8 - 4)^2$
    $= 40 + 4^2$
    $= 40 + 16$
    $= 56$
    The solution is 56.

23. 7

24. $12 + x = 90$
    $12 + x - 12 = 90 - 12$
    $x = 78$
    The number is 78.

25. $6 \cdot 5 \cdot 4 \cdot 3 \cdot 2 \cdot 1 = 720$

26. **Strategy** First, to find the total cost of the computer system, add the prices of the components. Then, to find the balance of the checking account, subtract the total cost of the computer system from $2,276.

    Solution $850 + 270 + 175 + 425 = 1720$
    $2,276 - 720 = 556$
    The balance is $556.

27. **Strategy**
    a. To find the perimeter of the square, substitute 24 for s in the formula below.

    b. To find the area of the square, substitute 24 for s in the formula below.

    Solution
    a. $P = 4s$
    $P = 4 \cdot 24$
    $P = 96$
    The perimeter is 96 cm.

    b. $A = s^2$
    $A = (24)^2$
    $A = 576$
    The area is 576 cm^2.

28. **Strategy** To find the data processor's take home pay, subtract the sum of his deductions, taxes (854), retirement (272), and insurance (108), from the total salary (5,690).

    Solution $5,690 - (854 + 272 + 108)$
    $= 5,690 - 1,234$
    $= 4,456$
    The data processor's take home pay is $4,456.

29. **Strategy**
    a. To determine between which two years the number of vehicles sold increased the most, for year there was an increase, find the increase. Then compare the numbers.
    b. The amount of increase is the largest amount.

    Solution
    1998 to 1999
    　55,000 − 25,000 = 30,000
    1999 to 2000
    　135,000 − 55,000 = 80,000
    2000 to 2001
    　175,000 − 135,000 = 40,000
    2001 to 2002
    　300,000 − 175,000 = 125,000
    　125,000 > 80,000 > 40,000 > 30,000
    a. The number of vehicles sold increased the most from 2001 to 2002.
    b. The amount of increase was 125,000 vehicles.

30. **Strategy** To find the commission earned, substitute 2 for U and 480 for R in the given formula and solve for C.

    Solution $C = U \cdot R$
    　　　　　$= 2 \cdot 480$
    　　　　　$= 960$
    The commission earned is $960.

31. **Strategy** To find the value per share of the fund, substitute 5,500,000 for C and 500,000 for S in the given formula and solve for V.

    Solution $V = \dfrac{C}{S} = \dfrac{5,5000,000}{500,000} = 11$
    The value per share is $11.

Chapter 2: Integers

Prep Test

1. 54 > 45
2. 4 units
3. 7654 + 8193 = 15,847
4. 6097 − 2318 = 3779
5. 472 × 56 = 26,432
6. 144 ÷ 24 = 6
7. $22 = y + 9$
   $22 - 9 = y + 9 - 9$
   $13 = y$
8. $12b = 60$
   $\dfrac{12b}{12} = \dfrac{60}{12}$
   $b = 5$
9. $P = C + M$
   $P = 129 + 43 = 172$
   The price is $172.
10. $(8-6)^2 + 12 \div 4 \cdot 3^2$
    $= 2^2 + 12 \div 4 \cdot 3^2$
    $= 4 + 12 \div 4 \cdot 9$
    $= 4 + 3 \cdot 9$
    $= 4 + 27$
    $= 31$

Go Figure

Group the numbers
$(2 \cdot 15)(4 \cdot 5)(10)(20) = (30)(20)(10)(20)$
This group of numbers will produce 4 zeros at the end of the number, when you multiply the first 20 natural numbers.

Section 2.1

Objective A Exercises

1.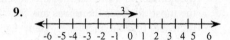

3.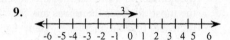

5.

7.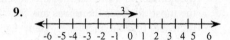

9.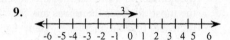

1 is 3 units to the right of −2.

11.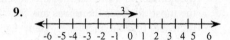

−1 is 4 units to the left of 3.

13.

3 is 6 units to the right of −3.

15.

A is −4, and C is −2.

17.

A is −7, and D is −4.

19. −2 > −5

21. 3 > −7

23. −42 < 27

25. 53 > −46

27. −51 < −20

29. −131 < 101

31. −7, −2, 0, 3

33. −5, −3, 1, 4

35. −4, 0, 5, 9

37. −10, −7, −5, 4, 12

39. −11, −7, −2, 5, 10

Objective B Exercises

41. −45

43. 88

45. −n

47. d

49. the opposite of negative thirteen

51. the opposite of negative p

53. five plus negative ten

55. negative fourteen minus negative three

57. negative thirteen minus eight

59. m plus negative n

61. −(−7) = 7

63. −(−61) = 61

65. −(46) = −46

67. −(−73) = 73

69. −(−z) = z

71. −(p) = −p

Objective C Exercises

73. $|-4| = 4$

75. $|9| = 9$

77. $|-11| = 11$

79. $|-12| = 12$

81. $|-23| = 23$

83. $-|27| = -27$

85. $|25| = 25$

87. $-|-41| = -41$

89. $-|-93| = -93$

91. $|x|$
    $|-10| = 10$

93. $|-x|$
    $|-8| = 8$

95. $|-y|$
    $|-(-6)| = |6| = 6$

97. $|-12| = 12, |8| = 8$
    $12 > 8$
    $|-12| > |8|$

99. $|6| = 6, |13| = 13$
    $6 < 13$
    $|6| < |13|$

101. $|-1| = 1, |-17| = 17$
     $1 < 17$
     $|-1| < |-17|$

103. $x = x$
     $|x| = |-x|$

105. $-|6| = -6, -(4) = -4, |-7| = 7, -(-9) = 9$
     $-|6|, -(4), |-7|, -(-9)$

107. $-|-7| = -7, -9 = -9, -(5) = -5, |4| = 4$
     $-9, -|-7|, -(5), |4|$

109. $-(-3) = 3, -|-8| = -8, |5| = 5,$
     $-|10| = -10, -(-2) = 2$
     $-|10|, -|-8|, -(-2), -(-3), |5|$

111. The absolute value of a number is the distance from zero to the number on the number line. If $|y| = 11$, then y must be a number that is 11 units from 0 on the number line. Therefore, y is −11 or 11.

113. x must be less than 7 and greater than −7.
     −6, −5, −4, −3, −2, −1, 0, 1, 2, 3, 4, 5, 6

Objective D Exercises

115. Strategy   To find the wind chill factor, use the given table.

      Solution   Find the number where the column with 10 and the row with 20 cross. Read the number −9.
The wind chill factor is −9°F.

117. **Strategy** To find the cooling power, use the given table.

    Solution Find the number where the column with −15 and the row with 10 cross. Read the number −35.
    The cooling power is −35°F.

119. **Strategy** To find the situation which feels colder:
    →From the given table, find the wind chill factor with a temperature of −30°F with a 5-mph wind and the wind chill factor with a temperature of −20°F with a 10-mph wind.
    →Compare the wind chill factors.

    Solution The wind chill factor with a temperature of −30°F and 5-mph wind is −46°F.
    The wind chill factor with a temperature of −20°F and 10-mph wind is −41°F.
    −46 < −41
    −30°F with a 5-mph wind feels colder.

121. **Strategy**
    a. To find the earnings per share in 2000, read the number in the bar graph below the bar corresponding to 2000.
    b. To find the earnings per share in 2002, read the number in the bar graph below the bar corresponding to 2002.

    Solution
    a. The earnings per share for 2000 were −27¢.
    b. The earnings per share for 2002 were −40¢.

123. **Strategy** To find a year in which Mycogen had a profit, use the bar graph to find a year in which earnings per share were positive.

    Solution The only recorded positive earnings per share were 11¢. Earnings per share were 11¢ in 1994.
    Mycogen did earn a profit during the years shown. Mycogen earned a profit in 2003.

125. **Strategy** To determine which stock showed the least net change, compare the absolute values of the numbers −1 and −2. The smaller number represents the least net change.

    Solution $|-1| = 1, |-2| = 2$
    $1 < 2$
    Stock B showed the least net change.

127. **Strategy** To determine which quarter has the greater loss, compare the absolute values of the numbers −26,800, and −24,900. The larger number corresponds to the quarter with the greater loss.

    Solution $|-26,800| = 26,800$
    $|-24,900| = 24,900$
    $26,800 > 24,900$
    The loss was greater during the third quarter.

Critical Thinking 2.1

129. a. Two numbers that are 4 units from 2 on the number line are −2 and 6.

    b. Two numbers that are 5 units from 3 on the number line are −2 and 8.

131. a. Enter the number 9. Press the +/– key.
   b. Enter the number 20. Press the +/– key.
   c. Enter the number 148. Press the +/– key.
   d. Enter the number 573. Press the +/– key.

133. Since x is an integer and $|x| < 10$, x must be less than 10 and greater than –10: –9, –8, –7, –6, –5, –4, –3, –2, –1, 0, 1, 2, 3, 4, 5, 6, 7, 8, 9.
Of these values of x, the numbers for which $|x| > 6$ are: –9, –8, –7, 7, 8, 9.

135. Since the first statement is false (It was either Student B or Student D), neither Student B nor Student D could have done it. Therefore, it was either Student A or Student C. Since the second statement is false (It was neither Student B nor Student C), either Student B or Student C did it. Since the first statement ensured that it was Student A or Student C, it must have been Student C.

Section 2.2

Objective A Exercises

1. a. To add two integers with the same sign, add the absolute values of the integers. Then attach the sign of the integers added.
    b. To add two integers with different signs, find the absolute value of each integer. Then subtract the smaller of these absolute values from the larger one. Then attach the sign of the integer with the larger absolute value.

3. $-3 + (-8) = -11$

5. $-8 + 3 = -5$

7. $-5 + 13 = 8$

9. $6 + (-10) = -4$

11. $3 + (-5) = -2$

13. $-4 + (-5) = -9$

15. $-6 + 7 = 1$

17. $(-5) + (-10) = -15$

19. $-7 + 7 = 0$

21. $(-15) + (-6) = -21$

23. $0 + (-14) = -14$

25. $73 + (-54) = 19$

27. $2 + (-3) + (-4) = -1 + (-4)$
    $= -5$

29. $-3 + (-12) + (-15) = -15 + (-15)$
    $= -30$

31. $-17 + (-3) + 29 = -20 + 29$
    $= 9$

33. $11 + (-22) + 4 + (-5)$
    $= -11 + 4 + (-5)$
    $= -7 + (-5)$
    $= -12$

35. $-22 + 10 + 2 + (-18)$
    $= -12 + 2 + (-18)$
    $= -10 + (-18)$
    $= -28$

37. $-25 + (-31) + 24 + 19$
    $= -56 + 24 + 19$
    $= -32 + 19$
    $= -13$

39. $3 + (-21) = -18$

41. $-5 + 16 = 11$

43. $(-3) + (-8) + 12 = -11 + 12$
    $= 1$

45. $x + (-7)$

47. a. $-57{,}931{,}000{,}000 + (-27{,}578{,}000{,}000) = -85{,}509{,}000{,}000$
The total of the U.S. balance of trade in Japan and Mexico is $-\$85{,}509{,}000{,}000$.

b. $-49{,}409{,}000{,}000 + (-27{,}578{,}000{,}000) = -76{,}987{,}000{,}000$
The total of the U.S. balance of trade with Canada and Mexico is $-\$76{,}987{,}000{,}000$.

c. $-57{,}931{,}000{,}000 + (-10{,}140{,}000{,}000) = -60{,}071{,}000{,}000$
The total of the U.S. balance of trade with Japan and France is $-\$60{,}071{,}000{,}000$.

49. $-a + b$
$-(-8) + (-3) = 8 + (-3)$
$= 5$

51. $-x + y$
$-(-5) + (-7) = 5 + (-7)$
$= -2$

53. $a + b + c$
$-10 + (-6) + 5 = -16 + 5$
$= -11$

55. $-x + (-y) + z$
$-(-2) + (-8) + (-11) = 2 + (-8) + (-11)$
$= -6 + (-11)$
$= -17$

57. The Addition Property of Zero

59. The Associative Property of Addition

61. $-13 + 0 = -13$

63. $18 + (-18) = 0$

65. $6 = -3 + z$
$\overline{6 \mid -3 + (-8)}$
$6 \neq -11$
No, -8 is not a solution of the equation $6 = -3 + z$.

67. $-7 + m = -15$
$\overline{-7 + (-8) \mid -15}$
$-15 = -15$
Yes, -8 is a solution of the equation $-7 + m = -15$.

69. $1 + z = z + 2$
$\overline{1 + (-4) \mid -4 + 2}$
$-3 \neq -2$
No, -4 is not a solution of the equation $1 + z = z + 2$.

Objective B Exercises

71. The word *minus* indicates the operation of subtraction. The word *negative* indicates a number less than zero.

73. $6 - 9 = 6 + (-9)$
$= -3$

75. $-9 - 4 = -9 + (-4)$
$= -13$

77. $3 - (-4) = 3 + 4$
$= 7$

79. $-4 - (-4) = -4 + 4$
$= 0$

81. $-10 - 7 = -10 + (-7)$
$= -17$

83. $(-7) - (-4) = -7 + 4$
$= -3$

85. $-4 - (-16) = -4 + 16$
$= 12$

87. $3 - (-24) = 3 + 24$
$= 27$

89. $(-41) - 65 = -41 + (-65)$
$= -106$

91. $-95 - (-28) = -95 + 28$
$= -67$

93. $-10 - (-4) = -10 + 4$
$= -6$

95. $-9 - 6 = -9 + (-6)$
$= -15$

97. $-t - r$

99. $49 - (-33) = 49 + 33$
$= 82$
The difference between the highest and lowest temperatures ever recorded in South America is 82°C.

101. $-4 - 3 - 2 = -4 + (-3) + (-2)$
$= -7 + (-2)$
$= -9$

103. $12 - (-7) - 8 = 12 + 7 + (-8)$
$= 19 + (-8)$
$= 11$

105. $4 - 12 - (-8) = 4 + (-12) + 8$
$= -8 + 8$
$= 0$

107. $-16 - 47 - 63 - 12$
$= -16 + (-47) + (-63) + (-12)$
$= -63 + (-63) + (-12)$
$= -126 + (-12)$
$= -138$

109. $12 - (-6) + 8 = 12 + 6 + 8$
$= 18 + 8$
$= 26$

111. $-8 - (-14) + 7 = -8 + 14 + 7$
$= 6 + 7$
$= 13$

113. $9 - 12 + 0 - 5 = 9 + (-12) + 0 + (-5)$
$= -3 + 0 + (-5)$
$= -3 + (-5)$
$= -8$

115. $5 + 4 - (-3) - 7 = 5 + 4 + 3 + (-7)$
$= 9 + 3 + (-7)$
$= 12 + (-7)$
$= 5$

117. $-13 + 9 - (-10) - 4 = -13 + 9 + 10 + (-4)$
$= -4 + 10 + (-4)$
$= 6 + (-4)$
$= 2$

119. $-x - y$
$-(-3) - 9 = 3 + (-9)$
$= -6$

121. $-x - (-y)$
$-(-3) - (-9) = 3 + 9$
$= 12$

123. $a - b - c$
$4 - (-2) - 9 = 4 + 2 + (-9)$
$= 6 + (-9)$
$= -3$

125. $x - y - (-z)$
$-9 - 3 - (-30) = -9 + (-3) + 30$
$= -12 + 30$
$= 18$

127.
$$\begin{array}{c|c} x - 7 = -10 \\ \hline -3 - 7 & -10 \\ -3 + (-7) & -10 \\ -10 = -10 \end{array}$$
Yes, -3 is a solution of the equation $x - 7 = -10$.

129.
$$\begin{array}{c|c} -5 - w = 7 \\ \hline -5 - (-2) & 7 \\ -5 + 2 & 7 \\ -3 \neq 7 \end{array}$$
No, -2 is not a solution of the equation $-5 - w = 7$.

131.
$$\begin{array}{c|c} -t - 5 = 7 + t \\ \hline -(-6) - 5 & 7 + (-6) \\ 6 - 5 & 7 + (-6) \\ 6 + (-5) & 7 + (-6) \\ 1 = 1 \end{array}$$
Yes, -6 is a solution of the equation $-t - 5 = 7 + t$.

Objective C Exercises

133. Strategy
a. To find the difference, subtract the elevation of Death Valley (-86) from the elevation of Mt. Aconcagua ($6{,}960$).

b. To find the difference, subtract the elevation of the Lake Assal (-156) from the elevation of Mt. Kilimangaro ($5{,}895$).

Solution
a. $6{,}960 - (-86) = 6{,}960 + 86$
$= 7{,}046$
The difference in elevation is 7,046 m.

b. $5{,}895 - (-156) = 5{,}895 + 156$
$= 6{,}051$
The difference in elevation is 6,051 m.

Chapter 2: Integers

135. Strategy — To determine for which continent the difference between the highest and lowest elevations is smallest:
→ Find the difference between the highest and lowest elevation for each continent.
→ Compare the differences.

Solution
Africa: $5,895 - (-156)$
$= 5,895 + 156 = 6,051$
Asia: $8,850 - (-411)$
$= 8,850 + 411 = 9,261$
Europe: $5,642 - (-28)$
$= 5,642 + 28 = 5,670$
America: $6,960 - (-86)$
$= 6,960 + 86 = 7,046$
$5,670 < 6,051 < 7,046 < 9,261$
The difference between the highest and lowest elevations is smallest in Europe.

137. Strategy — To find the difference, subtract the average temperature at 40,000 ft (-70) from the average temperature at 12,000 ft (16).

Solution
$16 - (-70) = 16 + 70 = 86$
The difference in temperature is 86°.

139. Strategy — To find how much colder, subtract the average temperature at 30,000 ft (-48) from the average temperature at 20,000 ft (-12).

Solution
$-12 - (-48) = -12 + 48 = 36$
The temperature is 36° colder.

141. Strategy — To find the golfer's score, substitute 49 for N and 52 for P in the given formula and solve for S.
$S = N - P$
$S = 49 - 52$
$S = 49 + (-52)$
$S = -3$
The golfer's score is -3.

143. Strategy — To find d, replace a by 7 and b by -12 in the given formula and solve for d.

Solution
$d = |a - b|$
$d = |7 - (-12)|$
$d = |7 + 12|$
$d = |19|$
$d = 19$
The distance between the two points is 19 units.

Critical Thinking 2.2

145. a. The opposite, or additive inverse, of 5 is -5. The difference between 5 and -5 is $5 - (-5) = 5 + 5 = 10$, which is not 0. The additive inverse of 0 is 0. The difference between 0 and 0 is $0 - 0 = 0$.
The statement is sometimes true.

b. When adding two numbers with the same sign, we add the absolute values of the numbers and then attach the sign of the addends. When adding two negative numbers, the signs of the addends are negative. Therefore, we would attach a negative sign on the sum. The statement is always true.

147. To simplify $-17 - (-18) + (-5)$ using a scientific calculator, enter 17, press the $+/-$ key, press the subtraction key $(-)$, enter 18, press the $+/-$ key, press the addition key $(+)$, enter 5, press the $+/-$ key, press the $=$ key. The result in the display should be -4.

Section 2.3

Objective A Exercises

1. To multiply two integers with the same sign, multiply the absolute values of the factors. The product is positive.
To multiply two integers with different signs, multiply the absolute values of the factors. The product is negative.

3. $-4 \cdot 6 = -24$

5. $-2(-3) = 6$

7. $(9)(2) = 18$

9. $5(-4) = -20$

11. $-8(2) = -16$

13. $(-5)(-5) = 25$

15. $(-7)(0) = 0$

17. $14(3) = 42$

19. $-32(4) = -128$

21. $(-8)(-26) = 208$

23. $9(-27) = -243$

25. $-5 \cdot (23) = -115$

27. $-7(-34) = 238$

29. $4 \cdot (-8) \cdot 3 = -32 \cdot 3$
    $= -96$

31. $(-6)(5)(7) = -30(7)$
    $= -210$

33. $-8(-7)(-4) = 56(-4)$
    $= -224$

35. $2(-20) = -40$

37. $-30(-6) = 180$

39. $-q(r) = -qr$

41. a. $-24{,}666{,}000(4) = -98{,}664{,}000$
    The annual net income for Midway Games would be $-\$98{,}664{,}000$.

    b. $-6{,}693{,}000(4) = -26{,}772{,}000$
    The annual net income for Pinnacle Entertainment would be $-\$26{,}772{,}000$.

    c. $-3{,}262{,}000(4) = -13{,}048{,}000$
    The annual net income for Granite Broadcasting would be $-\$13{,}048{,}000$.

43. The Multiplication Property of One

45. The Associative Property of Multiplication

47. $-6 \cdot (5 \cdot 10) = (-6 \cdot 5) \cdot 10$

49. $1(-14) = -14$

51. $-xy$
    $-(-3)(-8) = 3(-8)$
    $= -24$

53. $-xyz$
    $-(-6)(2)(-5) = 6(2)(-5)$
    $= 12(-5)$
    $= -60$

55. $-7n$
    $-7(-51) = 357$

57. $8ab$
    $8(7)(-1) = 56(-1)$
    $= -56$

59. $-5st$
    $-5(-40)(-8) = 200(-8)$
    $= -1{,}600$

61. $-5x = -15$
    $\overline{-5(-3) \mid -15}$
    $15 \neq -15$
    No, -3 is not a solution of the equation $-5x = 15$.

63. $-8 = -8a$
    $\overline{-8 \mid -8(0)}$
    $-8 \neq 0$
    No, 0 is not a solution of the equation $-8 = -8a$.

65. $-27 = -3c$
    $\overline{-27 \mid -3(9)}$
    $-27 = -27$
    Yes, 9 is a solution of the equation $-27 = -3c$.

Objective B Exercises

67. $18 \div (-3) = -6$

69. $(-64) \div (-8) = 8$

71. $-49 \div 1 = -49$

73. $-40 \div (-5) = 8$

75. $\dfrac{44}{-4} = -11$

77. $\dfrac{-98}{-7} = 14$

79. $-91 \div (-7) = 13$

81. $(-162) \div (-162) = 1$

83. $-130 \div (-5) = 26$

85. $(-92) \div (-4) = 23$

87. $\dfrac{550}{-5} = -110$

89. $\dfrac{-333}{-3} = 111$

91. $\dfrac{-9}{x}$

93. $-708{,}000 \div 3 = -236{,}000$
    The average monthly net income for Friendly Ice Cream was $-\$236{,}000$.

95. $-a \div b$
    $-(-36) \div (-4) = 36 \div (-4)$
    $= -9$

97. $(-a) \div (-b)$
    $-(-36) \div (-(-4)) = 36 \div 4$
    $= 9$

99. $\dfrac{-x}{y}$
    $\dfrac{-(-42)}{-7} = \dfrac{42}{-7}$
    $= -6$

101. $\dfrac{-x}{-y}$
    $\dfrac{-(-42)}{-(-7)} = \dfrac{42}{7}$
    $= 6$

103. $6 = \dfrac{-c}{-3}$
    $6 \;\Big|\; \dfrac{-18}{-3}$
    $6 = 6$
    Yes, 18 is a solution of the equation.

105. $\dfrac{21}{n} = 7$
    $\dfrac{21}{-3} \;\Big|\; 7$
    $-7 \ne 7$
    No, -3 is not a solution of the equation.

107. $\dfrac{m}{-4} = \dfrac{-16}{m}$
    $\dfrac{8}{-4} \;\Big|\; \dfrac{-16}{8}$
    $-2 = -2$
    Yes, 8 is a solution of the equation.

Objective C Exercises

109. Strategy   To find the average score, divide the combined scores (-12) by the number of golfers (4).

    Solution $-12 \div 4 = 3$
    The average golf score was -3.

111. Strategy   To find the average record low temperature for the first three months of the year:
    →Add the average temperatures for January (-70), February (-66), and March (-50).
    →Divide the sum by the number of months (3).

    Solution $-70 + (-66) + (-50) = -136 + (-50) = -186$
    $-186 \div 3 = -62$
    The average record low temperature for the first three months of the year is $-62°$F.

113. Strategy   To find the average daily low temperature for the week:
    →Add the seven temperature readings.
    →Divide by 7.

    Solution $4 + (-5) + 8 + (-1) + (-12) + (-14) + (-8) = -28$
    $-28 \div 7 = -4$
    The average daily low temperature for the week was $-4°$.

115. Strategy To find the wind chill factor, multiply the wind chill factor at 10°F with a 20 mph wind (–9) by 5.

Solution $-9 \cdot 5 = -45$
The wind chill factor is –45°F.

117. Strategy To find the next three numbers in the sequence:
→Find the multiplier by dividing the second number in the sequence (–4) by the first number (2).
→Use the multiplier to find the successive numbers in the sequence.

Solution $\frac{-4}{2} = -2$
$8 \cdot (-2) = -16$
$-16 \cdot (-2) = 32$
$32 \cdot (-2) = -64$
The next three numbers in the sequence are –16, 32, and –64.

119. Strategy To find the next three numbers in the sequence:
→Find the multiplier by dividing the second number in the sequence (–5) by the first number (–1).
→Use the multiplier to find the successive numbers in the sequence.

Solution $\frac{-5}{-1} = 5$
$-25 \cdot 5 = -125$
$-125 \cdot 5 = -625$
$-625 \cdot 5 = -3,125$
The next three numbers in the sequence are –125, –625, and –3,125.

Critical Thinking 2.3

121. a. We are looking for the largest possible product of two negative integers whose sum is –18. Find pairs of negative integers whose sum is –18 and look for a pattern.
$-17 + (-1) = -18; (-17)(-1) = 18$
$-16 + (-2) = -18; (-16)(-2) = 32$
$-15 + (-3) = -18; (-15)(-3) = 45$ The numbers are increasing.
⋮
$-10 + (-8) = -18; (-10)(-8) = 80$
$-9 + (-9) = -18; (-9)(-9) = 81$
$-8 + (-10) = -18; (-8)(-10) = 80$
$-7 + (-11) = -18; (-7)(-11) = 77$ The numbers are decreasing.
The largest possible product of two negative integers whose sum is –18 is 81.

b. We are looking for the smallest possible sum of two negative integers whose product is 16. List all the possible pairs of negative integers whose product is 16.
$-16(-1) = 16; -16 + (-1) = -17$
$-8(-2) = 16; -8 + (-2) = -10$
$-4(-4) = 16; -4 + (-4) = -8$
Of the numbers –17, –10, and –8, –17 is the smallest number.
The smallest possible sum of two negative integers whose product is 16 is –17.

123. By substituting negative integers for x in the inequality $1 - 3x < 12$, it can be shown that –3, –2, and –1 are the only negative integers that satisfy the inequality.

125. **a.** The product of two integers is positive with the signs of the two integers are the same. The product is negative when the signs of the two integers are different.

b. The quotient of two integers is positive with the signs of the two integers are the same. The quotient is negative when the signs of the two integers are different.

Section 2.4

Objective A Exercises

1. $x - 6 = 9$
   $x - 6 + 6 = 9 + 6$
   $x = 15$
   The solution is 15.

3. $8 = y - 3$
   $8 + 3 = y - 3 + 3$
   $11 = y$
   The solution is 11.

5. $x - 5 = -12$
   $x - 5 + 5 = -12 + 5$
   $x = -7$
   The solution is -7.

7. $-10 = z + 6$
   $-10 - 6 = z + 6 - 6$
   $-16 = z$
   The solution is -16.

9. $x + 12 = 4$
   $x + 12 - 12 = 4 - 12$
   $x = -8$
   The solution is -8.

11. $-12 = c - 12$
    $-12 + 12 = c - 12 + 12$
    $0 = c$
    The solution is 0.

13. $6 + x = 4$
    $6 + x - 6 = 4 - 6$
    $x = -2$
    The solution is -2.

15. $12 = n - 8$
    $12 + 8 = n - 8 + 8$
    $20 = n$
    The solution is 20.

17. $3m = -15$
    $\dfrac{3m}{3} = \dfrac{-15}{3}$
    $m = -5$
    The solution is -5.

19. $-10 = 5v$
    $\dfrac{-10}{5} = \dfrac{5v}{5}$
    $-2 = v$
    The solution is -2.

21. $-8x = -40$
    $\dfrac{-8x}{-8} = \dfrac{-40}{-8}$
    $x = 5$
    The solution is 5.

23. $-60 = -6v$
    $\dfrac{-60}{-6} = \dfrac{-6v}{-6}$
    $10 = v$
    The solution is 10.

25. $5x = -100$
    $\dfrac{5x}{5} = \dfrac{-100}{5}$
    $x = -20$
    The solution is -20.

27. $4x = 0$
    $\dfrac{4x}{4} = \dfrac{0}{4}$
    $x = 0$
    The solution is 0.

29. $-2r = 16$
    $\dfrac{-2r}{-2} = \dfrac{16}{-2}$
    $r = -8$
    The solution is -8.

31. $-72 = 18w$
    $\dfrac{-72}{18} = \dfrac{18w}{18}$
    $-4 = w$
    The solution is -4.

Objective B Exercises

33. The unknown number: n

    | ten less than a number | is | fifteen |

    $n - 10 = 15$
    $n - 10 + 10 = 15 + 10$
    $n = 25$
    The number is 25.

35. The unknown number: n

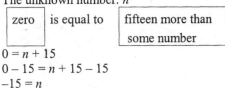

$0 = n + 15$
$0 - 15 = n + 15 - 15$
$-15 = n$
The number is -15.

37. The unknown number: n

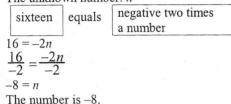

$16 = -2n$
$\dfrac{16}{-2} = \dfrac{-2n}{-2}$
$-8 = n$
The number is -8.

39. The unknown number: n

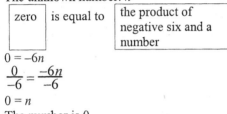

$0 = -6n$
$\dfrac{0}{-6} = \dfrac{-6n}{-6}$
$0 = n$
The number is 0.

41. Strategy To find the U.S. balance of trade in 1979, write and solve an equation using x to represent the U.S. balance of trade in 1979.

Solution

| the balance of trade in 1980 | was | $5158 million more than the balance of trade in 1979 |

$-19{,}407 = x + 5158$
$-19{,}407 - 5158 = x + 5158 - 5158$
$-24{,}565 = x$
The U.S. balance of trade in 1979 was $-\$24{,}565$ million.

43. Strategy To find this morning's temperature, write and solve an equation using x to represent this morning's temperature.

Solution

| the temperature now | is | 5° higher than it was in this morning |

$8 = x + 5$
$8 - 5 = x + 5 - 5$
$3 = x$
The temperature this morning was $3°C$.

45. Strategy To find the selling price of the car, replace P by 925 and C by 12,600 in the given formula and solve for S.

Solution $P = S - C$
$925 = S - 12{,}600$
$925 + 12{,}600 = S - 12{,}600 + 12{,}600$
$13{,}525 = S$
The selling price of the car should be $\$13{,}525$.

34 Chapter 2: Integers

47. **Strategy** To find the assets, replace N by 11 and L by 4 in the given formula and solve for A.

    Solution
    $N = A - L$
    $11 = A - 4$
    $11 + 4 = A - 4 + 4$
    $15 = A$
    The assets are $15 million.

Critical Thinking 2.4

49. a. False. For example, the solution of the equation $5x = 0$ is 0.

    b. False. For example, the solution of the equation $5x = -5$ is -1, a negative number.

    c. False. For example, the solution of the equation $-5x = -5$ is 1, a positive number.

51. In describing the addition, subtraction, and division properties of equations, students should be rephrasing the concept that the same number can be added to or subtracted from each side of an equation, or each side of an equation can be divided by the same nonzero number, without changing the solution of the equation. Be sure that students state that the same operation must be performed to <u>both sides</u> of the equation. Check to see if students remember to explain that the division property does not hold true for zero.

Section 2.5

Objective A Exercises

1. $3 - 12 \div 2 = 3 - 6$
   $= 3 + (-6)$
   $= -3$

3. $2(3 - 5) - 2 = 2(-2) - 2$
   $= -4 - 2$
   $= -4 + (-2)$
   $= -6$

5. $4 - (-3)^2 = 4 - 9$
   $= 4 + (-9)$
   $= -5$

7. $4 \cdot (2 - 4) - 4 = 4 \cdot (-2) - 4$
   $= -8 - 4$
   $= -8 + (-4)$
   $= -12$

9. $4 - (-2)^2 + (-3) = 4 - 4 + (-3)$
   $= 4 + (-4) + (-3)$
   $= 0 + (-3)$
   $= -3$

11. $3^3 - 4(2) = 27 - 4(2)$
    $= 27 - 8$
    $= 27 + (-8)$
    $= 19$

13. $3 \cdot (6 - 2) \div 6 = 3 \cdot 4 \div 6$
    $= 12 \div 6$
    $= 2$

15. $2^3 - (-3)^2 + 2 = 8 - 9 + 2$
    $= 8 + (-9) + 2$
    $= -1 + 2$
    $= 1$

17. $6 - 2(1 - 5) = 6 - 2(-4)$
    $= 6 - (-8)$
    $= 6 + 8$
    $= 14$

19. $6 - (-4)(-3)^2 = 6 - (-4)(9)$
    $= 6 - (-36)$
    $= 6 + 36$
    $= 42$

21. $4 \cdot 2 - 3 \cdot 7 = 8 - 3 \cdot 7$
    $= 8 - 21$
    $= 8 + (-21)$
    $= -13$

23. $(-2)^2 - 5(3) - 1 = 4 - 5(3) - 1$
    $= 4 - 15 - 1$
    $= 4 + (-15) + (-1)$
    $= -11 + (-1)$
    $= -12$

25. $3 \cdot 2^3 + 5 \cdot (3 + 2) - 17 = 3 \cdot 2^3 + 5 \cdot 5 - 17$
    $= 3 \cdot 8 + 5 \cdot 5 - 17$
    $= 24 + 5 \cdot 5 - 17$
    $= 24 + 25 - 17$
    $= 24 + 25 + (-17)$
    $= 49 + (-17)$
    $= 32$

27. $-12(6-8)+1^3 \cdot 3^2 \cdot 2 - 6(2)$
   $= -12(-2) + 1^3 \cdot 3^2 \cdot 2 - 6(2)$
   $= -12(-2) + 1 \cdot 9 \cdot 2 - 6(2)$
   $= 24 + 1 \cdot 9 \cdot 2 - 6(2)$
   $= 24 + 18 - 6(2)$
   $= 24 + 18 - 12$
   $= 24 + 18 + (-12)$
   $= 42 + (-12)$
   $= 30$

29. $-27 - (-3)^2 - 2 - 7 + 6 \cdot 3$
   $= -27 - 9 - 2 - 7 + 6 \cdot 3$
   $= -27 - 9 - 2 - 7 + 18$
   $= -27 + (-9) + (-2) + (-7) + 18$
   $= -36 + (-2) + (-7) + 18$
   $= -38 + (-7) + 18$
   $= -45 + 18$
   $= -27$

31. $16 - 4 \cdot 8 + 4^2 - (-18) - (-9)$
   $= 16 - 4 \cdot 8 + 16 - (-18) - (-9)$
   $= 16 - 32 + 16 - (-18) - (-9)$
   $= 16 + (-32) + 16 + 18 + 9$
   $= -16 + 16 + 18 + 9$
   $= 0 + 18 + 9$
   $= 18 + 9$
   $= 27$

33. $3a + 2b$
   $3(-2) + 2(4) = -6 + 2(4)$
   $= -6 + 8$
   $= 2$

35. $16 \div (ac)$
   $16 \div ((-2)(-1)) = 16 \div 2$
   $= 8$

37. $bc \div (2a)$
   $4(-1) \div ((2)(-2)) = 4(-1) \div (-4)$
   $= (-4) \div (-4)$
   $= 1$

39. $b^2 - c^2$
   $4^2 - (-1)^2 = 16 - 1$
   $= 16 + (-1)$
   $= 15$

41. $(b-a)^2 + 4c$
   $(4-(-2))^2 + 4(-1) = (4+2)^2 + 4(-1)$
   $= 6^2 + 4(-1)$
   $= 36 + 4(-1)$
   $= 36 + (-4)$
   $= 32$

43. $\dfrac{d-b}{c}$
   $\dfrac{3-4}{-1} = \dfrac{3+(-4)}{-1}$
   $= \dfrac{-1}{-1}$
   $= 1$

45. $\dfrac{b-d}{c-a}$
   $\dfrac{4-3}{-1-(-2)} = \dfrac{4+(-3)}{-1+2}$
   $= \dfrac{1}{1}$
   $= 1$

47. $(d-a)^2 \div 5$
   $(3-(-2))^2 \div 5 = (3+2)^2 \div 5$
   $= 5^2 \div 5$
   $= 25 \div 5$
   $= 5$

49. $(d-a)^2 - 3c$
   $(3-(-2))^2 - 3(-1)$
   $= (3+2)^2 - 3(-1)$
   $= 5^2 - 3(-1)$
   $= 25 - 3(-1)$
   $= 25 - (-3)$
   $= 25 + 3$
   $= 28$

Critical Thinking 2.5

51.
$$-2^2 - (-3)^2 + 5(4) \div 10 - (-6)$$
$$= -4 - 9 + 5(4) \div 10 - (-6)$$
$$= -4 - 9 + 20 \div 10 - (-6)$$
$$= -4 - 9 + 2 - (-6)$$
$$= -4 + (-9) + 2 + 6$$
$$= -13 + 2 + 6$$
$$= -11 + 6 = -5$$

We are looking for the smallest integer greater than –5.
The smallest integer greater than –5 is –4.
The smallest integer greater than $-2^2 - (-3)^2 + 5(4) \div 10 - (-6)$ is –4.

53. a.
$$x^2 - 2x - 8 = 0$$

$(-4)^2 - 2(-4) - 8$	0
$16 - 2(-4) - 8$	0
$16 + 8 - 8$	0
$24 - 8$	0
$16 \ne 0$	

No, –4 is not a solution of the equation.

b.
$$x^3 + 3x^2 - 5x - 15 = 0$$

$(-3)^3 + 3(-3)^2 - 5(-3) - 15$	0
$-27 + 3(9) - 5(-3) - 15$	0
$-27 + 27 + 15 - 15$	0
$0 + 15 - 15$	0
$15 - 15$	0
$0 = 0$	

Yes, –3 is not a solution of the equation.

55. $a \div bc$ when $a = 16$, $b = 2$, and $c = -4$ is:
$$16 \div (2)(-4) = 8(-4) = -32$$
$a \div (bc)$ when $a = 16$, $b = 2$, and $c = -4$ is:
$$16 \div [2(-4)] = 16 \div [-8] = -2$$
By Step 3 of the Order of Operations Agreement, multiplication and division are performed as they occur from left to right. Therefore, in the first solution above, the division $16 \div 2$ must be performed first; then the multiplication. By Step 1 of the Order of Operations Agreement, operations in parentheses must be performed first. Therefore, in the second solution above, the operation of multiplication inside the parentheses (bc) must be performed first; then the division.

Chapter Review Exercises

1. eight minus negative one

2. $-|-36| = -36$

3. $(-40)(-5) = 200$

4. $-a \div b$
$-(-27) \div (-3) = 27 \div (-3)$
$= -9$

5. $-28 + 14 = -14$

6. $-(-13) = 13$

7.

```
<-+--+--+--+--+--●--+--+--+--+--+--+-->
 -6 -5 -4 -3 -2 -1  0  1  2  3  4  5  6
```

8. $-24 = -6y$
$$\frac{-24}{-6} = \frac{-6y}{-6}$$
$4 = y$
The solution is 4.

9. $-51 \div (-3) = 17$

10. $\dfrac{840}{-4} = -210$

11. $-6 - (-7) - 15 - (-12)$
$= -6 + 7 + (-15) + 12$
$= 1 + (-15) + 12$
$= -14 + 12$
$= -2$

12. $-ab$
$-(-2)(-9) = 2(-9)$
$= -18$

13. $18 + (-13) + (-6) = 5 + (-6)$
$= -1$

14. $-18(4) = -72$

15. $(-2)^2 - (-3)^2 \div (1-4)^2 \cdot 2 - 6$
$= (-2)^2 - (-3)^2 \div (-3)^2 \cdot 2 - 6$
$= 4 - 9 \div 9 \cdot 2 - 6$
$= 4 - 1 \cdot 2 - 6$
$= 4 - 2 - 6$
$= 4 + (-2) + (-6)$
$= 2 + (-6)$
$= -4$

Chapter Review

16. $-x - y$
$-(-1) - 3 = 1 - 3$
$= 1 + (-3)$
$= -2$

17. $3 - (-8) = 3 + 8$
$= 11$
The difference between the number of strokes made by Diaz and the number by Sorenstam is 11 strokes.

18. $-15 - (-28) = -15 + 28$
$= 13$

19. The Commutative Property of Multiplication

20.
$$\begin{array}{c|c} -6 - t = 3 & \\ \hline -6 - (-9) & 3 \\ -6 + 9 & 3 \\ 3 = 3 & \end{array}$$
Yes, -9 is a solution of the equation $-6 - t = 3$.

21. $-9 + 16 - (-7) = -9 + 16 + 7$
$= 7 + 7$
$= 14$

22. $\dfrac{0}{-17} = 0$

23. $-5(2)(-6)(-1) = -10(-6)(-1)$
$= 60(-1)$
$= -60$

24. $3 + (-9) + 4 + (-10) = -6 + 4 + (-10)$
$= -2 + (-10)$
$= -12$

25. $(a - b)^2 - 2a$
$(-2 - (-3))^2 - 2(-2)$
$= (-2 + 3)^2 - 2(-2)$
$= 1^2 - 2(-2)$
$= 1 - 2(-2)$
$= 1 - (-4)$
$= 1 + 4$
$= 5$

26. $-8 > -10$

27. $-21 + 21 = 0$

28. $|-27| = 27$

29. The unknown number: n

| forty-eight | is | the product of negative six and some number |

$48 = -6n$
$\dfrac{48}{-6} = \dfrac{-6n}{-6}$
$-8 = n$
The number is -8.

30. Strategy To find the colder temperature, compare the numbers -4 and -12. The smaller number represents the colder temperature.

Solution $-4 > -12$
The colder temperature is $-12°C$.

31. Strategy To find the boiling point of neon:
→ Find the highest boiling point shown in the table.
→ Multiply the highest boiling point by 7.

Solution The highest boiling point shown in the table is $-34°$.
$-34(7) = -238$
The boiling point of neon is $-238°C$.

32. Strategy To find the temperature, add the increase (5) to the previous temperature (-8).

Solution $-8 + 5 = -3$
The temperature is $-3°C$.

33. Strategy To find d, replace a by 7 and b by -5 in the given formula and solve for d.

Solution $d = |a - b|$
$d = |7 - (-5)|$
$d = |7 + 5|$
$d = |12|$
$d = 12$
The distance between the two points is 12 units.

Chapter Test

1. negative three plus negative five

2. $-|-34| = -34$

3. $3 - (-15) = 3 + 15$
   $= 18$

4. $a + b$
   $(-11) + (-9) = -20$

5. $(-x)(-y)$
   $(-(-4)) \cdot (-(-6)) = 4 \cdot 6$
   $= 24$

6. The Commutative Property of Addition

7. $-360 \div -30 = 12$

8. $-3 + -6 + 11 = -9 + 11$
   $= 2$

9. $16 > -19$

10. $7 - (-3) - 12 = 7 + 3 - 12$
    $= 10 - 12$
    $= -2$

11. $a - b - c$
    $= 6 - (-2) - 11 = 6 + 2 - 11$
    $= 8 - 11$
    $= -3$

12. $-(-49) = 49$

13. $50 \cdot (-5) = -250$

14. $-|5|, -(3), |-9|, -(-11)$

15. $\begin{array}{c|c} 17 - x = 8 \\ \hline 17 - (-9) & 8 \\ 17 + 9 & 8 \\ 26 \neq 8 \end{array}$
    No, -9 is not a solution of the equation $17 - x = 8$.

16. [number line from -6 to 6 with arrow of length 2 ending at -3]
    -3 is 2 units to the right of -5.

17. Strategy To find the difference in scores, Woods' score (-12) from Daly's score (4).

    Solution $4 - (-12) = 4 + 12$
    $= 16$
    The difference in scores was 16 strokes.

18. $\dfrac{0}{-16} = 0$

19. $2bc - (c - a)^3$
    $(2 \cdot 4 \cdot (-1)) - (-1 + (-2))^3 = -8 - (-3)^3$
    $= -8 - (-27)$
    $= -8 + 27$
    $= 19$

20. -25

21. $c - 11 = 5$
    $c - 11 + 11 = 5 + 11$
    $c = 16$
    The solution is 16.

22. $0 - 11 = -11$

23. $-96 \div (-4) = 24$

24. $16 \div 4 - 12 \div (-2) = 4 - (-6)$
    $= 4 + 6$
    $= 10$

25. $\dfrac{-x}{y}$
    $\dfrac{-(-56)}{-8} = \dfrac{56}{-8}$
    $= -7$

26. $3xy$
    $3 \cdot (-2) \cdot (-10) = 3 \cdot 20$
    $= 60$

27. $-11w = 121$
    $\dfrac{-11w}{-11} = \dfrac{121}{-11}$
    $w = -11$
    The solution is -11.

28. $4 - 14 = -10$

29. Strategy To find the temperature, add the increase (11) to the previous temperature (–6).

Solution $-6 + 11 = 5$
The temperature is 5°C.

30. The unknown number: n

| the wind chill at –25°F with a 40 mph wind | is | Four times the wind chill factor at –25°F with a 40 mph wind |

$n = 4 \cdot (-16)$
$n = -64$
The wind chill factor is –64°F.

31. Strategy To find the temperature from yesterday, add the increase (8) to today's temperature (–13).

Solution $-13 + 8 = -5$
The temperature is –5°C.

32. $d = |a - b|$
$d = |4 - (-12)| = |4 + 12|$
$= |16|$
$= 16$
The solution is 16 units.

33. Strategy To find the assets, substitute 18 for N and 6 for L in the given formula and solve for A.

Solution $N = A - L$
$18 = A - 6$
$18 + 6 = A - 6 + 6$
$24 = A$
The assets are worth $24 million.

Cumulative Review Exercises

1. $-27 - (-32) = -27 + 32$
$= 5$

2. $439 \to 400$
$28 \to 30$
$400 \cdot 30 = 12,000$

3. $\begin{array}{r} 3{,}209 \\ 6\overline{)19{,}254} \\ \underline{-18} \\ 12 \\ \underline{-12} \\ 5 \\ \underline{-0} \\ 54 \\ \underline{-54} \\ 0 \end{array}$

4. $16 \div (3 + 5) \cdot 9 - 2^4 = 16 \div 8 \cdot 9 - 2^4$
$= 16 \div 8 \cdot 9 - 16$
$= 2 \cdot 9 - 16$
$= 18 - 16$
$= 18 + (-16)$
$= 2$

5. $-|-82| = -82$

6. 309,480

7. $5xy$
$5 \cdot 80 \cdot 6 = 400 \cdot 6$
$= 2,400$

8. $-294 \div (-14) = 21$

9. $-28 - (-17) = -28 + 17$
$= -11$

10. $-24 + 16 + (-32) = -8 + (-32)$
$= -40$

11. $44 \div 1 = 44$
$44 \div 2 = 22$
$44 \div 4 = 11$
$44 \div 11 = 4$
The factors of 44 are 1, 2, 4, 11, 22, and 44.

12. $x^4 y^2$
$2^4 \cdot 11^2 = (2 \cdot 2 \cdot 2 \cdot 2) \cdot (11 \cdot 11)$
$= 16 \cdot 121$
$= 1,936$

13. $62\underset{\uparrow}{9},874$ Given place value
$8 > 5$

629,874 rounded to the nearest thousand is 630,000.

14. $356 \to 400$
$481 \to 500$
$294 \to 300$
$117 \to +100$
$\overline{1{,}300}$

15. $-a - b$
   $-(-4) - (-5) = 4 + 5$
   $= 9$

16. $-100 \cdot 25 = -2{,}500$

17. 23
    $3\overline{)69}$
    $69 = 3 \cdot 23$

18. $3x = -48$
    $\dfrac{3x}{3} = \dfrac{-48}{3}$
    $x = -16$
    The solution is -16.

19. $(1-5)^2 \div (-6+4) + 8(-3)$
    $= (-4)^2 \div (-2) + 8(-3) = 16 \div (-2) + 8(-3)$
    $= -8 + 8(-3)$
    $= -8 + (-24)$
    $= -32$

20. $-c \div d$
    $-(-32) \div (-8) = 32 \div (-8)$
    $= -4$

21. $\dfrac{a}{b}$
    $\dfrac{39}{-13} = -3$

22. $-62 < 26$

23. $-18(-7) = 126$

24. $12 + p = 3$
    $12 - 12 + p = 3 - 12$
    $p = -9$
    The solution is -9.

25. $2 \cdot 2 \cdot 2 \cdot 2 \cdot 2 \cdot 7 \cdot 7 = 2^5 \cdot 7^2$

26. $4a + (a-b)^3$
    $4(5) + (5-2)^3 = 4(5) + 3^3$
    $= 4(5) + 27$
    $= 20 + 27$
    $= 47$

27. $5{,}971$
    482
    $+3{,}609$
    $\overline{10{,}062}$

28. $-21 - 5 = -21 + (-5)$
    $= -26$

29. $7{,}352 \rightarrow 7{,}000$
    $1{,}986 \rightarrow \underline{-\,2{,}000}$
    $\phantom{1{,}986 \rightarrow -\,}5{,}000$

30. $3^4 \cdot 5^2 = (3 \cdot 3 \cdot 3 \cdot 3) \cdot (5 \cdot 5)$
    $= 81 \cdot 25$
    $= 2{,}025$

31. **Strategy**   To find the land area, add the land area prior to the purchase (891,364) to the amount of land purchased (831,321).

    Solution
    $891{,}364$
    $+\,831{,}321$
    $\overline{1{,}722{,}685}$

    The land area of the United States after the Louisiana purchase was $1{,}722{,}685 \text{ mi}^2$.

32. **Strategy**   To find the age, subtract the year of the birth (1879) from the year of his death (1955).

    Solution
    1955
    $-\,1879$
    $\overline{76}$

    Albert Einstein was 76 years old when he died.

33. **Strategy**   To find the amount, subtract the down payment (3,550) from the cost (17,750).

    Solution
    $17{,}750$
    $-\,\,\,3{,}550$
    $\overline{14{,}200}$

    The amount to be paid is $14,200.

34. **Strategy** To find the cost of the land, multiply the number of acres (25) times the cost per acre (3,690).

    Solution
    $$\begin{array}{r} 3{,}690 \\ \times\ 25 \\ \hline 18\ 450 \\ 73\ 80 \\ \hline 92{,}250 \end{array}$$
    The cost of the land is $92,250.

35. **Strategy** To find the temperature, add the increase (7) to the original temperature (−12).

    Solution $-12 + 7 = -5$
    The temperature is −5°C.

36. **Strategy**
    a. To find the difference in Arizona, subtract the record low (−40) from the record high (128).

    b. To find the greatest difference, subtract the record lows from the record highs for each state.

    Solution
    a. $128 - (-40) = 128 + 40$
       $= 168$
       The difference in temperatures is 168°F.

    b. $112 - (-27) = 112 + 27 = 139$
       $100 - (-80) = 100 + 80 = 180$
       $128 - (-40) = 128 + 40 = 168$
       $120 - (-29) = 120 + 29 = 149$
       $180 > 168 > 149 > 139$
       The greatest difference in temperatures is in Alaska.

37. **Strategy** To find the amount:
    → Add the sales figures for the first three quarters (28,550 + 34,850 + 31,700).
    → Subtract the sum from the goal for the year (120,000).

    Solution
    $$\begin{array}{r} 28{,}550 \\ 34{,}850 \\ +\ 31{,}700 \\ \hline 95{,}100 \end{array}$$
    $$\begin{array}{r} 120{,}000 \\ -\ 95{,}100 \\ \hline 24{,}900 \end{array}$$
    You must sell $24,900 in the last quarter to meet the goal.

38. **Strategy** To find the score, substitute 198 for N and 206 for P in the given formula and solve for S.

    Solution $S = N - P$
    $S = 198 - 206$
    $S = -8$
    The golfer's score is −8.

Chapter 3: Fractions

Prep Test

1. $4 \times 5 = 20$
2. $2 \cdot 2 \cdot 2 \cdot 3 \cdot 5 = 120$
3. $9 \times 1 = 9$
4. $-6 + 4 = -2$
5. $-10 - 3 = -13$
6. $63 \div 30 = 2 \text{ r } 3$
7. $24 \div 8 = 3$
   $24 \div 12 = 2$
   Both 8 and 12 divide evenly into 24.
8. $16 \div 4 = 4$
   $20 \div 4 = 5$
   4 divides into both 16 and 20.
9. $8 \times 7 + 3 = 56 + 3 = 59$
10. $8 = ? + 1$
    $8 = 7 + 1$
11. $44 < 48$

Go Figure

There are 7 children in the family.
Maria has twice as many brothers as sisters. She has 4 brothers (including Pedro) and 2 sisters.
Pedro has as many brothers as sisters. He has 3 sisters (including Maria) and 3 brothers.

Section 3.1

Objective A Exercises

1. $4 = 2^2$
   $8 = 2^3$
   The LCM $= 2^3 = 8$.

3. $2 = 2$
   $7 = 7$
   The LCM $= 2 \cdot 7 = 14$.

5. $6 = 2 \cdot 3$
   $10 = 2 \cdot 5$
   The LCM $= 2 \cdot 3 \cdot 5 = 30$.

7. $9 = 3^2$
   $15 = 3 \cdot 5$
   The LCM $= 3^2 \cdot 5 = 45$.

9. $12 = 2^2 \cdot 3$
   $16 = 2^4$
   The LCM $= 2^4 \cdot 3 = 48$.

11. $4 = 4^2$
    $10 = 2 \cdot 5$
    The LCM $= 2^2 \cdot 5 = 20$.

13. $14 = 2 \cdot 7$
    $42 = 2 \cdot 3 \cdot 7$
    The LCM $= 2 \cdot 3 \cdot 7 = 42$.

15. $24 = 2^3 \cdot 3$
    $36 = 2^2 \cdot 3^2$
    The LCM $= 2^3 \cdot 3^2 = 72$.

17. $30 = 2 \cdot 3 \cdot 5$
    $40 = 2^3 \cdot 5$
    The LCM $= 2^3 \cdot 3 \cdot 5 = 120$.

19. $3 = 3$
    $5 = 5$
    $10 = 2 \cdot 5$
    The LCM $= 2 \cdot 3 \cdot 5 = 30$.

21. $4 = 2^2$
    $8 = 2^3$
    $12 = 2^2 \cdot 3$
    The LCM $= 2^3 \cdot 3 = 24$.

23. $9 = 3^2$
    $36 = 2^2 \cdot 3^2$
    $45 = 5 \cdot 3^2$
    The LCM $= 2^2 \cdot 3^2 \cdot 5 = 180$.

25. $6 = 2 \cdot 3$
    $9 = 3^2$
    $15 = 3 \cdot 5$
    The LCM $= 2 \cdot 3^2 \cdot 5 = 90$.

27. $13 = 13$
    $26 = 2 \cdot 13$
    $39 = 3 \cdot 13$
    The LCM $= 2 \cdot 3 \cdot 13 = 78$.

Objective B Exercises

29. $9 = 3^2$
    $12 = 2^2 \cdot 3$
    The GCF $= 3$.

31. $18 = 2 \cdot 3^2$
    $30 = 2 \cdot 3 \cdot 5$
    The GCF = $2 \cdot 3 = 6$.

33. $14 = 2 \cdot 7$
    $42 = 2 \cdot 3 \cdot 7$
    The GCF = $2 \cdot 7 = 14$.

35. $16 = 2^4$
    $80 = 2^4 \cdot 5$
    The GCF = $2^4 = 16$.

37. $21 = 3 \cdot 7$
    $55 = 5 \cdot 11$
    The GCF = 1.

39. $8 = 2^3$
    $36 = 2^2 \cdot 3^2$
    The GCF = $2^2 = 4$.

41. $12 = 2^2 \cdot 3$
    $76 = 2^2 \cdot 19$
    The GCF = $2^2 = 4$.

43. $24 = 2^3 \cdot 3$
    $30 = 2 \cdot 3 \cdot 5$
    The GCF = $2 \cdot 3 = 6$.

45. $24 = 2^3 \cdot 3$
    $36 = 2^2 \cdot 3^2$
    The GCF = $2^2 \cdot 3 = 12$.

47. $45 = 3^2 \cdot 5$
    $75 = 3 \cdot 5^2$
    The GCF = $3 \cdot 5 = 15$.

49. $6 = 2 \cdot 3$
    $10 = 2 \cdot 5$
    $12 = 2^2 \cdot 3$
    The GCF = 2.

51. $6 = 2 \cdot 3$
    $15 = 3 \cdot 5$
    $36 = 2^2 \cdot 3^2$
    The GCF = 3.

53. $21 = 3 \cdot 7$
    $63 = 3^2 \cdot 7$
    $84 = 2^2 \cdot 3 \cdot 7$
    The GCF = $3 \cdot 7 = 21$.

55. $24 = 2^3 \cdot 3$
    $36 = 2^2 \cdot 3^2$
    $60 = 2^2 \cdot 3 \cdot 5$
    The GCF = $2^2 \cdot 3 = 12$.

Objective C Exercises

57. Strategy   To find how often the machines are starting to fill a box at the same time, find the LCM of 2 and 3.

    Solution $2 = 2$
               $3 = 3$
               The LCM = $2 \cdot 3 = 6$.
               Every 6 min, the machines will start to fill a box at the same time.

59. Strategy   To find the number of copies to be packaged together, find the GCF of 75, 100, and 150.

    Solution $75 = 3 \cdot 5^2$
               $100 = 2^2 \cdot 5^2$
               $150 = 2 \cdot 3 \cdot 5^2$
               The GCF = $5^2 = 25$.
               Each package should contain 25 copies of the magazine.

61. Strategy To find the time when all sessions begin again at the same time.
→Add the time of the break (10 min) to the 30-minute and the 40-minute session to find the time for each session including the break.
→Find the LCM of the two sessions found in step 1.
→Add the time found in step 2 to 9:00 A.M.

Solution
$30 + 10 = 40$
$40 + 10 = 50$
$40 = 2^3 \cdot 5$
$50 = 2 \cdot 5^2$
The LCM = $2^3 \cdot 5^2 = 200$.
200 min = 3 h 20 min
9 A.M. + 3 h 20 min
= 12:20 P.M.
All sessions will start at the same time at 12:20 P.M. Lunch should be scheduled at 12:20 P.M. if all participants are to eat at the same time.

Critical Thinking 3.1

63. $x = x$
$2x = 2 \cdot x$
The LCM = $2 \cdot x = 2x$.
The GCF = x.

65. Two numbers whose greatest common factor is 1 are said to be relatively prime. Students might use examples of relatively prime factors from the textbook; for example, the numbers 14 and 27 in Example 2 on page 153 are relatively prime. Examples also appear in the exercises on page 155 (see Exercises 37 and 38). Or students might list examples they have come up with independently.

Section 3.2

Objective A Exercises

1. $\frac{4}{5}$

3. $\frac{1}{4}$

5. $\frac{4}{3}, 1\frac{1}{3}$

7. $\frac{13}{5}, 2\frac{3}{5}$

9. $4\overline{)13}$ $\quad \frac{13}{4} = 3\frac{1}{4}$
$\underline{-12}$
$\quad 1$

11. $5\overline{)20}$ $\quad \frac{20}{5} = 4$
$\underline{-20}$
$\quad 0$

13. $10\overline{)27}$ $\quad \frac{27}{10} = 2\frac{7}{10}$
$\underline{-20}$
$\quad 7$

15. $8\overline{)56}$ $\quad \frac{56}{8} = 7$
$\underline{-56}$
$\quad 0$

17. $9\overline{)17}$ $\quad \frac{17}{9} = 1\frac{8}{9}$
$\underline{-9}$
$\quad 8$

19. $5\overline{)12}$ $\quad \frac{12}{5} = 2\frac{2}{5}$
$\underline{-10}$
$\quad 2$

21. $1\overline{)18}$ $\quad \frac{18}{1} = 18$
$\underline{-1}$
$\quad 8$
$\underline{-8}$
$\quad 0$

23. $15\overline{)32}$ $\quad \frac{32}{15} = 2\frac{2}{15}$
$\underline{-30}$
$\quad 2$

25. $\frac{8}{8} = 1$

27. $3\overline{)28}$ $\quad \frac{28}{3} = 9\frac{1}{3}$
$\underline{-27}$
$\quad 1$

29. $2\frac{1}{4} = \frac{(4 \cdot 2)+1}{4} = \frac{8+1}{4} = \frac{9}{4}$

31. $5\frac{1}{2} = \frac{(2 \cdot 5)+1}{2} = \frac{10+1}{2} = \frac{11}{2}$

33. $2\frac{4}{5} = \frac{(5 \cdot 2)+4}{5} = \frac{10+4}{5} = \frac{14}{5}$

35. $7\frac{5}{6} = \frac{(6 \cdot 7)+5}{6} = \frac{42+5}{6} = \frac{47}{6}$

37. $7 = \frac{7}{1}$

39. $8\frac{1}{4} = \frac{(4 \cdot 8)+1}{4} = \frac{32+1}{4} = \frac{33}{4}$

41. $10\frac{1}{3} = \frac{(3 \cdot 10)+1}{3} = \frac{30+1}{3} = \frac{31}{3}$

43. $4\frac{7}{12} = \frac{(12 \cdot 4)+7}{12} = \frac{48+7}{12} = \frac{55}{12}$

45. $8 = \frac{8}{1}$

47. $12\frac{4}{5} = \frac{(5 \cdot 12)+4}{5} = \frac{60+4}{5} = \frac{64}{5}$

Objective B Exercises

49. $12 \div 2 = 6$
$\frac{1}{2} = \frac{1 \cdot 6}{2 \cdot 6} = \frac{6}{12}$
$\frac{6}{12}$ is equivalent to $\frac{1}{2}$.

51. $24 \div 8 = 3$
$\frac{3}{8} = \frac{3 \cdot 3}{8 \cdot 3} = \frac{9}{24}$
$\frac{9}{24}$ is equivalent to $\frac{3}{8}$.

53. $51 \div 17 = 3$
$\frac{2}{17} = \frac{2 \cdot 3}{17 \cdot 3} = \frac{6}{51}$
$\frac{6}{51}$ is equivalent to $\frac{2}{17}$.

55. $32 \div 4 = 8$
$\frac{3}{4} = \frac{3 \cdot 8}{4 \cdot 8} = \frac{24}{32}$
$\frac{24}{32}$ is equivalent to $\frac{3}{4}$.

57. $18 \div 1 = 18$
$6 = \frac{6}{1} = \frac{6 \cdot 18}{1 \cdot 18} = \frac{108}{18}$
$\frac{108}{18}$ is equivalent to 6.

59. $90 \div 3 = 30$
$\frac{1}{3} = \frac{1 \cdot 30}{3 \cdot 30} = \frac{30}{90}$
$\frac{30}{90}$ is equivalent to $\frac{1}{3}$.

61. $21 \div 3 = 7$
$\frac{2}{3} = \frac{2 \cdot 7}{3 \cdot 7} = \frac{14}{21}$
$\frac{14}{21}$ is equivalent to $\frac{2}{3}$.

63. $49 \div 7 = 7$
$\frac{6}{7} = \frac{6 \cdot 7}{7 \cdot 7} = \frac{42}{49}$
$\frac{42}{49}$ is equivalent to $\frac{6}{7}$.

65. $18 \div 9 = 2$
$\frac{4}{9} = \frac{4 \cdot 2}{9 \cdot 2} = \frac{8}{18}$
$\frac{8}{18}$ is equivalent to $\frac{4}{9}$.

67. $4 \div 1 = 4$
$7 = \frac{7}{1} = \frac{7 \cdot 4}{1 \cdot 4} = \frac{28}{4}$
$\frac{28}{4}$ is equivalent to 7.

69. $\frac{3}{12} = \frac{3}{2 \cdot 2 \cdot 3} = \frac{1}{4}$

71. $\frac{33}{44} = \frac{3 \cdot 11}{2 \cdot 2 \cdot 11} = \frac{3}{4}$

73. $\frac{4}{24} = \frac{2 \cdot 2}{2 \cdot 2 \cdot 2 \cdot 3} = \frac{1}{6}$

75. $\frac{8}{33} = \frac{2 \cdot 2 \cdot 2}{3 \cdot 11} = \frac{8}{33}$

77. $\frac{0}{8} = 0$

79. $\frac{42}{36} = \frac{2 \cdot 3 \cdot 7}{2 \cdot 2 \cdot 3 \cdot 3} = \frac{7}{6}$

81. $\frac{16}{16} = 1$

83. $\frac{21}{35} = \frac{3 \cdot 7}{5 \cdot 7} = \frac{3}{5}$

85. $\frac{16}{60} = \frac{2 \cdot 2 \cdot 2 \cdot 2}{2 \cdot 2 \cdot 3 \cdot 5} = \frac{4}{15}$

87. $\frac{12}{20} = \frac{2 \cdot 2 \cdot 3}{2 \cdot 2 \cdot 5} = \frac{3}{5}$

89. $\frac{12m}{18} = \frac{\overset{1}{2} \cdot \overset{1}{2} \cdot 3 \cdot m}{\underset{1}{2} \cdot \underset{1}{3} \cdot 3} = \frac{2m}{3}$

91. $\frac{4y}{8} = \frac{\overset{1}{2} \cdot \overset{1}{2} \cdot y}{\underset{1}{2} \cdot \underset{1}{2} \cdot 2} = \frac{y}{2}$

93. $\frac{24a}{36} = \frac{\overset{1}{2} \cdot \overset{1}{2} \cdot 2 \cdot \overset{1}{3} \cdot a}{\underset{1}{2} \cdot \underset{1}{2} \cdot \underset{1}{3} \cdot 3} = \frac{2a}{3}$

95. $\frac{8c}{8} = \frac{\overset{1}{2} \cdot \overset{1}{2} \cdot \overset{1}{2} \cdot c}{\underset{1}{2} \cdot \underset{1}{2} \cdot \underset{1}{2}} = c$

97. $\frac{18k}{3} = \frac{2 \cdot 3 \cdot \overset{1}{3} \cdot k}{\underset{1}{3}} = 6k$

Objective C Exercises

99. $\frac{3}{8} = \frac{15}{40} \qquad \frac{2}{5} = \frac{16}{40}$
    $\frac{15}{40} < \frac{16}{40}$
    $\frac{3}{8} < \frac{2}{5}$

101. $\frac{3}{4} = \frac{27}{36} \qquad \frac{7}{9} = \frac{28}{36}$
     $\frac{27}{36} < \frac{28}{36}$
     $\frac{3}{4} < \frac{7}{9}$

103. $\frac{2}{3} = \frac{22}{33} \qquad \frac{7}{11} = \frac{21}{33}$
     $\frac{22}{33} > \frac{21}{33}$
     $\frac{2}{3} > \frac{7}{11}$

105. $\frac{17}{24} = \frac{34}{48} \qquad \frac{11}{16} = \frac{33}{48}$
     $\frac{34}{48} > \frac{33}{48}$
     $\frac{17}{24} > \frac{11}{16}$

107. $\frac{7}{15} = \frac{28}{60} \qquad \frac{5}{12} = \frac{25}{60}$
     $\frac{28}{60} > \frac{25}{60}$
     $\frac{7}{15} > \frac{5}{12}$

109. $\frac{5}{9} = \frac{35}{63} \qquad \frac{11}{21} = \frac{33}{63}$
     $\frac{35}{63} > \frac{33}{63}$
     $\frac{5}{9} > \frac{11}{21}$

111. $\frac{7}{12} = \frac{21}{36} \qquad \frac{13}{18} = \frac{26}{36}$
     $\frac{21}{36} < \frac{26}{36}$
     $\frac{7}{12} < \frac{13}{18}$

113. $\frac{4}{5} = \frac{36}{45} \qquad \frac{7}{9} = \frac{35}{45}$
     $\frac{36}{45} > \frac{35}{45}$
     $\frac{4}{5} > \frac{7}{9}$

115. $\frac{9}{16} = \frac{81}{144} \qquad \frac{5}{9} = \frac{80}{144}$
     $\frac{81}{144} > \frac{80}{144}$
     $\frac{9}{16} > \frac{5}{9}$

117. $\frac{5}{8} = \frac{50}{80} \qquad \frac{13}{20} = \frac{52}{80}$
     $\frac{50}{80} < \frac{52}{80}$
     $\frac{5}{8} < \frac{13}{20}$

Objective D Exercises

119. **Strategy**   To find the fractions, write a fraction with 250 in the numerator and 2,000 in the denominator. Write the fraction in simplest form.

       Solution $\frac{250}{2,000} = \frac{2 \cdot 5 \cdot 5 \cdot 5}{2 \cdot 2 \cdot 2 \cdot 2 \cdot 5 \cdot 5 \cdot 5} = \frac{1}{8}$

       250 lb is $\frac{1}{8}$ of a ton.

Section 3.2   47

121. Strategy To find the fractions, write a fraction with 50 in the numerator and 60 in the denominator. Write the fraction in simplest form.

Solution $\dfrac{50}{60} = \dfrac{2 \cdot 5 \cdot 5}{2 \cdot 2 \cdot 3 \cdot 5} = \dfrac{5}{6}$

The history class is $\dfrac{5}{6}$ of an hour.

123. Strategy To find the fractions, write a fraction with 18 in the numerator and 24 in the denominator. Write the fraction in simplest form.

Solution $\dfrac{18}{24} = \dfrac{2 \cdot 3 \cdot 3}{2 \cdot 2 \cdot 2 \cdot 3} = \dfrac{3}{4}$

18 karat gold is $\dfrac{3}{4}$ pure gold.

125. Strategy To find which criterion was cited by most people,
→Rewrite each fraction as a fraction with a denominator of 100.
→Compare all the fractions.

Solution
$\dfrac{1}{4} = \dfrac{1 \cdot 25}{4 \cdot 25} = \dfrac{25}{100}$

$\dfrac{13}{50} = \dfrac{13 \cdot 2}{50 \cdot 2} = \dfrac{26}{100}$

$\dfrac{4}{25} = \dfrac{4 \cdot 4}{25 \cdot 4} = \dfrac{16}{100}$

$\dfrac{2}{25} = \dfrac{2 \cdot 4}{25 \cdot 4} = \dfrac{8}{100}$

$\dfrac{3}{25} = \dfrac{3 \cdot 4}{25 \cdot 4} = \dfrac{12}{100}$

$\dfrac{26}{100} > \dfrac{25}{100} > \dfrac{16}{100} > \dfrac{12}{100} > \dfrac{8}{100} > \dfrac{3}{100}$

The location was the most cited criterion.

127. Strategy To determine if you answered more or less than $\dfrac{8}{10}$ of the questions correctly:

→Write a fraction with the number of questions answered correctly (42) in the numerator and the total number of questions (50) in the denominator.
→Rewrite $\dfrac{8}{10}$ as a fraction with a denominator of 50.
→Compare the two fractions.

Solution
$\dfrac{42}{50}$

$\dfrac{8}{10} = \dfrac{8 \cdot 5}{10 \cdot 5} = \dfrac{40}{50}$

$\dfrac{42}{50} > \dfrac{40}{50}$

You answered more than $\dfrac{8}{10}$ of the questions correctly.

© Houghton Mifflin Company. All rights reserved.

Chapter 3: Fractions

129. Strategy   To find which is a more serious problem,
→Rewrite the job market fraction $\frac{9}{25}$ as a fraction with a denominator of 100.
→Rewrite the quality of their children's education fraction $\frac{7}{20}$ as a fraction with a denominator of 100.
→Compare the two fractions.

Solution
$$\frac{9}{25} = \frac{9 \cdot 4}{25 \cdot 4} = \frac{36}{100}$$
$$\frac{7}{20} = \frac{7 \cdot 5}{20 \cdot 5} = \frac{35}{100}$$
$$\frac{36}{100} > \frac{35}{100}$$

The job market was a more serious problem than the quality of their children's education.

131. Strategy   To find which concern was cited by most people,
→Rewrite each fraction as a fraction with a denominator of 100.
→Compare all the fractions.

Solution
$$\frac{9}{25} = \frac{9 \cdot 4}{25 \cdot 4} = \frac{36}{100}$$
$$\frac{13}{50} = \frac{13 \cdot 2}{50 \cdot 2} = \frac{26}{100}$$
$$\frac{1}{2} = \frac{1 \cdot 50}{2 \cdot 50} = \frac{50}{100}$$
$$\frac{23}{50} = \frac{23 \cdot 2}{50 \cdot 2} = \frac{46}{100}$$
$$\frac{9}{25} = \frac{9 \cdot 4}{25 \cdot 4} = \frac{36}{100}$$
$$\frac{7}{20} = \frac{7 \cdot 5}{20 \cdot 5} = \frac{35}{100}$$
$$\frac{2}{5} = \frac{2 \cdot 20}{5 \cdot 20} = \frac{40}{100}$$
$$\frac{50}{100} > \frac{46}{100} > \frac{40}{100} > \frac{36}{100} = \frac{36}{100} > \frac{35}{100} > \frac{26}{100} > \frac{21}{100}$$

Having enough money to pay bills was the most cited serious problem.

133. Strategy   To determine the fraction:
→Find the total production by the six states by adding the amount produced in Maine (175), Michigan (75), New York (200), Ohio (100), Vermont (375), and Wisconsin (75).
→Write a fraction with the production in New York (200) in the numerator and total production in the denominator. Write the fraction in simplest form.

Solution
$175 + 75 + 200 + 100 + 375 + 75 = 1000$
$$\frac{200}{1000} = \frac{5 \cdot 5 \cdot 2 \cdot 2 \cdot 2}{5 \cdot 5 \cdot 5 \cdot 2 \cdot 2 \cdot 2} = \frac{1}{5}$$

$\frac{1}{5}$ of the maple syrup produced was in New York.

Section 3

135. **Strategy**
   a. To determine the fraction:
      → Find the total of the monthly expenses.
      → Write a fraction with the amount spent on entertainment in the numerator and the total of the monthly expenses in the denominator. Write the fraction in simplest form.

   b. To determine the fraction:
      → Find the total of the monthly expenses.
      → Write the fraction with the amount spent on taxes in the numerator and the total of the monthly expenses in the denominator. Write the fraction in simplest form.

   Solution
   a. $325 + 150 + 100 + 700 + 550 + 175 = 2{,}000$
      $$\frac{150}{2{,}000} = \frac{2 \cdot 3 \cdot 5 \cdot 5}{2 \cdot 2 \cdot 2 \cdot 2 \cdot 5 \cdot 5 \cdot 5} = \frac{3}{40}$$
      $\frac{3}{40}$ of the total monthly expenses was spent on entertainment.

   b. $325 + 150 + 100 + 700 + 550 + 175 = 2{,}000$
      $$\frac{175}{2{,}000} = \frac{5 \cdot 5 \cdot 7}{2 \cdot 2 \cdot 2 \cdot 2 \cdot 5 \cdot 5 \cdot 5} = \frac{7}{80}$$
      $\frac{7}{80}$ of the total monthly expenses was spent on taxes.

Critical Thinking 3.2

137. The number of squares crossed by the diagonal for m rows of n squares is $m + (n - 1)$.

139. a. Sales in the Northeast in 2005: $\frac{1}{3}$

   Sales in the Northeast in 2006: $\frac{3}{10}$

   $$\frac{1}{3} = \frac{1 \cdot 10}{3 \cdot 10} = \frac{10}{30} \qquad \frac{3}{10} = \frac{3 \cdot 3}{10 \cdot 3} = \frac{9}{30}$$
   $$\frac{10}{30} > \frac{9}{30}$$
   $$\frac{1}{3} > \frac{3}{10}$$

   Sales in the Northeast were a greater fraction of total sales in 2005.

   b. Sales in the Northwest in 2005: $\frac{1}{6}$

   Sales in the Northwest in 2006: $\frac{1}{5}$

   $$\frac{1}{6} = \frac{1 \cdot 5}{6 \cdot 5} = \frac{5}{30} \qquad \frac{1}{5} = \frac{1 \cdot 6}{5 \cdot 6} = \frac{6}{30}$$
   $$\frac{6}{30} > \frac{5}{30}$$
   $$\frac{1}{5} > \frac{1}{6}$$

   Sales in the Northwest were a greater fraction of total sales in 2006.

141. We know that there are 1003 terms in this expression.
$$\frac{2}{1003} \cdot \frac{4}{1002} \cdot \frac{6}{1001} \cdots \frac{2002}{3} \cdot \frac{2004}{2} \cdot \frac{2006}{1}$$
Rewrite the expression so that the denominators are in increasing order
$$\frac{2}{1} \cdot \frac{4}{2} \cdot \frac{6}{3} \cdots \frac{2002}{1001} \cdot \frac{2004}{1002} \cdot \frac{2006}{1003}$$
Now it is easy to see that each term of the expression containing 1003 terms can be reduced to 2.
$$2 \cdot 2 \cdot 2 \cdots 2 \cdot 2 \cdot 2 = 2^{1003}$$

Section 3.3

Objective A Exercises

1. $\frac{4}{11} + \frac{5}{11} = \frac{4+5}{11} = \frac{9}{11}$

3. $\frac{2}{3} + \frac{1}{3} = \frac{2+1}{3}$
   $= \frac{3}{3} = 1$

5. $\frac{5}{6} + \frac{5}{6} = \frac{5+5}{6}$
   $= \frac{10}{6} = \frac{5}{3} = 1\frac{2}{3}$

7. $\frac{7}{18} + \frac{13}{18} + \frac{1}{18} = \frac{7+13+1}{18}$
   $= \frac{21}{18} = \frac{7}{6} = 1\frac{1}{6}$

9. $\frac{7}{b} + \frac{9}{b} = \frac{7+9}{b}$
   $= \frac{16}{b}$

11. $\frac{5}{c} + \frac{4}{c} = \frac{5+4}{c}$
    $= \frac{9}{c}$

13. $\frac{1}{x} + \frac{4}{x} + \frac{6}{x} = \frac{1+4+6}{x}$
    $= \frac{11}{x}$

15. $\frac{1}{4} + \frac{2}{3} = \frac{3}{12} + \frac{8}{12}$
    $= \frac{3+8}{12} = \frac{11}{12}$

17. $\frac{7}{15} + \frac{9}{20} = \frac{28}{60} + \frac{27}{60}$
    $= \frac{28+27}{60} = \frac{55}{60} = \frac{11}{12}$

19. $\frac{2}{3} + \frac{1}{12} + \frac{5}{6} = \frac{8}{12} + \frac{1}{12} + \frac{10}{12}$
    $= \frac{8+1+10}{12}$
    $= \frac{19}{12} = 1\frac{7}{12}$

21. $\frac{7}{12} + \frac{3}{4} + \frac{4}{5} = \frac{35}{60} + \frac{45}{60} + \frac{48}{60}$
    $= \frac{35+45+48}{60}$
    $= \frac{128}{60} = \frac{32}{15} = 2\frac{2}{15}$

23. $-\frac{3}{4} + \frac{2}{3} = \frac{-3}{4} + \frac{2}{3}$
    $= \frac{-9}{12} + \frac{8}{12}$
    $= \frac{-9+8}{12} = \frac{-1}{12} = -\frac{1}{12}$

25. $\frac{2}{5} + \left(-\frac{11}{15}\right) = \frac{2}{5} + \frac{-11}{15}$
    $= \frac{6}{15} + \frac{-11}{15}$
    $= \frac{6+(-11)}{15} = \frac{-5}{15} = -\frac{1}{3}$

27. $\frac{3}{8} + \left(-\frac{1}{2}\right) + \frac{7}{12} = \frac{3}{8} + \frac{-1}{2} + \frac{7}{12}$
    $= \frac{9}{24} + \frac{-12}{24} + \frac{14}{24}$
    $= \frac{9+(-12)+14}{24}$
    $= \frac{11}{24}$

29. $\frac{2}{3} + \left(-\frac{5}{6}\right) + \frac{1}{4} = \frac{2}{3} + \frac{-5}{6} + \frac{1}{4}$
    $= \frac{16}{24} + \frac{-20}{24} + \frac{6}{24}$
    $= \frac{16+(-20)+6}{24}$
    $= \frac{2}{24} = \frac{1}{12}$

31. $8 + 7\frac{2}{3} = 15\frac{2}{3}$

33. $2\frac{1}{6} + 3\frac{1}{2} = 2\frac{1}{6} + 3\frac{3}{6}$
    $= 5\frac{4}{6} = 5\frac{2}{3}$

35. $8\frac{3}{5} + 6\frac{9}{20} = 8\frac{12}{20} + 6\frac{9}{20}$
$= 14\frac{21}{20} = 15\frac{1}{20}$

37. $5\frac{5}{12} + 4\frac{7}{9} = 5\frac{15}{36} + 4\frac{28}{36}$
$= 9\frac{43}{36} = 10\frac{7}{36}$

39. $2\frac{1}{4} + 3\frac{1}{2} + 1\frac{2}{3}$
$= 2\frac{3}{12} + 3\frac{6}{12} + 1\frac{8}{12}$
$= 6\frac{17}{12} = 7\frac{5}{12}$

41. $-\frac{5}{6} + \frac{4}{9} = \frac{-5}{6} + \frac{4}{9}$
$= \frac{-15}{18} + \frac{8}{18}$
$= \frac{-15+8}{18} = \frac{-7}{18} = -\frac{7}{18}$

43. $\frac{2}{7} + \frac{3}{14} + \frac{1}{4} = \frac{8}{28} + \frac{6}{28} + \frac{7}{28}$
$= \frac{8+6+7}{28}$
$= \frac{21}{28} = \frac{3}{4}$

45. $-\frac{5}{6} + \left(-\frac{2}{3}\right) = \frac{-5}{6} + \frac{-2}{3}$
$= \frac{-5}{6} + \frac{-4}{6}$
$= \frac{-5+(-4)}{6} = \frac{-9}{6}$
$= -\frac{3}{2} = -1\frac{1}{2}$

47. $3\frac{7}{12} + 2\frac{5}{8} = 3\frac{14}{24} + 2\frac{15}{24}$
$= 5\frac{29}{24} = 6\frac{5}{24}$

49. $\frac{7}{8} + 1\frac{1}{3} = \frac{21}{24} + 1\frac{8}{24}$
$= 1\frac{29}{24} = 2\frac{5}{24}$

51. $x + y$
$\frac{3}{5} + \frac{4}{5} = \frac{3+4}{5}$
$= \frac{7}{5} = 1\frac{2}{5}$

53. $x + y$
$\frac{2}{3} + \left(-\frac{3}{4}\right) = \frac{2}{3} + \frac{-3}{4}$
$= \frac{8}{12} + \frac{-9}{12}$
$= \frac{8+(-9)}{12} = \frac{-1}{12} = -\frac{1}{12}$

55. $x + y$
$\frac{5}{6} + \frac{8}{9} = \frac{15}{18} + \frac{16}{18}$
$= \frac{15+16}{18} = \frac{31}{18} = 1\frac{13}{18}$

57. $x + y$
$-\frac{5}{8} + \left(-\frac{1}{6}\right) = \frac{-5}{8} + \frac{-1}{6}$
$= \frac{-15}{24} + \frac{-4}{24}$
$= \frac{-15+(-4)}{24}$
$= \frac{-19}{24} = -\frac{19}{24}$

59. $x + y + z$
$\frac{3}{8} + \frac{1}{4} + \frac{7}{12} = \frac{9}{24} + \frac{6}{24} + \frac{14}{24}$
$= \frac{9+6+14}{24}$
$= \frac{29}{24} = 1\frac{5}{24}$

61. $x + y + z$
$1\frac{1}{2} + 3\frac{3}{4} + 6\frac{5}{12} = 1\frac{6}{12} + 3\frac{9}{12} + 6\frac{5}{12}$
$= 10\frac{20}{12} = 10\frac{5}{3} = 11\frac{2}{3}$

63. $x + y + z$
$4\frac{3}{5} + 8\frac{7}{10} + 1\frac{9}{20}$
$= 4\frac{12}{20} + 8\frac{14}{20} + 1\frac{9}{20}$
$= 13\frac{35}{20} = 13\frac{7}{4} = 14\frac{3}{4}$

65.
$$z + \frac{1}{4} = -\frac{7}{20}$$

$-\frac{3}{5} + \frac{1}{4}$	$-\frac{7}{20}$
$-\frac{12}{20} + \frac{5}{20}$	$-\frac{7}{20}$
$\frac{-12+5}{20}$	$-\frac{7}{20}$
$-\frac{7}{20}$	$= -\frac{7}{20}$

Yes, $-\frac{3}{5}$ is a solution of the equation.

67. $\frac{1}{4} + x = -\frac{7}{12}$

$$\begin{array}{c|c} \frac{1}{4}+\left(-\frac{5}{6}\right) & -\frac{7}{12} \\ \frac{3}{12}+\left(-\frac{10}{12}\right) & -\frac{7}{12} \\ \frac{3+(-10)}{12} & -\frac{7}{12} \\ -\frac{7}{12} = -\frac{7}{12} \end{array}$$

Yes, $-\frac{5}{6}$ is a solution of the equation.

69. $\frac{19}{50} + \frac{6}{25} = \frac{19}{50} + \frac{12}{50} = \frac{31}{50}$

$\frac{31}{50}$ of money borrowed on home-equity loans are spent on debt consolidation and home improvement.

Objective B Exercises

71. $\frac{7}{12} - \frac{5}{12} = \frac{7-5}{12} = \frac{2}{12} = \frac{1}{6}$

73. $\frac{11}{24} - \frac{7}{24} = \frac{11-7}{24} = \frac{4}{24} = \frac{1}{6}$

75. $\frac{8}{d} - \frac{3}{d} = \frac{8-3}{d} = \frac{5}{d}$

77. $\frac{5}{n} - \frac{10}{n} = \frac{5-10}{n} = \frac{-5}{n} = -\frac{5}{n}$

79. $\frac{3}{7} - \frac{5}{14} = \frac{6}{14} - \frac{5}{14}$
$= \frac{6-5}{14} = \frac{1}{14}$

81. $\frac{2}{3} - \frac{1}{6} = \frac{4}{6} - \frac{1}{6}$
$= \frac{4-1}{6} = \frac{3}{6} = \frac{1}{2}$

83. $\frac{11}{12} - \frac{2}{3} = \frac{11}{12} - \frac{8}{12}$
$= \frac{11-8}{12} = \frac{3}{12} = \frac{1}{4}$

85. $-\frac{1}{2} - \frac{3}{8} = \frac{-1}{2} - \frac{3}{8}$
$= \frac{-4}{8} - \frac{3}{8}$
$= \frac{-4-3}{8} = \frac{-7}{8} = -\frac{7}{8}$

87. $-\frac{3}{10} - \frac{4}{5} = \frac{-3}{10} - \frac{4}{5}$
$= \frac{-3}{10} - \frac{8}{10}$
$= \frac{-3-8}{10} = \frac{-11}{10}$
$= -1\frac{1}{10}$

89. $-\frac{5}{12} - \left(-\frac{2}{3}\right) = \frac{-5}{12} - \frac{-2}{3}$
$= \frac{-5}{12} - \frac{-8}{12}$
$= \frac{-5-(-8)}{12}$
$= \frac{-5+8}{12} = \frac{3}{12} = \frac{1}{4}$

91. $-\frac{5}{9} - \left(-\frac{11}{12}\right) = \frac{-5}{9} - \frac{-11}{12}$
$= \frac{-20}{36} - \frac{-33}{36}$
$= \frac{-20-(-33)}{36}$
$= \frac{-20+33}{36} = \frac{13}{36}$

93. $4\frac{11}{18} - 2\frac{5}{18} = 2\frac{6}{18} = 2\frac{1}{3}$

95. $8\frac{3}{4} - 2 = 6\frac{3}{4}$

97. $8\frac{5}{6} - 7\frac{3}{4} = 8\frac{10}{12} - 7\frac{9}{12}$
$= 1\frac{1}{12}$

99. $7 - 3\frac{5}{8} = 6\frac{8}{8} - 3\frac{5}{8}$
$= 3\frac{3}{8}$

101. $10 - 4\frac{8}{9} = 9\frac{9}{9} - 4\frac{8}{9}$
$= 5\frac{1}{9}$

103. $7\frac{3}{8} - 4\frac{5}{8} = 6\frac{11}{8} - 4\frac{5}{8}$
$= 2\frac{6}{8} = 2\frac{3}{4}$

105. $12\frac{5}{12} - 10\frac{17}{24} = 12\frac{10}{24} - 10\frac{17}{24}$
$= 11\frac{34}{24} - 10\frac{17}{24}$
$= 1\frac{17}{24}$

Section 3.3

107. $6\frac{2}{3} - 1\frac{7}{8} = 6\frac{16}{24} - 1\frac{21}{24}$
$= 5\frac{40}{24} - 1\frac{21}{24}$
$= 4\frac{19}{24}$

109. $10\frac{2}{5} - 8\frac{7}{10} = 10\frac{4}{10} - 8\frac{7}{10}$
$= 9\frac{14}{10} - 8\frac{7}{10}$
$= 1\frac{7}{10}$

111. $-\frac{7}{12} - \frac{7}{9} = \frac{-7}{12} - \frac{7}{9}$
$= \frac{-21}{36} - \frac{28}{36}$
$= \frac{-21 - 28}{36} = \frac{-49}{36} = -1\frac{13}{36}$

113. $-\frac{7}{8} - \left(-\frac{2}{3}\right) = \frac{-7}{8} - \frac{-2}{3}$
$= \frac{-21}{24} - \frac{-16}{24}$
$= \frac{-21 - (-16)}{24}$
$= \frac{-21 + 16}{24} = \frac{-5}{24} = -\frac{5}{24}$

115. $8 - 1\frac{7}{12} = 7\frac{12}{12} - 1\frac{7}{12}$
$= 6\frac{5}{12}$

117. $x - y$
$\frac{8}{9} - \frac{5}{9} = \frac{8-5}{9} = \frac{3}{9} = \frac{1}{3}$

119. $x - y$
$-\frac{11}{12} - \frac{5}{12} = \frac{-11}{12} - \frac{5}{12}$
$= \frac{-11 - 5}{12}$
$= \frac{-16}{12} = \frac{-4}{3} = -1\frac{1}{3}$

121. $x - y$
$-\frac{2}{3} - \left(-\frac{3}{4}\right) = \frac{-2}{3} - \frac{-3}{4}$
$= \frac{-8}{12} - \frac{-9}{12}$
$= \frac{-8 - (-9)}{12}$
$= \frac{-8 + 9}{12} = \frac{1}{12}$

123. $x - y$
$-\frac{3}{10} - \left(-\frac{7}{15}\right) = \frac{-3}{10} - \frac{-7}{15}$
$= \frac{-9}{30} - \frac{-14}{30}$
$= \frac{-9 - (-14)}{30}$
$= \frac{-9 + 14}{30} = \frac{5}{30} = \frac{1}{6}$

125. $x - y$
$5\frac{7}{9} - 4\frac{2}{3} = 5\frac{7}{9} - 4\frac{6}{9}$
$= 1\frac{1}{9}$

127. $x - y$
$7\frac{9}{10} - 3\frac{1}{2} = 7\frac{9}{10} - 3\frac{5}{10}$
$= 4\frac{4}{10} = 4\frac{2}{5}$

129. $x - y$
$5 - 2\frac{7}{9} = 4\frac{9}{9} - 2\frac{7}{9}$
$= 2\frac{2}{9}$

131. $x - y$
$10\frac{1}{2} - 5\frac{7}{12} = 10\frac{6}{12} - 5\frac{7}{12}$
$= 9\frac{18}{12} - 5\frac{7}{12} = 4\frac{11}{12}$

133. $\frac{4}{5} = \frac{31}{20} - y$

$\frac{4}{5}$	$\frac{31}{20} - \left(-\frac{3}{4}\right)$
$\frac{4}{5}$	$\frac{31}{20} - \left(-\frac{15}{20}\right)$
$\frac{4}{5}$	$\frac{31 - (-15)}{20}$
$\frac{4}{5}$	$\frac{31 + 15}{20}$
$\frac{4}{5}$	$\frac{46}{20}$
$\frac{4}{5}$	$\frac{23}{10}$

$\frac{8}{10} \neq \frac{23}{10}$

No, $-\frac{3}{4}$ is not a solution of the equation.

135.
$$x - \frac{1}{4} = -\frac{17}{20}$$

$$\begin{array}{c|c} -\frac{3}{5} - \frac{1}{4} & -\frac{17}{20} \\ -\frac{12}{20} - \frac{5}{20} & -\frac{17}{20} \\ \frac{-12-5}{20} & -\frac{17}{20} \\ -\frac{17}{20} = -\frac{17}{20} \end{array}$$

Yes, $-\frac{3}{5}$ is a solution of the equation.

137. $\frac{12}{25} - \frac{8}{75} = \frac{36}{75} - \frac{8}{75} = \frac{28}{75}$

The difference between the fraction of personal income taxes and corporate income taxes is $\frac{28}{75}$.

Objective C Exercises

139. Strategy   To find the amount of property owned, subtract the amount sold $\left(1\frac{1}{2}\right)$ from the amount originally owned $\left(3\frac{1}{4}\right)$.

    Solution
$$3\frac{1}{4} - 1\frac{1}{2} = 3\frac{1}{4} - 1\frac{2}{4}$$
$$= 2\frac{5}{4} - 1\frac{2}{4}$$
$$= 1\frac{3}{4}$$

You now own $1\frac{3}{4}$ acres.

141. Strategy   To find the number of hours still required, subtract the number of hours already contributed $\left(12\frac{1}{4}\right)$ from the original amount required (20).

    Solution
$$20 - 12\frac{1}{4} = 19\frac{4}{4} - 12\frac{1}{4}$$
$$= 7\frac{3}{4}$$

$7\frac{3}{4}$ h of community service are still required.

143. Strategy   To find the amount of weight to gain:
→Add the amounts already gained $\left(4\frac{1}{2} + 3\frac{3}{4}\right)$.
→Subtract the amount gained from the goal amount (15).

    Solution
$$4\frac{1}{2} + 3\frac{3}{4} = 4\frac{2}{4} + 3\frac{3}{4} = 7\frac{5}{4} = 8\frac{1}{4}$$
$$15 - 8\frac{1}{4} = 14\frac{4}{4} - 8\frac{1}{4} = 6\frac{3}{4}$$

The boxer has $6\frac{3}{4}$ lb left to gain.

145. Strategy
    a. To find which response was given most frequently, compare all the responses.

    b. To find the fraction, add the fractions of those who cook $0\left(\frac{2}{25}\right)$, $1\left(\frac{1}{20}\right)$ and $2\left(\frac{1}{10}\right)$ dinners at home.

    c. To find the fraction, add the fractions of those who cook $5\left(\frac{21}{100}\right)$, $6\left(\frac{9}{100}\right)$, and $7\left(\frac{19}{100}\right)$ dinners at home. Then compare the result to $\left(\frac{1}{2}\right)$.

Solution

a.
$$\frac{2}{25} = \frac{8}{100}$$
$$\frac{1}{20} = \frac{5}{100}$$
$$\frac{1}{10} = \frac{10}{100}$$
$$\frac{3}{20} = \frac{15}{100}$$
$$\frac{21}{100} > \frac{19}{100} > \frac{15}{100} > \frac{13}{100} > \frac{10}{100} > \frac{9}{100} > \frac{8}{100} > \frac{5}{100}$$

The response that was given most frequently was 5 dinners at home per week.

b.
$$\frac{2}{25} + \frac{1}{20} + \frac{1}{10} = \frac{8}{100} + \frac{5}{100} + \frac{10}{100} = \frac{23}{100}$$

The fraction of the adult population cooking two or fewer dinners is $\frac{23}{100}$.

c.
$$\frac{21}{100} + \frac{9}{100} + \frac{19}{100} = \frac{49}{100}$$
$$\frac{1}{2} = \frac{50}{100}$$
$$\frac{49}{100} < \frac{50}{100}$$

The fraction of the adult population cooking five or more dinners is $\frac{49}{100}$. This is less than half of the people.

147. Strategy   To find the difference in heights, subtract the low 1980s $\left(\frac{5}{32}\right)$ from the high in 1950s $\left(\frac{1}{4}\right)$.

    Solution $\frac{1}{4} - \frac{5}{32} = \frac{8}{32} - \frac{5}{32} = \frac{3}{32}$

The difference in heights between 1950s and 1980s was $\frac{3}{32}$ in.

149. Strategy To find the feet of fencing, replace a, b, and c with $6\frac{1}{4}$, $10\frac{3}{4}$, and $12\frac{1}{2}$ in the given formula and solve for P.

Solution $P = a + b + c$
$$P = 6\frac{1}{4} + 10\frac{3}{4} + 12\frac{1}{2} = 6\frac{1}{4} + 10\frac{3}{4} + 12\frac{2}{4} = 29\frac{1}{2}$$
The number of feet of fencing needed is $29\frac{1}{2}$ ft.

151. Strategy To find length of wood beams, replace a, b, and c with $25\frac{3}{4}$, $12\frac{1}{2}$, and $17\frac{1}{2}$ in the given formula and solve for P.

Solution $P = a + b + c$
$$P = 25\frac{3}{4} + 12\frac{1}{2} + 17\frac{1}{2} = 25\frac{3}{4} + 12\frac{2}{4} + 17\frac{2}{4} = 55\frac{3}{4}$$
The total length of wood beams needed is $55\frac{3}{4}$ ft.

153. Strategy To find the difference, subtract Harold Osborn's jump from Charles Austin's jump.

Solution $7\frac{5}{6} - 6\frac{1}{2} = 7\frac{5}{6} - 6\frac{3}{6} = 1\frac{2}{6} = 1\frac{1}{3}$
The difference was $1\frac{1}{3}$ ft.

Critical Thinking 3.3

155. No, because the parts are not equal in size.

157. $\frac{1}{8}$ is 3 squares. $\frac{5}{6}$ is 20 squares.

No, $\frac{1}{8} + \frac{5}{6}$ of the rectangle is not 24 squares.

When the figure contains 48 squares, $\frac{1}{8}$ is 6 squares. $\frac{5}{6}$ is 40 squares. If there were 16 squares, $\frac{1}{8}$ and $\frac{5}{6}$ could not be illustrated (unless fractions of squares were used).

The possible number of squares that could be used to illustrate the sum of $\frac{1}{8}$ and $\frac{5}{6}$ include 24, 48, 72, 96, ... The list includes all the multiples of 24, or the common multiples of 6 and 8.

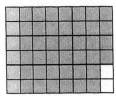

159. It is possible for the sum of the fractions to be greater than $\frac{3}{5}$ because there may be some households that have more than one type of pet, for example, a bird and a dog. Households with more than one type of pet would be represented in more than one of the fractions.

Section 3.4

Objective A Exercises

1. The idea of needing a common denominator when adding two fractions is similar to adding objects in the "real world." The sum of 3 apples and 5 oranges is 3 apples and 5 oranges. We cannot add unlike objects. The sum of 3 apples and 5 apples is 8 apples. We can add like objects. The sum of 3 ninths and 5 ninths is 8 ninths; we must have the same denominator in order to add the fractions.

   We don't need a common denominator when multiplying two fractions. Multiplication does not require that we have like objects. One-half of 6 apples is 3 apples. One-half of one-quarter is one-eighth of the pizza.

3. $\dfrac{2}{3} \cdot \dfrac{9}{10} = \dfrac{2 \cdot 9}{3 \cdot 10}$
   $= \dfrac{2 \cdot 3 \cdot 3}{3 \cdot 2 \cdot 5}$
   $= \dfrac{3}{5}$

5. $-\dfrac{6}{7} \cdot \dfrac{11}{12} = -\dfrac{6 \cdot 11}{7 \cdot 12}$
   $= -\dfrac{2 \cdot 3 \cdot 11}{7 \cdot 2 \cdot 2 \cdot 3}$
   $= -\dfrac{11}{14}$

7. $\dfrac{14}{15} \cdot \dfrac{6}{7} = \dfrac{14 \cdot 6}{15 \cdot 7}$
   $= \dfrac{2 \cdot 7 \cdot 2 \cdot 3}{3 \cdot 5 \cdot 7}$
   $= \dfrac{4}{5}$

9. $-\dfrac{6}{7} \cdot \dfrac{0}{10} = -\left(\dfrac{6}{7} \cdot \dfrac{0}{10}\right)$
   $= -\dfrac{6 \cdot 0}{7 \cdot 10}$
   $= -\dfrac{0}{70} = 0$

11. $\left(-\dfrac{4}{15}\right) \cdot \left(-\dfrac{3}{8}\right) = \dfrac{4}{15} \cdot \dfrac{3}{8}$
    $= \dfrac{4 \cdot 3}{15 \cdot 8}$
    $= \dfrac{2 \cdot 2 \cdot 3}{3 \cdot 5 \cdot 2 \cdot 2 \cdot 2}$
    $= \dfrac{1}{10}$

13. $-\dfrac{3}{4} \cdot \dfrac{1}{2} = -\left(\dfrac{3}{4} \cdot \dfrac{1}{2}\right)$
    $= -\dfrac{3 \cdot 1}{4 \cdot 2}$
    $= -\dfrac{3 \cdot 1}{2 \cdot 2 \cdot 2}$
    $= -\dfrac{3}{8}$

15. $\dfrac{9}{x} \cdot \dfrac{7}{y} = \dfrac{9 \cdot 7}{x \cdot y}$
    $= \dfrac{63}{xy}$

17. $-\dfrac{y}{5} \cdot \dfrac{z}{6} = -\left(\dfrac{y}{5} \cdot \dfrac{z}{6}\right)$
    $= -\dfrac{y \cdot z}{5 \cdot 6}$
    $= -\dfrac{yz}{30}$

19. $\dfrac{2}{3} \cdot \dfrac{3}{8} \cdot \dfrac{4}{9} = \dfrac{2 \cdot 3 \cdot 4}{3 \cdot 8 \cdot 9}$
    $= \dfrac{2 \cdot 3 \cdot 2 \cdot 2}{3 \cdot 2 \cdot 2 \cdot 2 \cdot 3 \cdot 3}$
    $= \dfrac{1}{9}$

21. $-\dfrac{7}{12} \cdot \dfrac{5}{8} \cdot \dfrac{16}{25} = -\left(\dfrac{7}{12} \cdot \dfrac{5}{8} \cdot \dfrac{16}{25}\right)$
    $= -\dfrac{7 \cdot 5 \cdot 16}{12 \cdot 8 \cdot 25}$
    $= -\dfrac{7 \cdot 5 \cdot 2 \cdot 2 \cdot 2 \cdot 2}{2 \cdot 2 \cdot 3 \cdot 2 \cdot 2 \cdot 2 \cdot 5 \cdot 5}$
    $= -\dfrac{7}{30}$

23. $\left(-\dfrac{3}{5}\right) \cdot \dfrac{1}{2} \cdot \left(-\dfrac{5}{8}\right) = \dfrac{3}{5} \cdot \dfrac{1}{2} \cdot \dfrac{5}{8}$
    $= \dfrac{3 \cdot 1 \cdot 5}{5 \cdot 2 \cdot 8}$
    $= \dfrac{3 \cdot 1 \cdot 5}{5 \cdot 2 \cdot 2 \cdot 2 \cdot 2}$
    $= \dfrac{3}{16}$

25. $6 \cdot \dfrac{1}{6} = \dfrac{6}{1} \cdot \dfrac{1}{6}$
    $= \dfrac{6 \cdot 1}{1 \cdot 6}$
    $= \dfrac{6}{6} = 1$

Chapter 3: Fractions

27. $\dfrac{3}{4} \cdot 8 = \dfrac{3}{4} \cdot \dfrac{8}{1}$
$= \dfrac{3 \cdot 8}{4 \cdot 1}$
$= \dfrac{3 \cdot 2 \cdot 2 \cdot 2}{2 \cdot 2 \cdot 1} = \dfrac{6}{1} = 6$

29. $12 \cdot \left(-\dfrac{5}{8}\right) = -\left(12 \cdot \dfrac{5}{8}\right)$
$= -\left(\dfrac{12}{1} \cdot \dfrac{5}{8}\right) = -\dfrac{12 \cdot 5}{1 \cdot 8}$
$= -\dfrac{2 \cdot 2 \cdot 3 \cdot 5}{1 \cdot 2 \cdot 2 \cdot 2}$
$= -\dfrac{15}{2} = -7\dfrac{1}{2}$

31. $-16 \cdot \dfrac{7}{30} = -\left(\dfrac{16}{1} \cdot \dfrac{7}{30}\right)$
$= -\dfrac{16 \cdot 7}{1 \cdot 30}$
$= -\dfrac{2 \cdot 2 \cdot 2 \cdot 2 \cdot 7}{1 \cdot 2 \cdot 3 \cdot 5}$
$= -\dfrac{56}{15} = -3\dfrac{11}{15}$

33. $\dfrac{6}{7} \cdot 0 = 0$

35. $\dfrac{5}{22} \cdot 2\dfrac{1}{5} = \dfrac{5}{22} \cdot \dfrac{11}{5}$
$= \dfrac{5 \cdot 11}{22 \cdot 5}$
$= \dfrac{5 \cdot 11}{2 \cdot 11 \cdot 5}$
$= \dfrac{1}{2}$

37. $3\dfrac{1}{2} \cdot 5\dfrac{3}{7} = \dfrac{7}{2} \cdot \dfrac{38}{7}$
$= \dfrac{7 \cdot 38}{2 \cdot 7}$
$= \dfrac{7 \cdot 2 \cdot 19}{2 \cdot 7}$
$= \dfrac{19}{1} = 19$

39. $3\dfrac{1}{3} \cdot \left(-\dfrac{7}{10}\right) = -\left(\dfrac{10}{3} \cdot \dfrac{7}{10}\right)$
$= -\dfrac{10 \cdot 7}{3 \cdot 10}$
$= -\dfrac{2 \cdot 5 \cdot 7}{3 \cdot 2 \cdot 5}$
$= -\dfrac{7}{3} = -2\dfrac{1}{3}$

41. $-1\dfrac{2}{3} \cdot \left(-\dfrac{3}{5}\right) = \dfrac{5}{3} \cdot \dfrac{3}{5}$
$= \dfrac{5 \cdot 3}{3 \cdot 5}$
$= 1$

43. $3\dfrac{1}{3} \cdot 2\dfrac{1}{3} = \dfrac{10}{3} \cdot \dfrac{7}{3}$
$= \dfrac{10 \cdot 7}{3 \cdot 3}$
$= \dfrac{2 \cdot 5 \cdot 7}{3 \cdot 3}$
$= \dfrac{70}{9} = 7\dfrac{7}{9}$

45. $3\dfrac{1}{3} \cdot (-9) = -\left(\dfrac{10}{3} \cdot \dfrac{9}{1}\right)$
$= -\dfrac{10 \cdot 9}{3 \cdot 1}$
$= -\dfrac{2 \cdot 5 \cdot 3 \cdot 3}{3 \cdot 1}$
$= -\dfrac{30}{1} = -30$

47. $8 \cdot 5\dfrac{1}{4} = \dfrac{8}{1} \cdot \dfrac{21}{4}$
$= \dfrac{8 \cdot 21}{1 \cdot 4}$
$= \dfrac{2 \cdot 2 \cdot 2 \cdot 3 \cdot 7}{1 \cdot 2 \cdot 2}$
$= \dfrac{42}{1} = 42$

49. $3\dfrac{1}{2} \cdot 1\dfrac{5}{7} \cdot \dfrac{11}{12} = \dfrac{7}{2} \cdot \dfrac{12}{7} \cdot \dfrac{11}{12}$
$= \dfrac{7 \cdot 12 \cdot 11}{2 \cdot 7 \cdot 12}$
$= \dfrac{7 \cdot 2 \cdot 2 \cdot 3 \cdot 11}{2 \cdot 7 \cdot 2 \cdot 2 \cdot 3}$
$= \dfrac{11}{2} = 5\dfrac{1}{2}$

51. $\dfrac{3}{4} \cdot \dfrac{14}{15} = \dfrac{3 \cdot 14}{4 \cdot 15}$
$= \dfrac{3 \cdot 2 \cdot 7}{2 \cdot 2 \cdot 3 \cdot 5}$
$= \dfrac{7}{10}$

53. $-\dfrac{9}{16} \cdot \dfrac{4}{27} = -\left(\dfrac{9 \cdot 4}{16 \cdot 27}\right)$
$= -\dfrac{3 \cdot 3 \cdot 2 \cdot 2}{2 \cdot 2 \cdot 2 \cdot 2 \cdot 3 \cdot 3 \cdot 3}$
$= -\dfrac{1}{12}$

55. $-\dfrac{7}{24}\cdot\dfrac{8}{21}\cdot\dfrac{3}{7}=-\left(\dfrac{7}{24}\cdot\dfrac{8}{21}\cdot\dfrac{3}{7}\right)$

$=-\dfrac{7\cdot 8\cdot 3}{24\cdot 21\cdot 7}$

$=-\dfrac{7\cdot 2\cdot 2\cdot 2\cdot 3}{2\cdot 2\cdot 2\cdot 3\cdot 3\cdot 7\cdot 7}$

$=-\dfrac{1}{21}$

57. $4\dfrac{4}{5}\cdot\dfrac{3}{8}=\dfrac{24}{5}\cdot\dfrac{3}{8}$

$=\dfrac{24\cdot 3}{5\cdot 8}$

$=\dfrac{2\cdot 2\cdot 2\cdot 3\cdot 3}{5\cdot 2\cdot 2\cdot 2}$

$=\dfrac{9}{5}=1\dfrac{4}{5}$

59. $-2\dfrac{2}{3}\cdot\left(-1\dfrac{11}{16}\right)=\dfrac{8}{3}\cdot\dfrac{27}{16}$

$=\dfrac{8\cdot 27}{3\cdot 16}$

$=\dfrac{2\cdot 2\cdot 2\cdot 3\cdot 3\cdot 3}{3\cdot 2\cdot 2\cdot 2\cdot 2}$

$=\dfrac{9}{2}=4\dfrac{1}{2}$

61. Strategy   To find the amount spent on housing, multiply the housing fraction $\left(\dfrac{13}{45}\right)$ by the income (45,000).

Solution $45{,}000\cdot\dfrac{13}{45}=13{,}000$

The typical household spends $13,000 per year on housing.

63. xy

$-\dfrac{5}{16}\cdot\dfrac{7}{15}=-\left(\dfrac{5\cdot 7}{16\cdot 15}\right)$

$=-\dfrac{5\cdot 7}{2\cdot 2\cdot 2\cdot 2\cdot 3\cdot 5}$

$=-\dfrac{7}{48}$

65. xy

$\dfrac{4}{7}\cdot 6\dfrac{1}{8}=\dfrac{4}{7}\cdot\dfrac{49}{8}$

$=\dfrac{4\cdot 49}{7\cdot 8}$

$=\dfrac{2\cdot 2\cdot 7\cdot 7}{7\cdot 2\cdot 2\cdot 2}$

$=\dfrac{7}{2}=3\dfrac{1}{2}$

67. xy

$-49\cdot\dfrac{5}{14}=-\left(\dfrac{49}{1}\cdot\dfrac{5}{14}\right)$

$=-\dfrac{49\cdot 5}{1\cdot 14}$

$=-\dfrac{7\cdot 7\cdot 5}{1\cdot 2\cdot 7}$

$=-\dfrac{35}{2}=-17\dfrac{1}{2}$

69. xy

$1\dfrac{3}{13}\cdot\left(-6\dfrac{1}{2}\right)=-\left(\dfrac{16}{13}\cdot\dfrac{13}{2}\right)$

$=-\dfrac{16\cdot 13}{13\cdot 2}$

$=-\dfrac{2\cdot 2\cdot 2\cdot 2\cdot 13}{13\cdot 2}$

$=-\dfrac{8}{1}=-8$

71. xyz

$\dfrac{3}{8}\cdot\dfrac{2}{3}\cdot\dfrac{4}{5}=\dfrac{3\cdot 2\cdot 4}{8\cdot 3\cdot 5}$

$=\dfrac{3\cdot 2\cdot 2\cdot 2}{2\cdot 2\cdot 2\cdot 3\cdot 5}$

$=\dfrac{1}{5}$

73. xyz

$2\dfrac{3}{8}\cdot\left(-\dfrac{3}{19}\right)\cdot\left(-\dfrac{4}{9}\right)=\dfrac{19}{8}\cdot\dfrac{3}{19}\cdot\dfrac{4}{9}$

$=\dfrac{19\cdot 3\cdot 4}{8\cdot 19\cdot 9}$

$=\dfrac{19\cdot 3\cdot 2\cdot 2}{2\cdot 2\cdot 2\cdot 19\cdot 3\cdot 3}$

$=\dfrac{1}{6}$

75. xyz

$\dfrac{5}{6}\cdot(-3)\cdot 1\dfrac{7}{15}=-\left(\dfrac{5}{6}\cdot\dfrac{3}{1}\cdot\dfrac{22}{15}\right)$

$=-\dfrac{5\cdot 3\cdot 22}{6\cdot 1\cdot 15}$

$=-\dfrac{5\cdot 3\cdot 2\cdot 11}{2\cdot 3\cdot 1\cdot 3\cdot 5}$

$=-\dfrac{11}{3}=-3\dfrac{2}{3}$

77.
$$\frac{3}{4}y = -\frac{1}{4}$$

$\frac{3}{4}\left(-\frac{1}{3}\right)$	$-\frac{1}{4}$
$-\frac{3\cdot 1}{4\cdot 3}$	$-\frac{1}{4}$
$-\frac{3\cdot 1}{2\cdot 2\cdot 3}$	$-\frac{1}{4}$
$-\frac{1}{4}$	$= -\frac{1}{4}$

Yes, $-\frac{1}{3}$ is a solution of the equation.

79.
$$\frac{4}{5}x = \frac{5}{3}$$

$\frac{4}{5}\cdot\frac{3}{4}$	$\frac{5}{3}$
$\frac{4\cdot 3}{5\cdot 4}$	$\frac{5}{3}$
$\frac{2\cdot 2\cdot 3}{5\cdot 2\cdot 2}$	$\frac{5}{3}$
$\frac{3}{5}$	$\neq \frac{5}{3}$

No, $\frac{3}{4}$ is not a solution of the equation.

81.
$$6x = 1$$

$6\left(-\frac{1}{6}\right)$	1
$\frac{6}{1}\left(-\frac{1}{6}\right)$	1
$-\frac{6\cdot 1}{1\cdot 6}$	1
$-\frac{2\cdot 3\cdot 1}{1\cdot 2\cdot 3}$	1
-1	$\neq 1$

No, $-\frac{1}{6}$ is not a solution of the equation.

Objective B Exercises

83. $\frac{5}{7} \div \frac{2}{5} = \frac{5}{7} \cdot \frac{5}{2}$
$= \frac{5\cdot 5}{7\cdot 2}$
$= \frac{25}{14} = 1\frac{11}{14}$

85. $\frac{4}{7} \div \left(-\frac{4}{7}\right) = -\left(\frac{4}{7} \div \frac{4}{7}\right)$
$= -\left(\frac{4}{7} \cdot \frac{7}{4}\right)$
$= -\frac{4\cdot 7}{7\cdot 4}$
$= -\frac{2\cdot 2\cdot 7}{7\cdot 2\cdot 2} = -\frac{1}{1} = -1$

87. $0 \div \frac{7}{9} = 0 \cdot \frac{9}{7} = 0$
Zero divided by a non-zero number is 0.

89. $\left(-\frac{1}{3}\right) \div \frac{1}{2} = -\left(\frac{1}{3} \div \frac{1}{2}\right)$
$= -\left(\frac{1}{3}\cdot\frac{2}{1}\right)$
$= -\frac{1\cdot 2}{3\cdot 1} = -\frac{2}{3}$

91. $-\frac{5}{16} \div \left(-\frac{3}{8}\right) = \frac{5}{16} \div \frac{3}{8}$
$= \frac{5}{16} \cdot \frac{8}{3}$
$= \frac{5\cdot 8}{16\cdot 3}$
$= \frac{5\cdot 2\cdot 2\cdot 2}{2\cdot 2\cdot 2\cdot 2\cdot 3}$
$= \frac{5}{6}$

93. $\frac{0}{1} \div \frac{1}{9} = \frac{0}{1} \cdot \frac{9}{1}$
$= \frac{0\cdot 9}{1\cdot 1}$
$= \frac{0}{1} = 0$

95. $6 \div \frac{3}{4} = \frac{6}{1} \cdot \frac{4}{3}$
$= \frac{6\cdot 4}{1\cdot 3}$
$= \frac{2\cdot 3\cdot 2\cdot 2}{1\cdot 3}$
$= \frac{8}{1} = 8$

97. $\frac{3}{4} \div (-6) = -\left(\frac{3}{4} \div \frac{6}{1}\right)$
$= -\left(\frac{3}{4}\cdot\frac{1}{6}\right)$
$= -\frac{3\cdot 1}{4\cdot 6}$
$= -\frac{3\cdot 1}{2\cdot 2\cdot 2\cdot 3}$
$= -\frac{1}{8}$

99. $\frac{9}{10} \div 0$
Division by zero is undefined.

Section 3.4

101. $\dfrac{5}{12} \div \left(-\dfrac{15}{32}\right) = -\left(\dfrac{5}{12} \div \dfrac{15}{32}\right)$

    $= -\left(\dfrac{5}{12} \cdot \dfrac{32}{15}\right)$

    $= -\dfrac{5 \cdot 32}{12 \cdot 15}$

    $= -\dfrac{5 \cdot 2 \cdot 2 \cdot 2 \cdot 2 \cdot 2}{2 \cdot 2 \cdot 3 \cdot 3 \cdot 5}$

    $= -\dfrac{8}{9}$

103. $\left(-\dfrac{2}{3}\right) \div (-4) = \dfrac{2}{3} \div \dfrac{4}{1}$

    $= \dfrac{2}{3} \cdot \dfrac{1}{4}$

    $= \dfrac{2 \cdot 1}{3 \cdot 4}$

    $= \dfrac{2 \cdot 1}{3 \cdot 2 \cdot 2}$

    $= \dfrac{1}{6}$

105. $\dfrac{8}{x} \div \left(-\dfrac{y}{4}\right) = -\left(\dfrac{8}{x} \div \dfrac{y}{4}\right)$

    $= -\left(\dfrac{8}{x} \cdot \dfrac{4}{y}\right)$

    $= -\dfrac{8 \cdot 4}{x \cdot y}$

    $= -\dfrac{32}{xy}$

107. $\dfrac{b}{6} \div \dfrac{5}{d} = \dfrac{b}{6} \cdot \dfrac{d}{5}$

    $= \dfrac{b \cdot d}{6 \cdot 5}$

    $= \dfrac{bd}{30}$

109. $3\dfrac{1}{3} \div \dfrac{5}{8} = \dfrac{10}{3} \cdot \dfrac{8}{5} = \dfrac{10 \cdot 8}{3 \cdot 5}$

    $= \dfrac{2 \cdot 5 \cdot 2 \cdot 2 \cdot 2}{3 \cdot 5}$

    $= \dfrac{16}{3} = 5\dfrac{1}{3}$

111. $5\dfrac{3}{5} \div \left(-\dfrac{7}{10}\right) = -\left(\dfrac{28}{5} \div \dfrac{7}{10}\right)$

    $= -\left(\dfrac{28}{5} \cdot \dfrac{10}{7}\right)$

    $= -\dfrac{28 \cdot 10}{5 \cdot 7}$

    $= -\dfrac{2 \cdot 2 \cdot 7 \cdot 2 \cdot 5}{5 \cdot 7}$

    $= -\dfrac{8}{1} = -8$

113. $-1\dfrac{1}{2} \div 1\dfrac{3}{4} = -\left(\dfrac{3}{2} \div \dfrac{7}{4}\right)$

    $= -\left(\dfrac{3}{2} \cdot \dfrac{4}{7}\right)$

    $= -\dfrac{3 \cdot 4}{2 \cdot 7}$

    $= -\dfrac{3 \cdot 2 \cdot 2}{2 \cdot 7}$

    $= -\dfrac{6}{7}$

115. $5\dfrac{1}{2} \div 11 = \dfrac{11}{2} \div \dfrac{11}{1}$

    $= \dfrac{11}{2} \cdot \dfrac{1}{11}$

    $= \dfrac{11 \cdot 1}{2 \cdot 11}$

    $= \dfrac{1}{2}$

117. $5\dfrac{2}{7} \div 1 = \dfrac{37}{7} \div \dfrac{1}{1}$

    $= \dfrac{37}{7} \cdot \dfrac{1}{1}$

    $= \dfrac{37 \cdot 1}{7 \cdot 1}$

    $= \dfrac{37}{7} = 5\dfrac{2}{7}$

119. $-16 \div 1\dfrac{1}{3} = -\left(\dfrac{16}{1} \div \dfrac{4}{3}\right)$

    $= -\left(\dfrac{16}{1} \cdot \dfrac{3}{4}\right)$

    $= -\dfrac{16 \cdot 3}{1 \cdot 4}$

    $= -\dfrac{2 \cdot 2 \cdot 2 \cdot 2 \cdot 3}{1 \cdot 2 \cdot 2}$

    $= -\dfrac{12}{1} = -12$

121. $2\dfrac{4}{13} \div 1\dfrac{5}{26} = \dfrac{30}{13} \div \dfrac{31}{26}$

    $= \dfrac{30}{13} \cdot \dfrac{26}{31}$

    $= \dfrac{30 \cdot 26}{13 \cdot 31}$

    $= \dfrac{2 \cdot 3 \cdot 5 \cdot 2 \cdot 13}{13 \cdot 31}$

    $= \dfrac{60}{31} = 1\dfrac{29}{31}$

123. $\dfrac{9}{10} \div \dfrac{3}{4} = \dfrac{9}{10} \cdot \dfrac{4}{3}$

    $= \dfrac{9 \cdot 4}{10 \cdot 3}$

    $= \dfrac{3 \cdot 3 \cdot 2 \cdot 2}{2 \cdot 5 \cdot 3}$

    $= \dfrac{6}{5} = 1\dfrac{1}{5}$

© Houghton Mifflin Company. All rights reserved.

Chapter 3: Fractions

125. $\left(-\dfrac{15}{24}\right) \div \dfrac{3}{5} = -\left(\dfrac{15}{24} \cdot \dfrac{5}{3}\right)$

$= -\dfrac{15 \cdot 5}{24 \cdot 3}$

$= -\dfrac{3 \cdot 5 \cdot 5}{2 \cdot 2 \cdot 2 \cdot 3 \cdot 3}$

$= -\dfrac{25}{24} = -1\dfrac{1}{24}$

127. $\dfrac{7}{8} \div 3\dfrac{1}{4} = \dfrac{7}{8} \div \dfrac{13}{4}$

$= \dfrac{7}{8} \cdot \dfrac{4}{13}$

$= \dfrac{7 \cdot 4}{8 \cdot 13}$

$= \dfrac{7 \cdot 2 \cdot 2}{2 \cdot 2 \cdot 2 \cdot 13}$

$= \dfrac{7}{26}$

129. $-3\dfrac{5}{11} \div 3\dfrac{4}{5} = -\left(\dfrac{38}{11} \div \dfrac{19}{5}\right)$

$= -\left(\dfrac{38}{11} \cdot \dfrac{5}{19}\right)$

$= -\dfrac{38 \cdot 5}{11 \cdot 19}$

$= -\dfrac{2 \cdot 19 \cdot 5}{11 \cdot 19}$

$= -\dfrac{10}{11}$

131. $x \div y$

$-\dfrac{5}{8} \div \left(-\dfrac{15}{2}\right) = \dfrac{5}{8} \div \dfrac{15}{2}$

$= \dfrac{5}{8} \cdot \dfrac{2}{15}$

$= \dfrac{5 \cdot 2}{8 \cdot 15}$

$= \dfrac{5 \cdot 2}{2 \cdot 2 \cdot 2 \cdot 3 \cdot 5}$

$= \dfrac{1}{12}$

133. $x \div y$

$\dfrac{1}{7} \div 0$

Division by zero is undefined.

135. $x \div y$

$-18 \div \dfrac{3}{8} = -\left(\dfrac{18}{1} \cdot \dfrac{8}{3}\right)$

$= -\dfrac{18 \cdot 8}{1 \cdot 3}$

$= -\dfrac{2 \cdot 3 \cdot 3 \cdot 2 \cdot 2 \cdot 2}{1 \cdot 3}$

$= -\dfrac{48}{1} = -48$

137. $x \div y$

$-\dfrac{1}{2} \div \left(-3\dfrac{5}{8}\right) = \dfrac{1}{2} \div \dfrac{29}{8}$

$= \dfrac{1}{2} \cdot \dfrac{8}{29}$

$= \dfrac{1 \cdot 8}{2 \cdot 29}$

$= \dfrac{1 \cdot 2 \cdot 2 \cdot 2}{2 \cdot 29}$

$= \dfrac{4}{29}$

139. $x \div y$

$6\dfrac{2}{5} \div (-4) = -\left(\dfrac{32}{5} \div \dfrac{4}{1}\right)$

$= -\left(\dfrac{32}{5} \cdot \dfrac{1}{4}\right)$

$= -\dfrac{32 \cdot 1}{5 \cdot 4}$

$= -\dfrac{2 \cdot 2 \cdot 2 \cdot 2 \cdot 2 \cdot 1}{5 \cdot 2 \cdot 2}$

$= -\dfrac{8}{5} = -1\dfrac{3}{5}$

141. $x \div y$

$-3\dfrac{2}{5} \div \left(-1\dfrac{7}{10}\right) = \dfrac{17}{5} \div \dfrac{17}{10}$

$= \dfrac{17}{5} \cdot \dfrac{10}{17}$

$= \dfrac{17 \cdot 10}{5 \cdot 17}$

$= \dfrac{17 \cdot 2 \cdot 5}{5 \cdot 17}$

$= \dfrac{2}{1} = 2$

143. Strategy — To find the number of servings, divide the net weight of Kellogg Honey Crunch Corn Flakes (24) by $\dfrac{3}{4}$.

Solution — $24 \div \dfrac{3}{4} = \dfrac{24}{1} \cdot \dfrac{4}{3} = \dfrac{24 \cdot 4}{1 \cdot 3} = 32$

There are 32 servings in a box of Kellogg Honey Crunch Corn Flakes.

Objective C Exercises

145. Strategy   To find the length of time, multiply the length of time for one chukker $\left(7\frac{1}{2}\right)$ by 4.

       Solution $7\frac{1}{2} \cdot 4 = \frac{15}{2} \cdot \frac{4}{1} = \frac{15 \cdot 4}{2 \cdot 1} = 30$

                 Four periods of play takes 30 min.

147. Strategy   →To find the number of feet in one rod, multiply the number of yards in one rod $\left(5\frac{1}{2}\right)$ by 3.

                 →To find the number of inches in one rod, multiply the number of yards in one rod $\left(5\frac{1}{2}\right)$ by 36.

       Solution $5\frac{1}{2} \cdot 3 = \frac{11}{2} \cdot \frac{3}{1} = \frac{11 \cdot 3}{2 \cdot 1} = 16\frac{1}{2}$

                    $5\frac{1}{2} \cdot 36 = \frac{11}{2} \cdot \frac{36}{1} = \frac{11 \cdot 36}{2 \cdot 1} = 198$

                 One rod is equivalent to $16\frac{1}{2}$ ft.

                 One rod is equivalent to 198 in.

149. Strategy   To find the amount of time, multiply the number of hours cleaning per week $\left(4\frac{1}{2}\right)$ by the number of weeks (52).

       Solution $4\frac{1}{2} \cdot 52 = \frac{9}{2} \cdot \frac{52}{1} = \frac{9 \cdot 52}{2 \cdot 1} = 234$

                 The average couple spends 234 h a year cleaning house.

151. Strategy   To find the number of house plots:

                 →Subtract the number of acres set aside (3) from the total number of acres $\left(25\frac{1}{2}\right)$ to find the number of acres to be sold in parcels.

                 →Divide the number of acres available by the number of acres in one parcel $\left(\frac{3}{4}\right)$.

       Solution $25\frac{1}{2} - 3 = 22\frac{1}{2}$

                    $22\frac{1}{2} \div \frac{3}{4} = \frac{45}{2} \cdot \frac{4}{3} = \frac{45 \cdot 4}{2 \cdot 3} = 30$

                 The developer plans to build 30 houses.

153. Strategy   To find the dimensions of the board, divide one side (14) by 2 and multiply the thickness $\left(\frac{7}{8}\right)$ by 2.

       Solution $14 \div 2 = 7$

                    $\frac{7}{8} \cdot 2 = \frac{7}{8} \cdot \frac{2}{1} = 1\frac{3}{4}$

                 The dimensions when the board is closed are 14 in. by 7 in. by $1\frac{3}{4}$ in.

155. **Strategy**

    a. To find the total per week, multiply the number of cans per day $\left(3\frac{1}{3}\right)$ by 7.

    b. To find the calories, multiply the cans per day $\left(3\frac{1}{3}\right)$ by 7 and 150.

    c. To find the difference, subtract the number of cans teenage girls drink $\left(2\frac{1}{3}\right)$ by the number of can teenage boys drink $\left(3\frac{1}{3}\right)$ and then multiply the difference by 7.

Solution

    a. $3\frac{1}{3} \cdot 7 = \frac{10}{3} \cdot \frac{7}{1} = \frac{70}{3} = 23\frac{1}{3}$

    The average teenage boy drinks $23\frac{1}{3}$ cans of soda a week.

    b. $3\frac{1}{3} \cdot 7 \cdot 150 = \frac{10}{3} \cdot \frac{7}{1} \cdot \frac{150}{1} = 3500$

    For one week, the number of calories is 3500.

    c. $3\frac{1}{3} - 2\frac{1}{3} = 1$
    $1 \cdot 7 = 7$

    The difference per week is 7 cans.

157. **Strategy**   To find the area, use the formula below, substitute $8\frac{1}{2}$ for L, 5 for W, and solve for A.

    Solution $A = LW$

$$A = 8\frac{1}{2} \cdot 5 = \frac{17}{2} \cdot \frac{5}{1}$$
$$= \frac{85}{2} = 42\frac{1}{2}$$

    The area of the rectangle is $42\frac{1}{2}$ yd^2.

159. **Strategy**   To find the area, use the formula below, substitute 12 for b, 16 for h, and solve for A.

    Solution $A = \frac{1}{2}bh$

$$A = \frac{1}{2} \cdot 12 \cdot 16 = \frac{1}{2} \cdot \frac{12}{1} \cdot \frac{16}{1} = 96$$

    The amount of canvas needed for the sail is 96 m^2.

161. Strategy To find the number of bags of seed:
→Find the area of the triangle, use the formula below, substitute 20 for b, 24 for h, and solve for A.
→Divide the area by 120.

Solution $A = \dfrac{1}{2}bh$

$A = \dfrac{1}{2} \cdot 20 \cdot 24 = \dfrac{1}{2} \cdot \dfrac{20}{1} \cdot \dfrac{24}{1} = 240$

$240 \div 120 = 2$

Two bags of grass seed should be purchased.

163. Strategy To find the rate, substitute $4\frac{2}{3}$ for d and $1\frac{1}{3}$ for t in the given formula and solve for r.

Solution $r = \dfrac{d}{t}$

$r = \dfrac{4\frac{2}{3}}{1\frac{1}{3}} = 4\frac{2}{3} \div 1\frac{1}{3}$

$r = \dfrac{14}{3} \div \dfrac{4}{3}$

$r = \dfrac{14}{3} \cdot \dfrac{3}{4} = \dfrac{14 \cdot 3}{3 \cdot 4} = 3\frac{1}{2}$

The rate of the hiker is $3\frac{1}{2}$ mph.

Critical Thinking 3.4

165. Divide $3\frac{1}{8}$ by $\frac{1}{8}$ to find the number of 50-mile segments between the two cities.

$3\frac{1}{8} \div \frac{1}{8} = \dfrac{25}{8} \div \dfrac{1}{8} = \dfrac{25}{8} \cdot \dfrac{8}{1} = 25$

Multiply 25 times 50 to find the number of miles.
$25 \cdot 50 = 1{,}250$
The distance between the two cities is 1,250 mi.

167. a. Commutative Property

$\dfrac{1}{2} \div \dfrac{1}{4} = \dfrac{1}{2} \cdot \dfrac{4}{1} = 2$

$\dfrac{1}{4} \div \dfrac{1}{2} = \dfrac{1}{4} \cdot \dfrac{2}{1} = \dfrac{1}{2}$

$\dfrac{1}{2} \div \dfrac{1}{4} \neq \dfrac{1}{4} \div \dfrac{1}{2}$

b. Associative Property

$\dfrac{1}{2} \div \left(\dfrac{1}{4} \div \dfrac{1}{8}\right) = \dfrac{1}{2} \div \left(\dfrac{1}{4} \cdot \dfrac{8}{1}\right) = \dfrac{1}{2} \div 2 = \dfrac{1}{2} \cdot \dfrac{1}{2} = \dfrac{1}{4}$

$\left(\dfrac{1}{2} \div \dfrac{1}{4}\right) \div \dfrac{1}{8} = \left(\dfrac{1}{2} \cdot \dfrac{4}{1}\right) \div \dfrac{1}{8} = 2 \div \dfrac{1}{8} = 2 \cdot 8 = 16$

$\dfrac{1}{2} \div \left(\dfrac{1}{4} \div \dfrac{1}{8}\right) \neq \left(\dfrac{1}{2} \div \dfrac{1}{4}\right) \div \dfrac{1}{8}$

c. Inverse Property

$7 \div \dfrac{1}{7} = 7 \cdot 7 = 49 \neq 1$

169. Our calendar year is 365 years. A complete orbit around the sun takes $365\frac{1}{4}$ days, so one solar year is $365\frac{1}{4}$ days. Therefore, after 365 days, the Earth is $\frac{1}{4}$ day short of a complete orbit around the sun. After two years, the earth is $\frac{1}{4}$ day $+\frac{1}{4}$ day $=\frac{1}{2}$ day short of a complete orbit around the sun. After 4 years, the earth is $\frac{1}{4}$ day $+\frac{1}{4}$ day $+\frac{1}{4}$ day $+\frac{1}{4}$ day $= 1$ day short of a complete orbit around the sun. Therefore, every four years, our calendar incorporates a leap year, which is a 366-day year, in order to make up for the one extra day in the four solar years.

Section 3.5

Objective A Exercises

1. $\frac{x}{4} = 9$
   $4 \cdot \frac{x}{4} = 4 \cdot 9$
   $x = 36$
   The solution is 36.

3. $-3 = \frac{m}{4}$
   $4(-3) = 4 \cdot \frac{m}{4}$
   $-12 = m$
   The solution is -12.

5. $\frac{2}{5}x = 10$
   $\frac{5}{2} \cdot \frac{2}{5}x = \frac{5}{2} \cdot 10$
   $x = 25$
   The solution is 25.

7. $-\frac{5}{6}w = 10$
   $-\frac{6}{5}\left(-\frac{5}{6}\right)w = -\frac{6}{5}(10)$
   $w = -12$
   The solution is -12.

9. $\frac{1}{4} + y = \frac{3}{4}$
   $\frac{1}{4} - \frac{1}{4} + y = \frac{3}{4} - \frac{1}{4}$
   $y = \frac{2}{4}$
   $y = \frac{1}{2}$
   The solution is $\frac{1}{2}$.

11. $x + \frac{1}{4} = \frac{5}{6}$
    $x = \frac{1}{4} - \frac{1}{4} = \frac{5}{6} - \frac{1}{4}$
    $x = \frac{10}{12} - \frac{3}{12}$
    $x = \frac{7}{12}$
    The solution is $\frac{7}{12}$.

13. $-\frac{2x}{3} = -\frac{1}{2}$
    $-\frac{3}{2}\left(-\frac{2}{3}x\right) = -\frac{3}{2}\left(-\frac{1}{2}\right)$
    $x = \frac{3}{4}$
    The solution is $\frac{3}{4}$.

15. $\frac{5n}{6} = -\frac{2}{3}$
    $\frac{6}{5}\left(\frac{5}{6}n\right) = \frac{6}{5}\left(-\frac{2}{3}\right)$
    $n = -\frac{4}{5}$
    The solution is $-\frac{4}{5}$.

17. $-\frac{3}{8}t = -\frac{1}{4}$
    $-\frac{8}{3}\left(-\frac{3}{8}t\right) = -\frac{8}{3}\left(-\frac{1}{4}\right)$
    $t = \frac{2}{3}$
    The solution is $\frac{2}{3}$.

Section 3.5   67

19. $4a = 6$
$\frac{4a}{4} = \frac{6}{4}$
$a = \frac{3}{2}$
$a = 1\frac{1}{2}$
The solution is $1\frac{1}{2}$.

21. $-9c = 12$
$\frac{-9c}{-9} = \frac{12}{-9}$
$c = -\frac{4}{3}$
$c = -1\frac{1}{3}$
The solution is $-1\frac{1}{3}$.

23. $-2x = \frac{8}{9}$
$\frac{-2x}{-2} = \frac{8}{9} \div (-2)$
$x = -\left(\frac{8}{9} \cdot \frac{1}{2}\right)$
$x = -\frac{4}{9}$
The solution is $-\frac{4}{9}$.

Objective B Exercises

25. The unknown number: n

   | a number minus one-third | equals | one-half |

   $n - \frac{1}{3} = \frac{1}{2}$
   $n - \frac{1}{3} + \frac{1}{3} = \frac{1}{2} + \frac{1}{3}$
   $n = \frac{3}{6} + \frac{2}{6}$
   $n = \frac{5}{6}$
   The number is $\frac{5}{6}$.

27. The unknown number: n

   | three-fifths times a number | is | nine-tenths |

   $\frac{3}{5}n = \frac{9}{10}$
   $\frac{5}{3} \cdot \frac{3}{5}n = \frac{5}{3} \cdot \frac{9}{10}$
   $n = \frac{3}{2}$
   $n = 1\frac{1}{2}$
   The number is $1\frac{1}{2}$.

29. The unknown number: n

   | the quotient of a number and negative four | is | three-fourths |

   $\frac{n}{-4} = \frac{3}{4}$
   $-4\left(\frac{n}{-4}\right) = -4\left(\frac{3}{4}\right)$
   $n = -3$
   The number is -3.

© Houghton Mifflin Company. All rights reserved.

31. The unknown number: n

[negative three-fourths of a number] is equal to [one-sixth]

$$-\frac{3}{4}n = \frac{1}{6}$$

$$-\frac{4}{3}\left(-\frac{3}{4}n\right) = -\frac{4}{3}\left(\frac{1}{6}\right)$$

$$n = -\frac{2}{9}$$

The number is $-\frac{2}{9}$.

33. Strategy   To find the salary after earning the master's degree, write and solve an equation using x to represent the salary after earning the master's degree.

Solution   [the salary prior to master's degree] is [one-half the salary after earning a master's degree]

$$40{,}000 = \frac{1}{2}x$$

$$2(40{,}000) = 2\left(\frac{1}{2}x\right)$$

$$80{,}000 = x$$

Her salary is $80,000 after earning the master's degree.

35. Strategy   To find the total number of quarts in the punch, write and solve an equation using x to represent the total number of quarts in the punch.

Solution   [the number of quarts of orange juice] is [three-fifths of the total number of quarts]

$$15 = \frac{3}{5}x$$

$$\frac{5}{3}(15) = \frac{5}{3}\left(\frac{3}{5}x\right)$$

$$25 = x$$

There is a total of 25 quarts in the punch.

37. Strategy   To find the mechanic's monthly income, write and solve an equation using x to represent the monthly income.

Solution   [the rent] is [two-fifths of the monthly income]

$$600 = \frac{2}{5}x$$

$$\frac{5}{2}(600) = \frac{5}{2}\left(\frac{2}{5}x\right)$$

$$1{,}500 = x$$

The mechanic's monthly income is $1,500.

39. Strategy To find the number of miles, replace a by 14 and g by 38 in the given formula and solve for m.

Solution
$$a = \frac{m}{g}$$
$$14 = \frac{m}{38}$$
$$38 \cdot 14 = 38 \cdot \frac{m}{38}$$
$$532 = m$$
The truck can travel 532 mi on 38 gal of diesel fuel.

Critical Thinking 3.5

41. $-\frac{x}{2} = \frac{2}{3}$

$$-2\left(-\frac{1}{2}x\right) = -2\left(\frac{2}{3}\right)$$
$$x = -\frac{4}{3}$$

$-9\left(-\frac{4}{3}\right) = 12$ and $12 > 10$, so $-9x > 10$, and **a** is true.

$-6\left(-\frac{4}{3}\right) = 8$, but 8 is not less than 8, so $-6x$ is not less than 8, and **b** is not true. Since **b** is not true, **c** is not true. Only **a** is true.

Section 3.6

Objective A Exercises

1. $\left(\frac{3}{4}\right)^2 = \frac{3}{4} \cdot \frac{3}{4}$
$= \frac{3 \cdot 3}{4 \cdot 4} = \frac{9}{16}$

3. $\left(-\frac{1}{6}\right)^3 = \left(-\frac{1}{6}\right)\left(-\frac{1}{6}\right)\left(-\frac{1}{6}\right)$
$= -\left(\frac{1}{6} \cdot \frac{1}{6} \cdot \frac{1}{6}\right)$
$= -\frac{1 \cdot 1 \cdot 1}{6 \cdot 6 \cdot 6} = -\frac{1}{216}$

5. $\left(2\frac{1}{4}\right)^2 = \left(\frac{9}{4}\right)^2$
$= \frac{9}{4} \cdot \frac{9}{4}$
$= \frac{9 \cdot 9}{4 \cdot 4} = \frac{81}{16} = 5\frac{1}{16}$

7. $\left(\frac{5}{8}\right)^3 \cdot \left(\frac{2}{5}\right)^2 = \frac{5}{8} \cdot \frac{5}{8} \cdot \frac{5}{8} \cdot \frac{2}{5} \cdot \frac{2}{5}$
$= \frac{5 \cdot 5 \cdot 5 \cdot 2 \cdot 2}{8 \cdot 8 \cdot 8 \cdot 5 \cdot 5}$
$= \frac{5}{128}$

9. $\left(\frac{18}{25}\right)^2 \cdot \left(\frac{5}{9}\right)^3$
$= \frac{18}{25} \cdot \frac{18}{25} \cdot \frac{5}{9} \cdot \frac{5}{9} \cdot \frac{5}{9}$
$= \frac{18 \cdot 18 \cdot 5 \cdot 5 \cdot 5}{25 \cdot 25 \cdot 9 \cdot 9 \cdot 9}$
$= \frac{4}{45}$

11. $\left(\frac{4}{5}\right)^4 \cdot \left(-\frac{5}{8}\right)^3$
$= \frac{4}{5} \cdot \frac{4}{5} \cdot \frac{4}{5} \cdot \frac{4}{5} \cdot \left(-\frac{5}{8}\right)\left(-\frac{5}{8}\right)\left(-\frac{5}{8}\right)$
$= -\left(\frac{4}{5} \cdot \frac{4}{5} \cdot \frac{4}{5} \cdot \frac{4}{5} \cdot \frac{5}{8} \cdot \frac{5}{8} \cdot \frac{5}{8}\right)$
$= -\frac{4 \cdot 4 \cdot 4 \cdot 4 \cdot 5 \cdot 5 \cdot 5}{5 \cdot 5 \cdot 5 \cdot 5 \cdot 8 \cdot 8 \cdot 8} = -\frac{1}{10}$

Chapter 3: Fractions

13. $7^2 \cdot \left(\dfrac{2}{7}\right)^3 = \dfrac{7}{1} \cdot \dfrac{7}{1} \cdot \dfrac{2}{7} \cdot \dfrac{2}{7} \cdot \dfrac{2}{7}$

    $= \dfrac{7 \cdot 7 \cdot 2 \cdot 2 \cdot 2}{1 \cdot 1 \cdot 7 \cdot 7 \cdot 7}$

    $= \dfrac{8}{7} = 1\dfrac{1}{7}$

15. $4 \cdot \left(\dfrac{4}{7}\right)^2 \cdot \left(-\dfrac{3}{4}\right)^3$

    $= \dfrac{4}{1} \cdot \dfrac{4}{7} \cdot \dfrac{4}{7} \cdot \left(-\dfrac{3}{4}\right) \cdot \left(-\dfrac{3}{4}\right) \cdot \left(-\dfrac{3}{4}\right)$

    $= -\left(\dfrac{4}{1} \cdot \dfrac{4}{7} \cdot \dfrac{4}{7} \cdot \dfrac{3}{4} \cdot \dfrac{3}{4} \cdot \dfrac{3}{4}\right)$

    $= -\dfrac{4 \cdot 4 \cdot 4 \cdot 3 \cdot 3 \cdot 3}{1 \cdot 7 \cdot 7 \cdot 4 \cdot 4 \cdot 4}$

    $= -\dfrac{27}{49}$

17. x^4

    $\left(\dfrac{2}{3}\right)^4 = \dfrac{2}{3} \cdot \dfrac{2}{3} \cdot \dfrac{2}{3} \cdot \dfrac{2}{3}$

    $= \dfrac{2 \cdot 2 \cdot 2 \cdot 2}{3 \cdot 3 \cdot 3 \cdot 3}$

    $= \dfrac{16}{81}$

19. $x^4 y^2$

    $\left(\dfrac{5}{6}\right)^4 \cdot \left(-\dfrac{3}{5}\right)^2 = \dfrac{5}{6} \cdot \dfrac{5}{6} \cdot \dfrac{5}{6} \cdot \dfrac{5}{6} \left(-\dfrac{3}{5}\right) \cdot \left(-\dfrac{3}{5}\right)$

    $= \dfrac{5}{6} \cdot \dfrac{5}{6} \cdot \dfrac{5}{6} \cdot \dfrac{5}{6} \cdot \dfrac{3}{5} \cdot \dfrac{3}{5}$

    $= \dfrac{5 \cdot 5 \cdot 5 \cdot 5 \cdot 3 \cdot 3}{6 \cdot 6 \cdot 6 \cdot 6 \cdot 5 \cdot 5}$

    $= \dfrac{25}{144}$

21. $x^3 y^2$

    $\left(\dfrac{2}{3}\right)^3 \cdot \left(1\dfrac{1}{2}\right)^2 = \dfrac{2}{3} \cdot \dfrac{2}{3} \cdot \dfrac{2}{3} \cdot \dfrac{3}{2} \cdot \dfrac{3}{2}$

    $= \dfrac{2 \cdot 2 \cdot 2 \cdot 3 \cdot 3}{3 \cdot 3 \cdot 3 \cdot 2 \cdot 2}$

    $= \dfrac{2}{3}$

Objective B Exercises

23. $\dfrac{\frac{9}{16}}{\frac{3}{4}} = \dfrac{9}{16} \div \dfrac{3}{4}$

    $= \dfrac{9}{16} \cdot \dfrac{4}{3} = \dfrac{3}{4}$

25. $\dfrac{-\frac{5}{6}}{\frac{15}{16}} = -\dfrac{5}{6} \div \dfrac{15}{16}$

    $= -\left(\dfrac{5}{6} \cdot \dfrac{16}{15}\right) = -\dfrac{8}{9}$

27. $\dfrac{\frac{2}{3} + \frac{1}{2}}{7} = \dfrac{\frac{7}{6}}{\frac{7}{1}}$

    $= \dfrac{7}{6} \div \dfrac{7}{1}$

    $= \dfrac{7}{6} \cdot \dfrac{1}{7} = \dfrac{1}{6}$

29. $\dfrac{2 + \frac{1}{4}}{\frac{3}{8}} = \dfrac{\frac{9}{4}}{\frac{3}{8}}$

    $= \dfrac{9}{4} \div \dfrac{3}{8}$

    $= \dfrac{9}{4} \cdot \dfrac{8}{3} = 6$

31. $\dfrac{\frac{9}{25}}{\frac{4}{5} - \frac{1}{10}} = \dfrac{\frac{9}{25}}{\frac{7}{10}}$

    $= \dfrac{9}{25} \div \dfrac{7}{10}$

    $= \dfrac{9}{25} \cdot \dfrac{10}{7} = \dfrac{18}{35}$

33. $\dfrac{\frac{1}{3} - \frac{3}{4}}{\frac{1}{6} + \frac{2}{3}} = \dfrac{-\frac{5}{12}}{\frac{5}{6}}$

    $= -\dfrac{5}{12} \div \dfrac{5}{6}$

    $= -\left(\dfrac{5}{12} \cdot \dfrac{6}{5}\right) = -\dfrac{1}{2}$

35. $\dfrac{3 + 2\frac{1}{3}}{5\frac{1}{6} - 1} = \dfrac{5\frac{1}{3}}{4\frac{1}{6}}$

    $= 5\dfrac{1}{3} \div 4\dfrac{1}{6}$

    $= \dfrac{16}{3} \div \dfrac{25}{6}$

    $= \dfrac{16}{3} \cdot \dfrac{6}{25} = \dfrac{32}{25} = 1\dfrac{7}{25}$

37. $\dfrac{5\frac{2}{3} - 1\frac{1}{6}}{3\frac{5}{8} - 2\frac{1}{4}} = \dfrac{4\frac{1}{2}}{1\frac{3}{8}}$

    $= 4\dfrac{1}{2} \div 1\dfrac{3}{8}$

    $= \dfrac{9}{2} \div \dfrac{11}{8}$

    $= \dfrac{9}{2} \cdot \dfrac{8}{11} = \dfrac{36}{11} = 3\dfrac{3}{11}$

Section 3.6

39. $\dfrac{x+y}{z}$

$\dfrac{\frac{2}{3}+\frac{3}{4}}{\frac{1}{12}} = \dfrac{\frac{17}{12}}{\frac{1}{12}}$

$= \dfrac{17}{12} \div \dfrac{1}{12}$

$= \dfrac{17}{12} \cdot \dfrac{12}{1} = 17$

41. $\dfrac{xy}{z}$

$\dfrac{\frac{3}{4}\cdot\left(-\frac{2}{3}\right)}{\frac{5}{8}} = \dfrac{-\frac{1}{2}}{\frac{5}{8}}$

$= -\dfrac{1}{2} \div \dfrac{5}{8}$

$= -\dfrac{1}{2} \cdot \dfrac{8}{5} = -\dfrac{4}{5}$

43. $\dfrac{x-y}{z}$

$\dfrac{2\frac{5}{8}-1\frac{1}{4}}{1\frac{3}{8}} = \dfrac{1\frac{3}{8}}{1\frac{3}{8}}$

$= 1\frac{3}{8} \div 1\frac{3}{8} = \dfrac{11}{8} \div \dfrac{11}{8}$

$= \dfrac{11}{8} \cdot \dfrac{8}{11} = 1$

45. $\dfrac{4x}{x+5} = -\dfrac{4}{3}$

$\dfrac{4\left(-\frac{3}{4}\right)}{-\frac{3}{4}+5}$	$-\dfrac{4}{3}$
$\dfrac{-3}{\frac{17}{4}}$	$-\dfrac{4}{3}$
$-3 \div \dfrac{17}{4}$	$-\dfrac{4}{3}$
$\dfrac{-3}{1} \cdot \dfrac{4}{17}$	$-\dfrac{4}{3}$
$-\dfrac{12}{17} \neq -\dfrac{4}{3}$	

No, $-\dfrac{3}{4}$ is not a solution of the equation.

Objective C Exercises

47. $\dfrac{3}{7}\cdot\dfrac{14}{15}+\dfrac{4}{5} = \dfrac{2}{5}+\dfrac{4}{5}$

$= \dfrac{6}{5} = 1\dfrac{1}{5}$

49. $\left(\dfrac{5}{6}\right)^2 - \dfrac{5}{9} = \dfrac{25}{36} - \dfrac{5}{9}$

$= \dfrac{5}{36}$

51. $\dfrac{3}{4}\cdot\left(\dfrac{11}{12}-\dfrac{7}{8}\right)+\dfrac{5}{16} = \dfrac{3}{4}\cdot\dfrac{1}{24}+\dfrac{5}{16}$

$= \dfrac{1}{32}+\dfrac{5}{16}$

$= \dfrac{11}{32}$

53. $\dfrac{11}{16}-\left(\dfrac{3}{4}\right)^2 + \dfrac{7}{8} = \dfrac{11}{16}-\dfrac{9}{16}+\dfrac{7}{8}$

$= \dfrac{1}{8}+\dfrac{7}{8}$

$= 1$

55. $\left(1\dfrac{1}{3}-\dfrac{5}{6}\right)+\dfrac{7}{8}\div\left(-\dfrac{1}{2}\right)^2$

$= \dfrac{1}{2}+\dfrac{7}{8}\div\left(-\dfrac{1}{2}\right)^2$

$= \dfrac{1}{2}+\dfrac{7}{8}\div\dfrac{1}{4}$

$= \dfrac{1}{2}+\dfrac{7}{8}\cdot\dfrac{4}{1}$

$= \dfrac{1}{2}+\dfrac{7}{2} = 4$

57. $\left(\dfrac{2}{3}\right)^2 + \dfrac{8-7}{3-9}\div\dfrac{3}{8} = \left(\dfrac{2}{3}\right)^2 + \left(-\dfrac{1}{6}\right)\div\dfrac{3}{8}$

$= \dfrac{4}{9}+\left(-\dfrac{1}{6}\right)\div\dfrac{3}{8}$

$= \dfrac{4}{9}+\left(-\dfrac{1}{6}\right)\cdot\dfrac{8}{3}$

$= \dfrac{4}{9}+\left(\dfrac{-4}{9}\right) = 0$

59. $\dfrac{1}{2}+\dfrac{\frac{13}{25}}{4-\frac{3}{4}}\div\dfrac{1}{5} = \dfrac{1}{2}+\dfrac{\frac{13}{25}}{\frac{13}{4}}\div\dfrac{1}{5}$

$= \dfrac{1}{2}+\dfrac{13}{25}\div\dfrac{13}{4}\div\dfrac{1}{5}$

$= \dfrac{1}{2}+\dfrac{13}{25}\cdot\dfrac{4}{13}\div\dfrac{1}{5}$

$= \dfrac{1}{2}+\dfrac{4}{25}\div\dfrac{1}{5} = \dfrac{1}{2}+\dfrac{4}{25}\cdot\dfrac{5}{1}$

$= \dfrac{1}{2}+\dfrac{4}{5} = \dfrac{13}{10} = 1\dfrac{3}{10}$

61. $\left(\dfrac{2}{3}\right)^2 + \dfrac{\frac{5}{8}-\frac{1}{4}}{\frac{2}{3}-\frac{1}{6}}\cdot\dfrac{8}{9} = \left(\dfrac{2}{3}\right)^2 + \dfrac{\frac{3}{8}}{\frac{1}{2}}\cdot\dfrac{8}{9}$

$= \left(\dfrac{2}{3}\right)^2 + \dfrac{3}{8}\div\dfrac{1}{2}\cdot\dfrac{8}{9}$

$= \frac{4}{9} + \frac{3}{8} \div \frac{1}{2} \cdot \frac{8}{9}$

$= \frac{4}{9} + \frac{3}{8} \cdot \frac{2}{1} \cdot \frac{8}{9}$

$= \frac{4}{9} + \frac{3}{4} \cdot \frac{8}{9}$

$= \frac{4}{9} + \frac{2}{3} = \frac{10}{9} = 1\frac{1}{9}$

63. $\dfrac{x}{y} - z^2$

$\dfrac{\frac{5}{6}}{\frac{1}{3}} - \left(-\frac{3}{4}\right)^2 = \frac{5}{6} \div \frac{1}{3} - \left(-\frac{3}{4}\right)^2$

$= \frac{5}{6} \div \frac{1}{3} - \frac{9}{16}$

$= \frac{5}{6} \cdot \frac{3}{1} - \frac{9}{16}$

$= \frac{5}{2} - \frac{9}{16}$

$= \frac{31}{16} = 1\frac{15}{16}$

65. $xy^3 + z$

$\frac{9}{10} \cdot \left(\frac{1}{3}\right)^3 + \frac{7}{15} = \frac{9}{10} \cdot \frac{1}{27} + \frac{7}{15}$

$= \frac{1}{30} + \frac{7}{15}$

$= \frac{1}{2}$

67. $\dfrac{w}{xy} - z$

$\dfrac{2\frac{1}{2}}{4 \cdot \frac{3}{8}} - \frac{2}{3} = \dfrac{\frac{5}{2}}{\frac{3}{2}} - \frac{2}{3}$

$= \frac{5}{2} \div \frac{3}{2} - \frac{2}{3}$

$= \frac{5}{2} \cdot \frac{2}{3} - \frac{2}{3}$

$\frac{5}{3} - \frac{2}{3} = \frac{3}{3} = 1$

69. $\dfrac{12w}{\frac{1}{6} - w} = -7$

$\begin{array}{c|c} \dfrac{12\left(-\frac{1}{3}\right)}{\frac{1}{6} - \left(-\frac{1}{3}\right)} & -7 \\ \dfrac{-4}{\frac{1}{2}} & -7 \\ -4 \div \frac{1}{2} & -7 \\ -4 \cdot 2 & -7 \\ -8 \neq -7 \end{array}$

No, $-\frac{1}{3}$ is not a solution of the equation.

Critical Thinking 3.6

71. $1^5 = 1$

$\dfrac{9}{10} < 1$

73. $\left(-1\frac{1}{10}\right)^2 = \left(-\frac{11}{10}\right)\left(-\frac{11}{10}\right) = 1.21$

$(0.9)^2 = (0.9)(0.9) = 0.81$

$1.21 > 0.81$

$\left(-1\frac{1}{10}\right)^2 > (0.9)^2$

75. The expression will be a minimum when x has its smallest value. The smallest whole number is 0.

$\left(\frac{3}{4}\right)^2 + x^5 \div \frac{7}{8} = \frac{9}{16} + 0^5 \div \frac{7}{8}$

$= \frac{9}{16} + 0 \cdot \frac{8}{7}$

$= \frac{9}{16}$

The minimum value is $\dfrac{9}{16}$.

77. The reason the "puzzle" works as it does is because the sum of the fractions $\dfrac{1}{2}, \dfrac{1}{3}$ and $\dfrac{1}{9}$ is $\dfrac{17}{18}$, not 1.

Chapter Review Exercises

1. $\begin{array}{r} 9 \\ 2\overline{)19} \\ \underline{-18} \\ 1 \end{array}$ $\quad \dfrac{19}{2} = 9\dfrac{1}{2}$

2. $6\dfrac{2}{9} - 3\dfrac{7}{18} = 6\dfrac{4}{18} - 3\dfrac{7}{18}$
   $= 5\dfrac{22}{18} - 3\dfrac{7}{18}$
   $= 2\dfrac{15}{18} = 2\dfrac{5}{6}$

3. $x \div y$
   $2\dfrac{5}{8} \div 1\dfrac{3}{4} = \dfrac{21}{8} \div \dfrac{7}{4}$
   $= \dfrac{21}{8} \cdot \dfrac{4}{7}$
   $= \dfrac{3}{2} = 1\dfrac{1}{2}$

4. $\left(-2\dfrac{1}{3}\right) \cdot \dfrac{3}{7} = -\left(\dfrac{7}{3} \cdot \dfrac{3}{7}\right)$
   $= -\dfrac{7 \cdot 3}{3 \cdot 7}$
   $= -1$

5. $3\dfrac{3}{4} \div 1\dfrac{7}{8} = \dfrac{15}{4} \div \dfrac{15}{8}$
   $= \dfrac{15}{4} \cdot \dfrac{8}{15}$
   $= \dfrac{15 \cdot 8}{4 \cdot 15}$
   $= \dfrac{3 \cdot 5 \cdot 2 \cdot 2 \cdot 2}{2 \cdot 2 \cdot 3 \cdot 5} = 2$

6. $3 \cdot \dfrac{8}{9} = \dfrac{3}{1} \cdot \dfrac{8}{9}$
   $= \dfrac{3 \cdot 8}{1 \cdot 9}$
   $= \dfrac{3 \cdot 2 \cdot 2 \cdot 2}{1 \cdot 3 \cdot 3}$
   $= \dfrac{8}{3} = 2\dfrac{2}{3}$

7. $\dfrac{x}{y+z}$
   $\dfrac{\dfrac{7}{8}}{\dfrac{4}{5} + \left(-\dfrac{1}{2}\right)} = \dfrac{\dfrac{7}{8}}{\dfrac{3}{10}}$
   $= \dfrac{7}{8} \div \dfrac{3}{10}$
   $= \dfrac{7}{8} \cdot \dfrac{10}{3} = \dfrac{35}{12} = 2\dfrac{11}{12}$

8. $\dfrac{3}{5} = \dfrac{9}{15}$ $\quad \dfrac{7}{15} = \dfrac{7}{15}$
   $\dfrac{9}{15} > \dfrac{7}{15}$
   $\dfrac{3}{5} > \dfrac{7}{15}$

9. $50 = 2 \cdot 5^2$
   $75 = 3 \cdot 5^2$
   $\text{LCM} = 2 \cdot 3 \cdot 5^2 = 150$

10. $6\dfrac{11}{15} + 4\dfrac{7}{10} = 6\dfrac{22}{30} + 4\dfrac{21}{30}$
    $= 10\dfrac{43}{30}$
    $= 11\dfrac{13}{30}$

11. xy
    $8 \cdot \dfrac{5}{12} = \dfrac{8}{1} \cdot \dfrac{5}{12}$
    $= \dfrac{8 \cdot 5}{1 \cdot 12}$
    $= \dfrac{2 \cdot 2 \cdot 2 \cdot 5}{1 \cdot 2 \cdot 2 \cdot 3}$
    $= \dfrac{10}{3} = 3\dfrac{1}{3}$

12. $\dfrac{10}{7}, 1\dfrac{3}{7}$

13. $\dfrac{7}{8} = \dfrac{35}{40}$ $\quad \dfrac{17}{20} = \dfrac{34}{40}$
    $\dfrac{35}{49} > \dfrac{34}{40}$
    $\dfrac{7}{8} > \dfrac{17}{20}$

14. $\dfrac{\dfrac{5}{8} - \dfrac{1}{4}}{\dfrac{1}{2} + \dfrac{1}{8}} = \dfrac{\dfrac{3}{8}}{\dfrac{5}{8}}$
    $= \dfrac{3}{8} \div \dfrac{5}{8}$
    $= \dfrac{3}{8} \cdot \dfrac{8}{5}$
    $= \dfrac{3 \cdot 8}{8 \cdot 5} = \dfrac{3}{5}$

15. $72 \div 9 = 8$
    $\dfrac{4}{9} = \dfrac{4 \cdot 8}{9 \cdot 8} = \dfrac{32}{72}$
    $\dfrac{32}{72}$ is equivalent to $\dfrac{4}{9}$.

16. x^2y^3

$$\left(\frac{8}{9}\right)^2 \cdot \left(-\frac{3}{4}\right)^3$$
$$= \frac{8}{9} \cdot \frac{8}{9} \cdot \left(-\frac{3}{4}\right) \cdot \left(-\frac{3}{4}\right) \cdot \left(-\frac{3}{4}\right)$$
$$= -\left(\frac{8}{9} \cdot \frac{8}{9} \cdot \frac{3}{4} \cdot \frac{3}{4} \cdot \frac{3}{4}\right)$$
$$= -\frac{8 \cdot 8 \cdot 3 \cdot 3 \cdot 3}{9 \cdot 9 \cdot 4 \cdot 4 \cdot 4} = -\frac{1}{3}$$

17. $ab^2 - c$

$$4 \cdot \left(\frac{1}{2}\right)^2 - \frac{5}{7} = 4 \cdot \frac{1}{4} - \frac{5}{7}$$
$$= \frac{4}{1} \cdot \frac{1}{4} - \frac{5}{7} = 1 - \frac{5}{7}$$
$$= \frac{7}{7} - \frac{5}{7}$$
$$= \frac{2}{7}$$

18. $42 = 2 \cdot 3 \cdot 7$
$63 = 3 \cdot 3 \cdot 7$
GCF $= 3 \cdot 7 = 21$

19. $2\frac{5}{14} = \frac{(14 \cdot 2) + 5}{14}$
$= \frac{28 + 5}{14} = \frac{33}{14}$

20. $x + y + z$

$$\frac{5}{8} + \left(-\frac{3}{4}\right) + \frac{1}{2} = \frac{5}{8} + \frac{-3}{4} + \frac{1}{2}$$
$$= \frac{5}{8} + \frac{-6}{8} + \frac{4}{8}$$
$$= \frac{5 + (-6) + 4}{8}$$
$$= \frac{3}{8}$$

21. $\frac{5}{9} \div \left(-\frac{2}{3}\right) = -\left(\frac{5}{9} \div \frac{2}{3}\right)$
$= -\left(\frac{5}{9} \cdot \frac{3}{2}\right)$
$= -\frac{5 \cdot 3}{9 \cdot 2}$
$= -\frac{5 \cdot 3}{3 \cdot 3 \cdot 2} = -\frac{5}{6}$

22. $\frac{2}{5} \div \frac{4}{7} + \frac{3}{8} = \frac{2}{5} \cdot \frac{7}{4} + \frac{3}{8}$
$= \frac{7}{10} + \frac{3}{8}$
$= \frac{28}{40} + \frac{15}{40}$
$= \frac{28 + 15}{40} = \frac{43}{40} = 1\frac{3}{40}$

23. $5\frac{1}{4} \cdot \frac{8}{9} \cdot (-3) = -\left(\frac{21}{4} \cdot \frac{8}{9} \cdot \frac{3}{1}\right)$
$= -\frac{21 \cdot 8 \cdot 3}{4 \cdot 9 \cdot 1}$
$= -\frac{3 \cdot 7 \cdot 2 \cdot 2 \cdot 2 \cdot 3}{2 \cdot 2 \cdot 3 \cdot 3 \cdot 1}$
$= -14$

24. $\frac{2}{3} - \frac{11}{18} = \frac{12}{18} - \frac{11}{18}$
$= \frac{12 - 11}{18}$
$= \frac{1}{18}$

25. $\frac{7}{8} - \left(-\frac{5}{6}\right) = \frac{7}{8} - \left(-\frac{5}{6}\right)$
$= \frac{21}{24} - \frac{-20}{24}$
$= \frac{21 - (-20)}{24} = \frac{21 + 20}{24}$
$= \frac{41}{24} = 1\frac{17}{24}$

26. $\left(-\frac{3}{8}\right)^2 \cdot 4^2 = \frac{9}{64} \cdot 16$
$= \frac{9}{64} \cdot \frac{16}{1}$
$= \frac{9}{4} = 2\frac{1}{4}$

27. $3\frac{7}{12} + 5\frac{1}{2} = 3\frac{7}{12} + 5\frac{6}{12}$
$= 8\frac{13}{12} = 9\frac{1}{12}$

28. $\frac{30}{105} = \frac{2 \cdot 3 \cdot 5}{3 \cdot 5 \cdot 7} = \frac{2}{7}$

29. $a - b$
$7 - 2\frac{3}{10} = 6\frac{10}{10} - 2\frac{3}{10}$
$= 4\frac{7}{10}$

30. $-\dfrac{5}{9} = \dfrac{1}{6} + p$

$-\dfrac{5}{9} - \dfrac{1}{6} = \dfrac{1}{6} - \dfrac{1}{6} + p$

$-\dfrac{10}{18} - \dfrac{3}{18} = p$

$-\dfrac{13}{18} = p$

The solution is $-\dfrac{13}{18}$.

31. **Strategy** To find the fraction, write a fraction with 40 in the numerator and the number of minutes in one hour (60) in the denominator. Write the fraction in simplest form.

Solution $\dfrac{40}{60} = \dfrac{2 \cdot 2 \cdot 2 \cdot 5}{2 \cdot 2 \cdot 3 \cdot 5} = \dfrac{2}{3}$

40 min is $\dfrac{2}{3}$ of an hour.

32. **Strategy** To find the entire length, substitute $12\dfrac{1}{12}$, $29\dfrac{1}{3}$, and $26\dfrac{3}{4}$ for a, b, and c in the given formula and solve for P.

Solution $P = a + b + c$

$P = 12\dfrac{1}{12} + 29\dfrac{1}{3} + 26\dfrac{3}{4}$

$= 12\dfrac{1}{12} + 29\dfrac{4}{12} + 26\dfrac{9}{12}$

$= 68\dfrac{2}{12} = 68\dfrac{1}{6}$

The entire length is $68\dfrac{1}{6}$ yd.

33. **Strategy** To find the amount of weight to gain:
→Add the amounts already gained $\left(3\dfrac{1}{2} + 2\dfrac{1}{4}\right)$
→Subtract the amount gained from the goal amount (12).

Solution $3\dfrac{1}{2} + 2\dfrac{1}{4} = 3\dfrac{2}{4} + 2\dfrac{1}{4}$

$= 5\dfrac{3}{4}$

$12 - 5\dfrac{3}{4} = 11\dfrac{4}{4} - 5\dfrac{3}{4} = 6\dfrac{1}{4}$

The wrestler has $6\dfrac{1}{4}$ lb left to gain.

34. **Strategy** To find the number of units:
→Find the number of minutes in 8 h.
→Divide the number of minutes worked by the time to assemble one unit.

Solution $8 \cdot 60 = 480$

$480 \div 2\dfrac{1}{2} = 480 \div \dfrac{5}{2}$

$= \dfrac{480}{1} \cdot \dfrac{2}{5}$

$= 192$

35. **Strategy** To find the overtime pay, substitute $6\dfrac{1}{4}$ for H and 24 for R in the given formula and solve for P.

Solution $P = RH$

$P = 24 \cdot 6\dfrac{1}{4}$

$P = 24 \cdot \dfrac{25}{4}$

$P = \dfrac{24}{1} \cdot \dfrac{25}{4} = 150$

The employee is due $150 in overtime pay.

Chapter 3: *Fractions*

36. Strategy To find the final veolcity, substitute 0 for S and $15\frac{1}{2}$ for t in the given formula and solve for t.

Solution
$V = S + 32t$
$V = 0 + 32 \cdot 15\frac{1}{2}$
$V = 32 \cdot \frac{31}{2}$
$V = \frac{32}{1} \cdot \frac{31}{2} = 496$
The final velocity is 496 ft/s.

Chapter Test

1. $7\overline{)18}$ $\quad \frac{18}{7} = 2\frac{4}{7}$
   $\underline{-14}$
   $\quad 4$

2. $7\frac{3}{4} - 3\frac{5}{6} = 7\frac{9}{12} - 3\frac{10}{12}$
   $= 6\frac{21}{12} - 3\frac{10}{12}$
   $= 3\frac{11}{12}$

3. xy
   $6\frac{3}{7} \cdot 3\frac{1}{2} = \frac{45}{7} \cdot \frac{7}{2}$
   $= \frac{45}{2}$
   $= 22\frac{1}{2}$

4. $-\frac{2}{3} \cdot \left(\frac{-7}{8}\right) = \frac{2}{3} \cdot \frac{7}{8}$
   $= \frac{2 \cdot 7}{3 \cdot 8}$
   $= \frac{7}{12}$

5. $30 = 2 \cdot 3 \cdot 5$
   $45 = 3 \cdot 3 \cdot 5$
   $\text{LCM} = 2 \cdot 3^2 \cdot 5 = 90$

6. $\frac{11}{12} + \left(\frac{-3}{8}\right) = \frac{22}{24} + \frac{-9}{24}$
   $= \frac{22 - 9}{24}$
   $= \frac{13}{24}$

7. $x^3 y^2$
   $\left(1\frac{1}{2}\right)^3 \cdot \left(\frac{5}{6}\right)^2 = \left(\frac{3}{2}\right)^3 \cdot \left(\frac{5}{6}\right)^2$
   $= \frac{3}{2} \cdot \frac{3}{2} \cdot \frac{3}{2} \cdot \frac{5}{6} \cdot \frac{5}{6}$
   $= \frac{3 \cdot 3 \cdot 3 \cdot 5 \cdot 5}{2 \cdot 2 \cdot 2 \cdot 6 \cdot 6}$
   $= \frac{75}{32} = 2\frac{11}{32}$

8. $3\frac{4}{5} = \frac{(5 \cdot 3) + 4}{5}$
   $= \frac{15 + 4}{5} = \frac{19}{5}$

9. $-\frac{7}{12} \div \left(\frac{-3}{4}\right) = \frac{7}{12} \cdot \frac{4}{3}$
   $= \frac{7 \cdot 4}{12 \cdot 3} = \frac{7}{9}$

10. $\frac{2}{7} \div \frac{3}{14} + \frac{2}{3} = \frac{2}{7} \cdot \frac{14}{3} + \frac{2}{3}$
    $= \frac{4}{3} + \frac{2}{3} = 2$

11. $\frac{x}{yz}$
    $\frac{\frac{7}{20}}{\frac{2}{15} \cdot \frac{3}{8}} = \frac{\frac{7}{20}}{\frac{1}{20}}$
    $= \frac{7}{20} \div \frac{1}{20}$
    $= \frac{7}{20} \cdot \frac{20}{1} = 7$

12. $18 = 2 \cdot 3^2$
    $54 = 2 \cdot 3^3$
    $\text{GCF} = 2 \cdot 3^2 = 18$

13. $\dfrac{13}{14} - \dfrac{16}{21} = \dfrac{39}{42} - \dfrac{32}{42}$

$= \dfrac{7}{42} = \dfrac{1}{6}$

14. $\dfrac{60}{75} = \dfrac{2 \cdot 2 \cdot 3 \cdot 5}{3 \cdot 5 \cdot 5} = \dfrac{4}{5}$

15. $x + y + z$

$1\dfrac{3}{8} + \dfrac{1}{2} + \dfrac{5}{6} = 1\dfrac{9}{24} + \dfrac{12}{24} + \dfrac{20}{24}$

$= 2\dfrac{17}{24}$

16. $\dfrac{5}{6} = \dfrac{25}{30} \qquad \dfrac{11}{15} = \dfrac{22}{30}$

$\dfrac{25}{30} > \dfrac{22}{30}$

$\dfrac{5}{6} > \dfrac{11}{15}$

17. $a^2 b - c^2$

$\left(\dfrac{2}{3}\right)^2 \cdot 9 - \left(\dfrac{3}{5}\right)^2 = \dfrac{2}{3} \cdot \dfrac{2}{3} \cdot 9 - \dfrac{3}{5} \cdot \dfrac{3}{5}$

$= 4 - \dfrac{9}{25}$

$= \dfrac{100}{25} - \dfrac{9}{25}$

$= \dfrac{91}{25} = 3\dfrac{16}{25}$

18. $\dfrac{\dfrac{3}{4} - \dfrac{1}{3}}{\dfrac{1}{6} + \dfrac{1}{3}} = \dfrac{\dfrac{9}{12} - \dfrac{4}{12}}{\dfrac{1}{6} + \dfrac{2}{6}} = \dfrac{\dfrac{5}{12}}{\dfrac{3}{6}}$

$= \dfrac{5}{12} \div \dfrac{3}{6}$

$= \dfrac{5}{12} \cdot \dfrac{6}{3} = \dfrac{5}{6}$

19. $\dfrac{x - y}{z^3}$

$\dfrac{\dfrac{4}{9} - \dfrac{10}{27}}{\left(\dfrac{1}{3}\right)^3} = \dfrac{\dfrac{12}{27} - \dfrac{10}{27}}{\dfrac{2}{3} \cdot \dfrac{2}{3} \cdot \dfrac{2}{3}} = \dfrac{\dfrac{2}{27}}{\dfrac{8}{27}}$

$= \dfrac{2}{27} \div \dfrac{8}{27}$

$= \dfrac{2}{27} \cdot \dfrac{27}{8} = \dfrac{1}{4}$

20. $x \div y$

$-\dfrac{8}{9} \div \dfrac{16}{27} = \dfrac{-8}{9} \cdot \dfrac{27}{16}$

$= -\dfrac{3}{2} = -1\dfrac{1}{2}$

21. $\dfrac{3x}{5} = -\dfrac{3}{10}$

$\dfrac{5}{3} \cdot \dfrac{3}{5} x = \dfrac{5}{3} \cdot \left(-\dfrac{3}{10}\right)$

$x = -\dfrac{1}{2}$

22. $z + \dfrac{1}{5} = \dfrac{11}{20}$

$\begin{array}{c|c} -\dfrac{3}{4} + \dfrac{1}{5} & \dfrac{11}{20} \\ -\dfrac{15}{20} + \dfrac{4}{20} & \dfrac{11}{20} \\ -\dfrac{11}{20} & \dfrac{11}{20} \end{array}$

$-\dfrac{11}{20} \neq \dfrac{11}{20}$

No, $-\dfrac{3}{4}$ is not a solution of the equation.

23. $2\dfrac{7}{8} \cdot \dfrac{2}{11} \cdot 4 = \dfrac{23}{8} \cdot \dfrac{2}{11} \cdot \dfrac{4}{1}$

$= \dfrac{23}{11} = 2\dfrac{1}{11}$

24. $x + \dfrac{1}{3} = \dfrac{5}{6}$

$x + \dfrac{1}{3} - \dfrac{1}{3} = \dfrac{5}{6} - \dfrac{1}{3}$

$x = \dfrac{5}{6} - \dfrac{2}{6}$

$x = \dfrac{3}{6} = \dfrac{1}{2}$

25. $28 \div 7 = 4$

$\dfrac{3}{7} = \dfrac{3 \cdot 4}{7 \cdot 4} = \dfrac{12}{28}$

$\dfrac{12}{28}$ is equivalent to $\dfrac{3}{7}$

78   Chapter 3: Fractions

26. The unknown number: n

[a number minus one-half] equals [one-third]

$n - \dfrac{1}{2} = \dfrac{1}{3}$

$n - \dfrac{1}{2} + \dfrac{1}{2} = \dfrac{1}{3} + \dfrac{1}{2}$

$n = \dfrac{2}{6} + \dfrac{3}{6}$

$n = \dfrac{5}{6}$

The number is $\dfrac{5}{6}$.

27. $\dfrac{10}{24} = \dfrac{5}{12}$

28. Strategy   To find the amount of weight to lose:
→Add the amounts already lost $\left(11\dfrac{1}{6} + 8\dfrac{5}{8}\right)$.
→Subtract the amount lost from the goal amount (30).

Solution $11\dfrac{1}{6} + 8\dfrac{5}{8} = 11\dfrac{4}{24} + 8\dfrac{15}{24}$

$= 19\dfrac{19}{24}$

$30 - 19\dfrac{19}{24} = 29\dfrac{24}{24} - 19\dfrac{19}{24}$

$= 10\dfrac{5}{24}$

The patient has $10\dfrac{5}{24}$ lbs left to lose.

29. Strategy   To find the amount of hamburger meat, multiply the number of hamburgers per person (2) by the weight of each hamburger $\left(\dfrac{1}{4}\right)$ by the number of guests (35).

Solution $2 \cdot \dfrac{1}{4} \cdot 35 = \dfrac{2}{1} \cdot \dfrac{1}{4} \cdot \dfrac{35}{1} = 17\dfrac{1}{2}$

You should buy $17\dfrac{1}{2}$ lbs of hamburger.

30. Strategy   To find the amount of felt needed, substitute 20 for b and 12 for h in the given formula and solve for A.

Solution $A = \dfrac{1}{2}bh$

$A = \dfrac{1}{2} \cdot 20 \cdot 12 = 120$

The amount of felt needed is 120 in^2.

31. Strategy   To find the amount of hours still required:
→Add the hours already contributed $\left(7\dfrac{1}{4} + 2\dfrac{3}{4}\right)$.
→Subtract the amount already contributed from the total hours required (20).

Solution $7\dfrac{1}{4} + 2\dfrac{3}{4} = 10$

$20 - 10 = 10$

You must still contribute 10 h of community service.

32. Strategy To find the number of units:
→ Multiply the number of hours (6) by 60 minutes.
→ Divide the minutes by the time to assemble one unit $\left(4\frac{1}{2}\right)$.

Solution $6 \cdot 60 = 360$

$$360 \div 4\frac{1}{2} = \frac{360}{1} \div \frac{9}{2}$$
$$= \frac{360}{1} \cdot \frac{2}{9} = 80$$

The employee can assemble 80 units in 6 h.

33. Strategy To find the cost, substitute price per share $\left(12\frac{3}{4}\right)$ for S and number of shares (400) for N in the given formula and solve for C.

Solution $C = SN$

$$C = 12\frac{3}{4} \cdot 400 = \frac{51}{4} \cdot \frac{400}{1}$$
$$= 5,100$$

The cost is $5,100.

Cumulative Review Exercises

1. $3a + (a-b)^3$
$3 \cdot 4 + (4-1)^3 = 3 \cdot 4 + 3^3$
$= 3 \cdot 4 + 27$
$= 12 + 27$
$= 39$

2. $4 \cdot \frac{7}{8} = \frac{4}{1} \cdot \frac{7}{8}$
$= \frac{4 \cdot 7}{1 \cdot 8}$
$= \frac{2 \cdot 2 \cdot 7}{1 \cdot 2 \cdot 2 \cdot 2}$
$= \frac{7}{2} = 3\frac{1}{2}$

3. $4\frac{7}{9} + 3\frac{5}{6} = 4\frac{14}{18} + 3\frac{15}{18}$
$= 7\frac{29}{18}$
$= 8\frac{11}{18}$

4. $-42 - (-27) = -42 + 27$
$= -15$

5. $72 = 2 \cdot 2 \cdot 2 \cdot 3 \cdot 3$
$108 = 2 \cdot 2 \cdot 3 \cdot 3 \cdot 3$
GCF $= 2 \cdot 2 \cdot 3 \cdot 3 = 36$

6. $3\frac{1}{13} \cdot 5\frac{1}{5} = \frac{40}{13} \cdot \frac{26}{5} = \frac{40 \cdot 26}{13 \cdot 5}$
$= \frac{2 \cdot 2 \cdot 2 \cdot 5 \cdot 2 \cdot 13}{13 \cdot 5}$
$= 16$

7. $\frac{8}{9} \div \left(-\frac{4}{5}\right) = -\left(\frac{8}{9} \div \frac{4}{5}\right)$
$= -\left(\frac{8}{9} \cdot \frac{5}{4}\right)$
$= -\frac{8 \cdot 5}{9 \cdot 4}$
$= -\frac{2 \cdot 2 \cdot 2 \cdot 5}{3 \cdot 3 \cdot 2 \cdot 2}$
$= -\frac{10}{9} = -1\frac{1}{9}$

8. $-\frac{2}{3} - \left(-\frac{2}{5}\right) = \frac{-2}{3} - \left(\frac{-2}{5}\right)$
$= \frac{-10}{15} - \frac{-6}{15} = \frac{-10 - (-6)}{15}$
$= \frac{-10 + 6}{15} = \frac{-4}{15} = -\frac{4}{15}$

9. $\dfrac{\frac{1}{5} + \frac{1}{4}}{\frac{1}{4} - \frac{1}{5}} = \dfrac{\frac{9}{20}}{\frac{1}{20}}$
$= \frac{9}{20} \div \frac{1}{20}$
$= \frac{9}{20} \cdot \frac{20}{1} = \frac{9 \cdot 20}{20 \cdot 1}$
$= 9$

10. $\frac{7}{11} = \frac{35}{55}$ $\frac{4}{5} = \frac{44}{55}$
$\frac{35}{55} < \frac{44}{55}$
$\frac{7}{11} < \frac{4}{5}$

11. $-2\dfrac{1}{3} \div 1\dfrac{2}{7} = -\left(\dfrac{7}{3} \div \dfrac{9}{7}\right)$
$= -\left(\dfrac{7}{3} \cdot \dfrac{7}{9}\right)$
$= -\dfrac{7 \cdot 7}{3 \cdot 9}$
$= -\dfrac{7 \cdot 7}{3 \cdot 3 \cdot 3} = -\dfrac{49}{27}$
$= -1\dfrac{22}{27}$

12. $-\dfrac{3}{8} \cdot \dfrac{2}{5} \cdot \left(-\dfrac{4}{9}\right) = \dfrac{3}{8} \cdot \dfrac{2}{5} \cdot \dfrac{4}{9}$
$= \dfrac{3 \cdot 2 \cdot 4}{8 \cdot 5 \cdot 9}$
$= \dfrac{3 \cdot 2 \cdot 2 \cdot 2}{2 \cdot 2 \cdot 2 \cdot 5 \cdot 3 \cdot 3}$
$= \dfrac{1}{15}$

13. abc
$\dfrac{4}{7} \cdot 1\dfrac{1}{6} \cdot 3 = \dfrac{4}{7} \cdot \dfrac{7}{6} \cdot \dfrac{3}{1}$
$= \dfrac{4 \cdot 7 \cdot 3}{7 \cdot 6 \cdot 1}$
$= \dfrac{2 \cdot 2 \cdot 7 \cdot 3}{7 \cdot 2 \cdot 3 \cdot 1} = 2$

14. $8\dfrac{3}{4} - 1\dfrac{5}{7} = 8\dfrac{21}{28} - 1\dfrac{20}{28}$
$= 7\dfrac{1}{28}$

15. $\dfrac{7}{12} - \left(-\dfrac{3}{8}\right) = \dfrac{7}{12} - \dfrac{-3}{8}$
$= \dfrac{14}{24} - \dfrac{-9}{24} = \dfrac{14 - (-9)}{24}$
$= \dfrac{14 + 9}{24} = \dfrac{23}{24}$

16. $\dfrac{2}{5} \div \dfrac{9-6}{3+7} + \left(-\dfrac{1}{2}\right)^2 = \dfrac{2}{5} \div \dfrac{3}{10} + \left(-\dfrac{1}{2}\right)^2$
$= \dfrac{2}{5} \div \dfrac{3}{10} + \dfrac{1}{4}$
$= \dfrac{2}{5} \cdot \dfrac{10}{3} + \dfrac{1}{4}$
$= \dfrac{4}{3} + \dfrac{1}{4} = \dfrac{19}{12}$
$= 1\dfrac{7}{12}$

17. $a - b$
$\dfrac{3}{4} - \left(-\dfrac{7}{8}\right) = \dfrac{3}{4} - \dfrac{-7}{8}$
$= \dfrac{6}{8} - \dfrac{-7}{8}$
$= \dfrac{6 - (-7)}{8}$
$= \dfrac{6 + 7}{8}$
$= \dfrac{13}{8} = 1\dfrac{5}{8}$

18. $1\dfrac{9}{16} + 4\dfrac{5}{8} = 1\dfrac{9}{16} + 4\dfrac{10}{16}$
$= 5\dfrac{19}{16} = 6\dfrac{3}{16}$

19. $28 = -7y$
$\dfrac{28}{-7} = \dfrac{-7y}{-7}$
$-4 = y$
The solution is -4.

20. $9 \overline{)\,41\,}$ $\dfrac{41}{9} = 4\dfrac{5}{9}$
    $\underline{-36}$
       5

21. $\dfrac{5}{14} - \dfrac{9}{42} = \dfrac{15}{42} - \dfrac{9}{42}$
$= \dfrac{6}{42} = \dfrac{1}{7}$

22. $x^3 y^4$
$\left(\dfrac{7}{12}\right)^3 \left(\dfrac{6}{7}\right)^4 = \dfrac{7}{12} \cdot \dfrac{7}{12} \cdot \dfrac{7}{12} \cdot \dfrac{6}{7} \cdot \dfrac{6}{7} \cdot \dfrac{6}{7} \cdot \dfrac{6}{7}$
$= \dfrac{7 \cdot 7 \cdot 7 \cdot 6 \cdot 6 \cdot 6 \cdot 6}{12 \cdot 12 \cdot 12 \cdot 7 \cdot 7 \cdot 7 \cdot 7}$
$= \dfrac{3}{28}$

23. $2a - (b - a)^2$
$2 \cdot 2 - (-3 - 2)^2 = 2 \cdot 2 - (-5)^2$
$= 2 \cdot 2 - 25$
$= 4 - 25 = 4 + (-25)$
$= -21$

24.   6,847
     3,501
   $\underline{+\ 924}$
   11,272

Cumulative Review

25. $(x-y)^3 + 5x$
    $(8-6)^3 + 5 \cdot 8 = 2^3 + 5 \cdot 8$
    $= 8 + 5 \cdot 8$
    $= 8 + 40$
    $= 48$

26. $x + \dfrac{4}{5} = \dfrac{1}{4}$
    $x + \dfrac{4}{5} - \dfrac{4}{5} = \dfrac{1}{4} - \dfrac{4}{5}$
    $x = \dfrac{5}{20} - \dfrac{16}{20}$
    $x = \dfrac{5 - 16}{20}$
    $x = -\dfrac{11}{20}$
    The solution is $-\dfrac{11}{20}$.

27. $89,357 \rightarrow 90,000$
    $66,042 \rightarrow -70,000$
    $20,000$

28. $-8 - (-12) - (-15) - 32$
    $= -8 + 12 + 15 + (-32)$
    $= 4 + 15 + (-32)$
    $= 19 + (-32)$
    $= -13$

29. $7\dfrac{3}{4} = \dfrac{(4 \cdot 7) + 3}{4}$
    $= \dfrac{28 + 3}{4}$
    $= \dfrac{31}{4}$

30. $5\overline{)35}^{7}$
    $2\overline{)70}$
    $2\overline{)140}$
    $140 = 2 \cdot 2 \cdot 5 \cdot 7 = 2^2 \cdot 5 \cdot 7$

31. **Strategy** To determine how many more calories you would burn:
    →Calculate the number of calories burned by bicycling at 12 mph for 4 h.
    →Calculate the number of calories burned by walking at a rate of 3 mph for 5 h.
    →Find the difference between the two calculations.

    Solution Bicycling: $4 \cdot 410 = 1,640$
    Walking: $320 \cdot 5 = 1,600$
    $1,640 - 1,600 = 40$
    You would burn 40 more calories by bicycling at 12 mph for 4 h.

32. **Strategy** To find the projected increase, subtract the population in 1990 (13,581,000) from the projected population in 2025 (15,321,000).

    Solution $15,321,000 - 13,581,000 = 1,740,000$
    The projected increase in population is 1,740,000 people.

33. **Strategy** To find the average life span of a $100 bill, multiply the average life span of a $1 $\left(1\dfrac{1}{2}\text{ years}\right)$ by 6.

    Solution $1\dfrac{1}{2} \cdot 6 = \dfrac{3}{2} \cdot \dfrac{6}{1} = 9$
    The average life span of a $100 bill is 9 years.

34. Strategy To find the length of fencing, substitute $16\frac{1}{2}$ for s in the given formula and solve for P.

Solution $P = 4s$

$P = 4 \cdot 16\frac{1}{2} = \frac{4}{1} \cdot \frac{33}{2}$

$P = \frac{132}{2} = 66$

The length of fencing needed is 66 ft.

35. Strategy To find the distance traveled, substitute $5\frac{1}{2}$ for r and $\frac{3}{4}$ for t in the given formula and solve for d.

Solution $d = rt$

$d = 5\frac{1}{2} \cdot \frac{3}{4}$

$d = \frac{11}{2} \cdot \frac{3}{4}$

$d = \frac{33}{8} = 4\frac{1}{8}$

The distance traveled is $4\frac{1}{8}$ miles.

36. Strategy To find the pressure, substitute $14\frac{3}{4}$ for D in the given formula and solve for P.

Solution $P = 15 + \frac{1}{2}D$

$P = 15 + \frac{1}{2} \cdot 14\frac{3}{4}$

$P = 15 + \frac{1}{2} \cdot \frac{59}{4}$

$P = 15 + 7\frac{3}{8} = 22\frac{3}{8}$

The pressure is $22\frac{3}{8}$ pounds per square inch.

Chapter 4: Decimals and Real Numbers

Prep Test

1. $\frac{3}{10}$

2. 36,900

3. four thousand seven hundred ninety-one

4. 6842

5. [number line with point at -3, marks from -6 to 6]

6. $-37 + 8892 + 465 = 8855 + 465 = 9320$

7. $2403 - (-765) = 2403 + 765 = 3168$

8. $-844(-91) = 76,804$

9.
```
     278 r18
23)6412
    -46
    ‾‾‾
    181
   -161
   ‾‾‾‾
    202
   -184
   ‾‾‾‾
     18
```

10. $8^2 = (8)(8) = 64$

Go Figure

If it takes one loaf 30 minutes to fill the oven, doubling in volume each minute, then at 29 minutes, the oven would be half filled. At 28 minutes, the oven would be one quarter filled with one loaf. So for the oven to be half filled with two loaves of bread, would be at the 28-minute mark.

Section 4.1

Objective A Exercises

1. The digit 5 is in the thousandths' place.

3. The digit 5 is in the ten-thousandths' place.

5. The digit 5 is in the hundredths' place.

7. $\frac{3}{10} = 0.3$ [three tenths]

9. $\frac{21}{100} = 0.21$ [twenty-one hundredths]

11. $\frac{461}{1,000} = 0.461$ [four hundred sixty-one thousandths]

13. $\frac{93}{1,000} = 0.093$ [ninety-three thousandths]

15. $0.1 = \frac{1}{10}$ [one tenth]

17. $0.47 = \frac{47}{100}$ [forty-seven hundredths]

19. $0.289 = \frac{289}{1,000}$ [two hundred eighty-nine thousandths]

21. $0.09 = \frac{9}{100}$ [nine hundredths]

23. thirty-seven hundredths

25. nine and four tenths

27. fifty-three ten-thousandths

29. forty-five thousandths

31. twenty-six and four hundredths

33. 3.0806

35. 407.03

37. 246.024

39. 73.02684

Objective B Exercises

41. $0.7 = 0.70$
    $0.70 > 0.56$
    $0.7 > 0.56$

43. $3.605 > 3.065$

45. $9.04 = 9.040$
    $9.004 < 9.040$
    $9.004 < 9.04$

47. $9.31 = 9.310$
    $9.310 > 9.031$
    $9.31 > 9.031$

49. $4.6 < 40.6$

51. $0.07046 > 0.07036$

53. 0.609, 0.660, 0.696, 0.699
    0.609, 0.66, 0.696, 0.699

55. 1.237, 1.327, 1.372, 1.732

57. 21.780, 21.805, 21.870, 21.875
21.78, 21.805, 21.87, 21.875

Objective C Exercises

59. 5.3̄9̄8 *Given place value*
    9 > 5
    5.398 rounded to the nearest tenth is 5.4.

61. 30.0̄0̄92 *Given place value*
    0 < 5
    30.0092 rounded to the nearest tenth is 30.0.

63. 413.5̄9̄72 *Given place value*
    7 > 5
    413.5972 rounded to the nearest hundredth is 413.60.

65. 6.061̄7̄45 *Given place value*
    7 > 5
    6.061745 rounded to the nearest thousandth is 6.062.

67. 96.8̄027 *Given place value*
    8 > 5
    96.8027 rounded to the nearest whole number is 97.

69. 5,439.8̄3 *Given place value*
    8 > 5
    5,439.83 rounded to the nearest whole number is 5,440.

71. 0.0235̄9̄1 *Given place value*
    9 > 5
    0.023591 rounded to the nearest ten-thousandth is 0.0236.

Objective D Exercises

73. Strategy To find the weight, round 0.1763668 to the nearest hundredth.

    Solution 0.1763668 rounded to the nearest hundredth is 0.18. The weight of a nickel to the nearest hundredth is 0.18 oz.

75. Strategy To find the distance, round 26.21875 to the nearest tenth.

    Solution 26.21875 rounded to the nearest tenth is 26.2. To the nearest tenth, the Boston Marathon is 26.2 mi.

77. Strategy To determine who had the greatest average yards per carry, compare all the averages and find the greatest.

    Solution 4.99 is greatest
    Barry Sanders had the greatest average carry.

79. Strategy To find the length, round 42.195 to the nearest tenth of a kilometer.

    Solution 42.195 rounded to the nearest tenth of a kilometer is 42.2. To the nearest tenth of a kilometer, the marathon was 42.2 km.

81. Strategy
    To find the shipping and handling charges for each order, compare each amount ordered with the amounts in the table and read the corresponding shipping and handling charges from the table.

    Solution
    a. $10.01 < $12.42 < $20.00
    The shipping and handling charge is $2.40.

    b. $20.01 < $23.56 < $30.00
    The shipping and handling charge is $3.60.

    c. $40.01 < $47.80 < $50.00
    The shipping and handling charge is $6.00.

    d. $66.91 > $50.01
    The shipping and handling charge is $7.00.

    e. $30.01 < $35.75 < $40.00
    The shipping and handling charge is $4.70.

f. $20.00 = $20.00
The shipping and handling charge is $2.40.

g. $10.01 < $18.25 < $20.00
The shipping and handling charge is $2.40.

Critical Thinking 4.1

83. a. Answers will vary.
For example, 0.11, 0.12, 0.13, 0.14, 0.15, 0.16, 0.17, 0.18, and 0.19 are numbers between 0.1 and 0.2. But any number of digits can be attached to 0.1, and the number will be between 0.1 and 0.2. For example, 0.123456789 is a number between 0.1 and 0.2.

b. Answers will vary.
For example, 1.01, 1.02, 1.03, 1.04, 1.05, 1.06, 1.07, 1.08, and 1.09 are numbers between 1 and 1.1. But any number of digits can be attached to 1.0, and the number will be between 1 and 1.1. For example, 1.0123456789 is a number between 1 and 1.1.

c. Answers will vary.
For example, 0.001, 0.002, 0.003, and 0.004 are numbers between 0 and 0.005. But any number of digits can be attached to 0.001, 0.002, 0.003, or 0.004, and the number will be between 0 and 0.005. For example, 0.00123456789 is a number between 0 and 0.005.

85. A sales tax or meal tax is always rounded up to the nearest cent. Also, if a product is priced at "3 for $1.00," a customer is charged $.34 for one item.
Generally interest paid by an institution is rounded down, for example, interest paid by a bank on a NOW account.

Section 4.2

Objective A Exercises

1. 111
    1.864
    39
    $+25.0781$
    65.9421

3. $2\,1$
    35.9
    8.217
    $+146.74$
    190.857

5. 36.47
    -15.21
    21.26

7. $7\,9\,10$
    $28.\cancel{0}\cancel{0}$
    -6.74
    21.26

9. $59\,11\,10$
    $6.\cancel{0}\cancel{2}\cancel{0}$
    -3.252
    2.768

11. $-42.1 - 8.6 = -42.1 + (-8.6)$
    $= -50.7$

13. $5.73 - 9.042 = 5.73 + (-9.042)$
    $= -3.312$

15. $-9.37 + 3.465 = -5.905$

17. $-19 - (-2.65) = -19 + 2.65$
    $= -16.35$

19. $-12.3 - 4.07 + 6.82$
    $= -12.3 + (-4.07) + 6.82$
    $= -16.37 + 6.82$
    $= -9.55$

21. $-5.6 - (-3.82) - 17.409$
    $= -5.6 + 3.82 + (-17.409)$
    $= -1.78 + (-17.409)$
    $= -19.189$

23. $6.24 + 8.573 + 19.06 + 22.488$
    $= 14.813 + 19.06 + 22.488$
    $= 33.873 + 22.488$
    $= 56.361$

25. $62.57 - 8.9 = 62.57 + (-8.9)$
    $= 53.67$

27. $-65.47 + (-32.91) = -98.38$

29. $-138.72 - 510.64$
    $= -138.72 + (-510.64) = -649.36$

31. $-31 - (-62.09) = -31 + 62.09$
    $= 31.09$

Chapter 4: *Decimals and Real Numbers*

33. $\begin{array}{r} 6.408 \\ +5.917 \\ \hline 12.325 \end{array} \rightarrow \begin{array}{r} 6 \\ +6 \\ \hline 12 \end{array}$

35. $\begin{array}{r} 56.87 \\ -23.24 \\ \hline 33.63 \end{array} \rightarrow \begin{array}{r} 60 \\ -20 \\ \hline 40 \end{array}$

37. $\begin{array}{r} 0.931 \\ -0.628 \\ \hline 0.303 \end{array} \rightarrow \begin{array}{r} 0.9 \\ -0.6 \\ \hline 0.3 \end{array}$

39. $\begin{array}{r} 87.65 \\ -49.032 \\ \hline 38.618 \end{array} \rightarrow \begin{array}{r} 90 \\ -50 \\ \hline 40 \end{array}$

41. **a.** $46.353 + 5.863 + 1.23 = 53.446$
    The total children is 53.446 million.

    b. $46.353 - 5.863 = 40.49$
    There are 40.49 million more children in Public School.

43. $x + y$
    $5.904 + (-7.063) = -1.159$

45. $x + y$
    $-6.175 + (-19.49) = -25.665$

47. $x + y + z$
    $-6.059 + 3.884 + 15.71$
    $= -2.175 + 15.71$
    $= 13.535$

49. $x + y + z$
    $-16.219 + 47 + (-2.3885)$
    $= 30.781 + (-2.3885)$
    $= 28.3925$

51. $x - y$
    $6.029 - (-4.708) = 6.029 + 4.708$
    $= 10.737$

53. $x - y$
    $-21.073 - 6.48 = -21.073 + (-6.48)$
    $= -27.553$

55. $x - y$
    $-8.21 - (-6.798) = -8.21 + 6.798$
    $= -1.412$

57. $\dfrac{0.8 - p = 3.6}{\begin{array}{c|c} 0.8 - (-2.8) & 3.6 \\ 0.8 + 2.8 & 3.6 \\ 3.6 = 3.6 \end{array}}$
    Yes, -2.8 is a solution of the equation.

59. $\dfrac{27.4 = y - 9.4}{\begin{array}{c|c} 27.4 & 36.8 - 9.4 \\ 27.4 = 27.4 \end{array}}$
    Yes, 36.8 is a solution of the equation.

Objective B Exercises

61. $\begin{array}{r} 3.4 \\ \times 0.5 \\ \hline 1.70 \end{array}$

63. $\begin{array}{r} 8.29 \\ \times 0.004 \\ \hline 0.03316 \end{array}$

65. $(-6.3)(-2.4) = 15.12$

67. $-1.3(4.2) = -5.46$

69. $1.31(-0.006) = -0.00786$

71. $(-100)(4.73) = -473$

73. $1{,}000 \cdot 4.25 = 4{,}250$

75. $6.71 \cdot 10^4 = 67{,}100$

77. $(0.06)(-0.4)(-1.5) = (-0.024)(-1.5)$
    $= 0.036$

79. $\begin{array}{r} 9.81 \rightarrow 10 \\ 0.77 \rightarrow \times 0.8 \\ \hline 8.0 \end{array}$
    $9.81 \cdot 0.77 = 7.5537$

81. $\begin{array}{r} 6.88 \rightarrow 7 \\ 9.97 \rightarrow \times 10 \\ \hline 70 \end{array}$
    $6.88 \cdot 9.97 = 68.5936$

83. $\begin{array}{r} 28.45 \rightarrow 30 \\ 1.13 \rightarrow \times 1 \\ \hline 30 \end{array}$
    $28.45 \cdot 1.13 = 32.1485$

85. $20{,}000(0.6336) = 12{,}672$
    12,672 British pounds would be exchanged for 20,000 U.S. dollars.

87. ab
    $6.27 \cdot 8 = 50.16$

89. $10t$
    $10(-4.8) = -48$

91. ab
    $(0.379)(-0.22) = -0.08338$

93. cd
$(-2.537)(-9.1) = 23.0867$

95. $1.6 = -0.2z$
$\begin{array}{c|c} 1.6 & -0.2(-8) \\ 1.6 & = 1.6 \end{array}$
Yes, -8 is a solution of the equation.

97. $-83.25r = 8.325$
$\begin{array}{c|c} -83.25(-10) & 8.325 \\ 832.5 & \neq 8.325 \end{array}$
No, -10 is not a solution of the equation.

Objective C Exercises

99. $\begin{array}{r} 32.3 \\ 0.5 \overline{)16.15} \\ \underline{-15} \\ 11 \\ \underline{-10} \\ 15 \\ \underline{-15} \\ 0 \end{array}$

101. $27.08 \div (-0.4) = -67.7$

103. $(-3.312) \div (-0.8) = 4.14$

105. $-2.501 \div 0.41 = -6.1$

107. $55.63 \div 8.8 \approx 6.3$

109. $(-52.8) \div (-9.1) \approx 5.8$

111. $6.457 \div 8 \approx 0.81$

113. $0.0416 \div (-0.53) \approx -0.08$

115. $52.78 \div 10 = 5.278$

117. $48.05 \div 10^2 = 0.4805$

119. $-19.04 \div 0.75 \approx -25.4$

121. $27.735 \div (-60.3) \approx -0.5$

123. $42.43 \rightarrow 40$
$3.8 \rightarrow 4$
$40 \div 4 = 10$
$42.43 \div 3.8 \approx 11.17$

125. $6.398 \rightarrow 6$
$5.5 \rightarrow 6$
$6 \div 6 = 1$
$6.398 \div 5.5 \approx 1.16$

127. $1.237 \rightarrow 1$
$0.021 \rightarrow 0.02$
$1 \div 0.02 = 50$
$1.237 \div 0.021 \approx 58.90$

129. $33.14 \rightarrow 30$
$4.6 \rightarrow 5$
$30 \div 5 = 6$
$33.14 \div 4.6 \approx 7.20$

131. $9.9 \div 4.0 \approx 2.5$
DSL's market is 2.5 times greater in 2003 than 2001.

133. $\dfrac{x}{y}$
$\dfrac{3.542}{0.7} = 3.542 \div 0.7 = 5.06$

135. $\dfrac{x}{y}$
$\dfrac{0.648}{-2.7} = 0.648 \div (-2.7) = -0.24$

137. $\dfrac{x}{y}$
$\dfrac{-8.034}{-3.9} = (-8.034) \div (-3.9) = 2.06$

139. $\dfrac{x}{y}$
$\dfrac{-2.501}{0.41} = -2.501 \div 0.41 = -6.1$

141. $\dfrac{q}{-8} = -3.1$
$\begin{array}{c|c} \dfrac{24.8}{-8} & -3.1 \\ -3.1 & = -3.1 \end{array}$
Yes, 24.8 is a solution of the equation.

143. $21 = \dfrac{t}{0.4}$
$\begin{array}{c|c} 21 & \dfrac{-8.4}{0.4} \\ 21 & \approx -21 \end{array}$
No, -8.4 is not a solution of the equation.

Objective D Exercises

145. $\begin{array}{r} 0.375 \\ 8\overline{)3.000} \end{array}$
$\dfrac{3}{8} = 0.375$

147. $11\overline{)8.0000}^{0.7272}$
$\frac{8}{11} = 0.\overline{72}$

149. $12\overline{)7.00000}^{0.58333}$
$\frac{7}{12} = 0.58\overline{3}$

151. $4\overline{)7.00}^{1.75}$
$\frac{7}{4} = 1.75$

153. Write $\frac{1}{2}$ as a decimal.
$2\overline{)1.0}^{0.5}$
$1\frac{1}{2} = 1.5$

155. Write $\frac{1}{6}$ as a decimal.
$6\overline{)1.0000}^{0.1666} = 0.1\overline{6}$
$4\frac{1}{6} = 4.1\overline{6}$

157. Write $\frac{1}{4}$ as a decimal.
$4\overline{)1.00}^{0.25}$
$2\frac{1}{4} = 2.25$

159. Write $\frac{8}{9}$ as a decimal.
$9\overline{)8.000}^{0.888} = 0.\overline{8}$
$3\frac{8}{9} = 3.\overline{8}$

161. $0.2 = \frac{2}{10} = \frac{1}{5}$

163. $0.75 = \frac{75}{100} = \frac{3}{4}$

165. $0.125 = \frac{125}{1,000} = \frac{1}{8}$

167. $2.5 = 2\frac{5}{10} = 2\frac{1}{2}$

169. $4.55 = 4\frac{55}{100} = 4\frac{11}{20}$

171. $1.72 = 1\frac{72}{100} = 1\frac{18}{25}$

173. $0.045 = \frac{45}{1,000} = \frac{9}{200}$

175. $\frac{9}{10} = 0.90$
$0.90 > 0.89$
$\frac{9}{10} > 0.89$

177. $\frac{4}{5} = 0.800$
$0.800 < 0.803$
$\frac{4}{5} < 0.803$

179. $\frac{4}{9} = 0.\overline{444}$
$0.444 < 0.\overline{444}$
$0.444 < \frac{4}{9}$

181. $\frac{3}{25} = 0.12$
$0.13 > 0.12$
$0.13 > \frac{3}{25}$

183. $\frac{5}{16} = 0.3125$
$0.3125 > 0.3120$
$\frac{5}{16} > 0.312$

185. $\frac{10}{11} = 0.\overline{90}9$
$0.\overline{90}9 > 0.909$
$\frac{10}{11} > 0.909$

Objective E Exercises

187. Strategy To find your monthly salary, divide your annual salary (47,619) by 12.

Solution $47{,}619 \div 12 = 3{,}968.25$
Your monthly salary is $3,968.25.

189. Strategy To find the temperature fall during the 27-minute period, subtract the final temperature (−20) from the initial temperature (12.22).

Solution $12.22 - (-20) = 12.22 + 20 = 32.22$
The temperature dropped 32.22°C during the 27-minute period.

191. Strategy To find the cost per can, divide the total cost (8.89) by the number of cans (24).

Solution $8.89 \div 24 \approx 0.37$
A diet cola costs approximately $0.37 per can.

193. Strategy To find the cost to operate the motor, multiply the number of hours operating the motor (90) times the cost per hour (0.038).

Solution $90 \cdot 0.038 = 3.42$
The cost to run the motor is $3.42.

195. Strategy To find the amount deductible, multiply number of miles (11,842) by the standard deduction per mile for 2002 tax return (0.365).

Solution $11{,}842 \cdot 0.365 \approx 4322.33$
The amount deductible was $4322.33.

197. Strategy
  a. To find which 10-year period had the greatest increase, subtract consumption from 1950 from 1960, subtract consumption from 1960 from 1970, subtract consumption from 1970 from 1980, subtract consumption from 1980 from 1990, subtract 1990 from 2000, and compare results.

  b. To find which 10-year period had the least increase subtract consumption from 1950 from 1960, subtract consumption from 1960 from 1970, subtract consumption from 1970 from 1980, subtract consumption from 1980 from 1990, subtract 1990 from 2000, and compare results.

Solution
  a. 1950 to 1960, $2.2 - 1.7 = 0.5$
  1960 to 1970, $3.1 - 2.2 = 0.9$
  1970 to 1980, $4.4 - 3.1 = 1.3$
  1980 to 1990, $5.4 - 4.4 = 1.0$
  1990 to 2000, $5.5 - 5.4 = 0.1$
  The greatest increase was from 1970 to 1980.

  b. Use the Solution from part **a**.
  The increase was the least from 1990 to 2000.

199. Strategy To find the amount remaining in the budget:
  →Add the amounts already spent.
  →Subtract the amount already spent from the budget (1620).

Solution
```
    62.78
   164.93
    35.50
   560
 + 291.62
  -------
  1114.83

  1620.00
 -1114.83
  -------
   505.17
```
You have $505.17 remaining in the budget.

201. Strategy To find the bookkeeper's total income:
→ Find the amount of overtime pay by multiplying the hours of overtime (6) by the overtime rate (24.75).
→ Add the overtime pay to the regular salary (660).

Solution $6 \cdot 24.75 = 148.50$
$148.50 + 660 = 808.50$
The bookkeeper's total income for the week is $808.50.

203. Strategy To estimate the bill, estimate the following items and then add the estimated values: soup (5.75), cheese sticks (6.25), swordfish (26.95), chicken divan (24.95), and carrot cake (7.25).

Solution
$5.75 \rightarrow 6.00$
$6.25 \rightarrow 6.00$
$26.95 \rightarrow 30.00$
$24.95 \rightarrow 20.00$
$7.25 \rightarrow +7.00$
69.00

The bill is approximately $69.00.

205. Strategy
a. To determine if life expectancy has increased for both males and females between each 10-year period shown in the graph, see if the numbers increase from left to right along each of the lines in the graph.

b. To determine whether males or females had a longer life expectancy, compare the life expectancy for males in 2000 with the life expectancy for females in 2000. To find the difference in the life expectancy, subtract the smaller of the two numbers from the larger.

c. To determine during which year the difference between male and female life expectancy was greatest, subtract the lower life expectancy from the higher for each year shown in the graph. Find the largest difference.

Solution
a. The numbers along each of the lines in the graph increase from left to right. Life expectancy has increased for both males and females between each 10-year period shown in the graph. Yes.

b. $79.4 > 73.6$
Females had a longer life expectancy.
$79.4 - 73.6 = 5.8$
Female life expectancy was longer by 5.8 years.

c. 1900: $49.6 - 49.1 = 0.5$
1910: $53.7 - 50.2 = 3.5$
1920: $56.3 - 54.6 = 1.7$
1930: $61.4 - 58.0 = 3.4$
1940: $65.3 - 60.9 = 4.4$
1950: $70.9 - 65.3 = 5.6$
1960: $73.2 - 66.6 = 6.6$
1970: $74.8 - 67.1 = 7.7$
1980: $77.5 - 69.9 = 7.6$
1990: $78.6 - 71.8 = 6.8$
2000: $79.4 - 73.6 = 5.6$
7.7 is the largest difference.
The difference between male and female life expectancy was greatest in 1970.

207. Strategy To find the total cost for parts **a, b,** and **c**.
→Multiply the weight of each item by the quantity.
→Multiply the cost of each item by the quantity.
→Find the sum of the weights.
→Find the sum of the costs.
→Convert the weights from ounces to pounds.
→Find the cost to the zone.
→Add the cost of candy to shipping charges

Solution
a. For 2 of code 116,
$2 \cdot 8 = 16$ oz
$2 \cdot 2.90 = \$5.80$
For 1 of code 130,
$1 \cdot 7 = 7$ oz
$1 \cdot 5.25 = \$5.25$
For 3 of code 149,
$3 \cdot 8 = 24$ oz
$3 \cdot 6.25 = \$18.75$
For 4 of code 182,
$4 \cdot 8 = 32$ oz
$4 \cdot 3.70 = \$14.80$
$16 + 7 + 24 + 32 = 79$oz
$5.80 + 5.25 + 18.75 + 14.80 = \$44.60$
$79/16 = 4.94$ lbs
To send 4.94 lbs to zone 4 will cost $8.30.
The total cost is $44.60 + 8.30 = \$52.90$.

b. For 1 of code 112,
$1 \cdot 16 = 16$ oz
$1 \cdot 4.75 = \$4.75$
For 4 of code 117,
$4 \cdot 16 = 64$ oz
$4 \cdot 5.50 = \$22$
For 2 of code 131,
$2 \cdot 16 = 32$ oz
$2 \cdot 9.95 = \$19.90$
For 3 of code 160,
$3 \cdot 8 = 24$ oz
$3 \cdot 1.95 = \$5.85$
For 5 of code 182,
$5 \cdot 8 = 40$ oz
$5 \cdot 3.70 = \$18.50$
$16 + 64 + 32 + 24 + 40 = 176$oz
$4.75 + 22 + 19.90 + 5.85 + 18.50 = \$71$
$176/16 = 11$ lbs
Sending 11 lbs to zone 3 costs $8.60.
The total cost is $71 + 8.60 = \$79.60$.

c. For 3 of code 117,
$3 \cdot 16 = 48$ oz
$3 \cdot 5.50 = \$16.50$
For 1 of code 131,
$1 \cdot 16 = 16$ oz
$1 \cdot 9.95 = \$9.95$
For 2 of code 155,
$2 \cdot 8 = 16$ oz
$2 \cdot 4.80 = \$9.60$
For 4 of code 160,
$4 \cdot 8 = 32$ oz
$4 \cdot 1.95 = \$7.80$
For 1 of code 182,
$1 \cdot 8 = 8$ oz
$1 \cdot 3.70 = \$3.70$
For 3 of code 199,
$3 \cdot 8 = 24$ oz
$3 \cdot 1.90 = \$5.70$
$48 + 16 + 16 + 32 + 8 + 24 = 136$ oz
$16.5 + 9.95 + 9.60 + 7.8 + 3.70 + 5.7 = \$53.25$
$136/16 = 8.5$ lbs
To send 8.5 lbs to zone 2 will cost $7.80.
$53.25 + 7.80 = \$61.05$.
The total cost is $61.05.

209. Strategy To find the perimeter, substitute 4.5 for L and 3.25 for W in the formula below and solve for P.

Solution $P = 2L + 2W$
$P = 2 \cdot 4.5 + 2 \cdot 3.25$
$= 9 + 6.5 = 15.5$
The perimeter is 15.5 in.

211. Strategy To find the area, substitute 4.5 for L and 3.25 for W in the formula below and solve for A.

Solution $A = LW$
$A = 4.5 \cdot 3.25 = 14.625$
The area is 14.625 in^2.

213. Strategy To find the perimeter, substitute 2.8, 4.75, and 6.4 for a, b, and c in the formula below and solve for P.

Solution $P = a + b + c$
$P = 2.8 + 4.75 + 6.4 = 13.95$
The perimeter is 13.95 m.

215. Strategy To find the perimeter, substitute 3.5 for s in the formula below and solve for P.

Solution
$P = 4s$
$P = 4 \cdot 3.5 = 14$
The perimeter is 14 ft.

217. Strategy To find the employee's federal earnings, substitute 694.89 for E and 132.69 for W in the given formula and solve for F.

Solution
$F = E - W$
$F = 694.89 - 132.69$
$F = 562.20$
The federal earnings are $562.20.

219. Strategy To find the force on the falling object, substitute 4.25 for m and -9.80 for g in the given formula and solve for F.

Solution
$F = ma$
$F = 4.25(-9.80)$
$F = -41.65$
The force on the object is -41.65 newtons.

221. Strategy To find the equity, replace V by 225,000 and L by 167,853.25 in the given formula and solve for E.

Solution
$E = V - L$
$E = 225,000 - 167,853.25$
$E = 57,146.75$
The equity on the home is $57,146.75.

Critical Thinking 4.2

223. $31.93 = 3193¢
Find the whole number factors of 3193. The prime factorization of 3193 is $31 \cdot 103$. The number of cents charged per pen must be less than 50, since the price was reduced. The factor 103 is not less than 50. Therefore, the price of the pen was reduced to 31¢.

225. The largest amount by which the estimate of the sum could differ from the exact sum is 0.098. For example, 0.149 is the largest three-place decimal that would be rounded down to 0.1. The exact sum of 0.149 and 0.149 is 0.298. To estimate the sum of 0.149 and 0.149, we would round each number to 0.1, and $0.1 + 0.1 = 0.2$.
The difference between 0.298 and 0.2 is 0.098. 0.151 is the smallest three-place decimal that would be rounded up to 0.2. The exact sum of 0.151 and 0.151 is 0.302. To estimate the sum of 0.151 and 0.151, we would round each number to 0.2, and $0.2 + 0.2 = 0.4$.
The difference between 0.4 and 0.302 is 0.098. If one addend is rounded up and the other rounded down, the difference will be less than 0.098.

Section 4.3

Objective A Exercises

1. $y + 3.96 = 8.45$
$y + 3.96 - 3.96 = 8.45 - 3.96$
$y = 4.49$
The solution is 4.49.

3. $-9.3 = c - 15$
$-9.3 + 15 = c - 15 + 15$
$5.7 = c$
The solution is 5.7.

5. $7.3 = -\dfrac{n}{1.1}$
$-1.1(7.3) = -1.1\left(-\dfrac{n}{1.1}\right)$
$-8.03 = n$
The solution is -8.03.

7. $-7x = 8.4$
$\dfrac{-7x}{-7} = \dfrac{8.4}{-7}$
$x = -1.2$
The solution is -1.2.

9. $y - 0.234 = -0.09$
$y - 0.234 + 0.234 = -0.09 + 0.234$
$y = 0.144$
The solution is 0.144.

11. $6.21r = -1.863$
$\dfrac{6.21r}{6.21} = \dfrac{-1.863}{6.21}$
$r = -0.3$

13. $-0.001 = x + 0.009$
$-0.001 - 0.009 = x + 0.009 - 0.009$
$-0.01 = x$
The solution is -0.01.

15. $\frac{x}{2} = -0.93$
$2 \cdot \frac{x}{2} = 2(-0.93)$
$x = -1.86$
The solution is -1.86.

17. $-9.85y = 2.0685$
$\frac{-9.85y}{-9.85} = \frac{2.0685}{-9.85}$
$y = -0.21$
The solution is -0.21.

19. $-6v = 15$
$\frac{-6v}{-6} = \frac{15}{-6}$
$v = -2.5$
The solution is -2.5.

21. $0.908 = 2.913 + x$
$0.908 - 2.913 = 2.913 - 2.913 + x$
$-2.005 = x$
The solution is -2.005.

23. $\frac{t}{-2.1} = -7.8$
$-2.1\left(\frac{t}{-2.1}\right) = -2.1(-7.8)$
$t = 16.38$
The solution is 16.38.

Objective B Exercises

25. Strategy — To find the selling price of the calendar, replace P by 8.50 and C by 15.23 in the given formula and solve for S.

Solution
$P = S - C$
$8.50 = S - 15.23$
$8.50 + 15.23 = S - 15.23 + 15.23$
$23.73 = S$
The selling price of the calendar should be $23.73.

27. Strategy — To find the velocity, replace a by 16 and t by 6.3 in the given formula and solve for v.

Solution
$a = \frac{v}{t}$
$16 = \frac{v}{6.3}$
$6.3(16) = 6.3 \cdot \frac{v}{6.3}$
$100.8 = v$
The velocity is 100.8 ft/s.

29. Strategy — To find the cost per kilowatt-hour, replace c by 0.01, w by 1,000 and t by 4 in the given formula and solve for k.

Solution
$c = 0.001wtk$
$0.01 = 0.001 \cdot 1{,}000 \cdot 4 \cdot k$
$0.01 = 4k$
$\frac{0.01}{4} = \frac{4k}{4}$
$0.0025 = k$
The cost per kilowatt-hour is $0.0025.

31. Strategy — To find the monthly lease payment, write and solve an equation using x to represent the monthly lease payment.

Solution

the total of the monthly payments	is	the product of the number of months of the lease and the monthly lease payment

$15{,}387 = 60x$
$\frac{15{,}387}{60} = \frac{60x}{60}$
$256.45 = x$
The monthly lease payment is $256.45.

94 Chapter 4: Decimals and Real Numbers

33. Strategy To find the width, substitute 225 for A and 18 for L in the formula below and solve for W.

Solution
$$A = LW$$
$$225 = 18 \cdot W$$
$$\frac{225}{18} = \frac{18W}{18}$$
$$W = 12.5$$
The width is 12.5 ft.

35. Strategy To find the length, substitute 30 for P in the formula below and solve for s.

Solution
$$P = 4s$$
$$30 = 4 \cdot s$$
$$\frac{30}{4} = \frac{4s}{4}$$
$$s = 7.5$$
The length of each edge is 7.5 ft.

Critical Thinking 4.3

37. a. For example, $x - 0.04 = -1$.

b. For example, $-3x = -6.3$.

39. Since x is divided by a, we can solve the equation for x by multiplying each side of the equation by a.
$$12 = \frac{x}{a}$$
$$a \cdot 12 = a \cdot \frac{x}{a}$$
$$a \cdot 12 = x$$

Let a be any positive number. Then find the corresponding value of x. We will use 1, 3, 5, 10, 100, and 1,000.

$a \cdot 12 = x$	$a \cdot 12 = x$	$a \cdot 12 = x$
$1 \cdot 12 = x$	$3 \cdot 12 = x$	$5 \cdot 12 = x$
$12 = x$	$36 = x$	$60 = x$

$a \cdot 12 = x$	$a \cdot 12 = x$	$a \cdot 12 = x$
$10 \cdot 12 = x$	$100 \cdot 12 = x$	$1{,}000 \cdot 12 = x$
$120 = x$	$1{,}200 = x$	$12{,}000 = x$

Students should note that as the value of a increases, the value of x increases.

Section 4.4

Objective A Exercises

1. a. In their descriptions, students should paraphrase the definition of square root on page 267 (a square root of a positive number x is a number whose square is x) and then relate it to finding the square root of a perfect square. A definition of a perfect square would be appropriate within the discussion.

b. A student's description should include the idea of finding a factor of the radicand that is a perfect square. Be sure that the student indicates that the *largest* perfect square factor must be found. For example, 4 is a perfect square factor of 32, but it is not the largest perfect square factor. Therefore, if 32 is written as the product of 4 and 8, then $\sqrt{32} = \sqrt{4 \cdot 8} = 2\sqrt{8}$, which is not in simplest form. The largest square factor of 32 is 16, so 32 is written as the product of 16 and 2: $\sqrt{32} = \sqrt{16 \cdot 2} = 4\sqrt{2}$, which is in simplest form.

3. Since $6^2 = 36$, $\sqrt{36} = 6$.

5. Since $3^2 = 9$, $-\sqrt{9} = -3$.

7. Since $13^2 = 169$, $\sqrt{169} = 13$.

9. Since $15^2 = 225$, $\sqrt{225} = 15$.

11. Since $5^2 = 25$, $-\sqrt{25} = -5$.

13. Since $10^2 = 100$, $-\sqrt{100} = -10$.

15. $\sqrt{8+17} = \sqrt{25}$
    $= 5$

17. $\sqrt{49} + \sqrt{9} = 7 + 3$
    $= 10$

19. $\sqrt{121} - \sqrt{4} = 11 - 2$
    $= 9$

21. $3\sqrt{81} = 3 \cdot 9$
    $= 27$

22. $8\sqrt{36} = 8 \cdot 6$
    $= 48$

23. $-2\sqrt{49} = -2 \cdot 7$
    $= -14$

25. $5\sqrt{16} - 4 = 5 \cdot 4 - 4$
    $= 20 - 4$
    $= 16$

27. $3 + 10\sqrt{1} = 3 + 10 \cdot 1$
    $= 3 + 10$
    $= 13$

29. $\sqrt{4} - 2\sqrt{16} = 2 - 2 \cdot 4$
    $= 2 - 8$
    $= 2 + (-8)$
    $= -6$

31. $5\sqrt{25} + \sqrt{49} = 5 \cdot 5 + 7$
    $= 25 + 7$
    $= 32$

33. $\sqrt{\frac{1}{100}} = \frac{1}{10}$

35. $\sqrt{\frac{9}{16}} = \frac{3}{4}$

37. $\sqrt{\frac{1}{4}} + \sqrt{\frac{1}{64}} = \frac{1}{2} + \frac{1}{8}$
    $= \frac{4}{8} + \frac{1}{8} = \frac{5}{8}$

39. $-4\sqrt{xy}$
    $-4\sqrt{3 \cdot 12} = -4\sqrt{36}$
    $= -4 \cdot 6$
    $= -24$

41. $8\sqrt{x+y}$
    $8\sqrt{19+6} = 8\sqrt{25}$
    $= 8 \cdot 5$
    $= 40$

43. $5 + 2\sqrt{ab}$
    $5 + 2\sqrt{27 \cdot 3} = 5 + 2\sqrt{81}$
    $= 5 + 2 \cdot 9$
    $= 5 + 18$
    $= 23$

45. $\sqrt{a^2 + b^2}$
    $\sqrt{3^2 + 4^2} = \sqrt{9 + 16}$
    $= \sqrt{25}$
    $= 5$

47. $\sqrt{c^2 - b^2}$
    $\sqrt{13^2 - 12^2} = \sqrt{169 - 144}$
    $= \sqrt{25}$
    $= 5$

49. $5 + \sqrt{9} = 5 + 3$
    $= 8$

51. $6 - \sqrt{25} = 6 - 5$
    $= 1$

53. $-4\sqrt{81} = -4 \cdot 9$
    $= -36$

Objective B Exercises

55. $\sqrt{3} \approx 1.7321$

57. $\sqrt{10} \approx 3.1623$

59. $2\sqrt{6} \approx 4.8990$

61. $3\sqrt{14} \approx 11.2250$

63. $-4\sqrt{2} \approx -5.6569$

65. $-8\sqrt{30} \approx -43.8178$

67. 23 is between the perfect squares 16 and 25.
    $\sqrt{16} = 4$ and $\sqrt{25} = 5$
    $4 < \sqrt{23} < 5$

69. 29 is between the perfect squares 25 and 36.
$\sqrt{25} = 5$ and $\sqrt{36} = 6$
$5 < \sqrt{29} < 6$

71. 62 is between the perfect squares 49 and 64.
$\sqrt{49} = 7$ and $\sqrt{64} = 8$
$7 < \sqrt{62} < 8$

73. 130 is between the perfect squares 121 and 144.
$\sqrt{121} = 11$ and $\sqrt{144} = 12$
$11 < \sqrt{130} < 12$

75. $\sqrt{8} = \sqrt{4 \cdot 2}$
$= \sqrt{4} \cdot \sqrt{2}$
$= 2\sqrt{2}$

77. $\sqrt{45} = \sqrt{9 \cdot 5}$
$= \sqrt{9} \cdot \sqrt{5}$
$= 3\sqrt{5}$

79. $\sqrt{20} = \sqrt{4 \cdot 5}$
$= \sqrt{4} \cdot \sqrt{5}$
$= 2\sqrt{5}$

81. $\sqrt{27} = \sqrt{9 \cdot 3}$
$= \sqrt{9} \cdot \sqrt{3}$
$= 3\sqrt{3}$

83. $\sqrt{48} = \sqrt{16 \cdot 3}$
$= \sqrt{16} \cdot \sqrt{3}$
$= 4\sqrt{3}$

85. $\sqrt{75} = \sqrt{25 \cdot 3}$
$= \sqrt{25} \cdot \sqrt{3}$
$= 5\sqrt{3}$

87. $\sqrt{63} = \sqrt{9 \cdot 7}$
$= \sqrt{9} \cdot \sqrt{7}$
$= 3\sqrt{7}$

89. $\sqrt{98} = \sqrt{49 \cdot 2}$
$= \sqrt{49} \cdot \sqrt{2}$
$= 7\sqrt{2}$

91. $\sqrt{112} = \sqrt{16 \cdot 7}$
$= \sqrt{16} \cdot \sqrt{7}$
$= 4\sqrt{7}$

93. $\sqrt{175} = \sqrt{25 \cdot 7}$
$= \sqrt{25} \cdot \sqrt{7}$
$= 5\sqrt{7}$

Objective C Exercises

95. Strategy To find the velocity of the tsunami wave, substitute 100 for d in the given formula and solve for v.

    Solution $v = 3\sqrt{d}$
    $v = 3\sqrt{100}$
    $v = 3 \cdot 10 = 30$
    The tsunami has a velocity of 30 ft per second.

97. Strategy To find the time for the object to fall, substitute 144 for d in the given equation and solve for t.

    Solution $t = \sqrt{\dfrac{d}{16}}$
    $t = \sqrt{\dfrac{144}{16}}$
    $t = \sqrt{9} = 3$
    It takes 3 s for the object to fall 144 ft.

99. Strategy To find the distance, substitute 144 for E and 36 for S in the given formula and solve for d.

    Solution $d = 4{,}000\sqrt{\dfrac{E}{S}} - 4{,}000$
    $d = 4{,}000\sqrt{\dfrac{144}{36}} - 4{,}000$
    $d = 4{,}000\sqrt{4} - 4{,}000$
    $d = 4{,}000 \cdot 2 - 4{,}000$
    $d = 8{,}000 - 4{,}000$
    $d = 4{,}000$
    The space explorer is 4,000 mi above the surface.

Critical Thinking 4.4

101. $\sqrt{4} = 2, \sqrt{100} = 10$
The whole numbers between 2 and 10 are 3, 4, 5, 6, 7, 8, and 9.
The whole numbers between $\sqrt{4}$ and $\sqrt{100}$ are 3, 4, 5, 6, 7, 8, and 9.

103. $\sqrt{\frac{1}{4} + \frac{1}{8}} = \sqrt{\frac{2}{8} + \frac{1}{8}} = \sqrt{\frac{3}{8}} \approx 0.61237$

$\sqrt{\frac{1}{3} + \frac{1}{9}} = \sqrt{\frac{3}{9} + \frac{1}{9}} = \sqrt{\frac{4}{9}} = \frac{2}{3} \approx 0.66667$

$\sqrt{\frac{1}{5} + \frac{1}{6}} = \sqrt{\frac{6}{30} + \frac{5}{30}} = \sqrt{\frac{11}{30}} \approx 0.60553$

$\sqrt{\frac{1}{5} + \frac{1}{6}} < \sqrt{\frac{1}{4} + \frac{1}{8}} < \sqrt{\frac{1}{3} + \frac{1}{9}}$

105. a. Since $36 = 6^2$, 36 is a perfect square. 36 has 9 factors (1, 2, 3, 4, 6, 9, 12, 18, 36).
The number 36 is a two-digit perfect square that has exactly nine factors.

b. 14 and 4 have a difference of 10 $(14 - 4 = 10)$.
4, the smaller number, is a perfect square $(2^2 = 4)$.
14, the larger number, is 2 less than a perfect square $(14 + 2 = 16$, and $4^2 = 16)$.
The numbers are 4 and 14.

107. The expression $3\sqrt{16}$ means 3 times the square root of 16. Students must recognize this as multiplication or a product. Some students may go on to say that since $\sqrt{16} = 4$, the expression $3\sqrt{16}$ can be rewritten as 3(4), which equals 12.

Section 4.5

Objective A Exercises

1. Draw a solid dot one half unit to the right of two on the number line.

3. Draw solid dot one half unit to the left of negative three on the number line.

5. Draw a solid dot one half unit to the left of negative 4 on the number line.

7. Draw a solid dot one half unit to the right of one on the number line.

9. The real numbers greater than 6 are to the right of 6 on the number line. Draw a parenthesis at 6. Draw a heavy line to the right of 6. Draw an arrow at the right of the line.

11. The real numbers less than 0 are to the left of 0 on the number line. Draw a parenthesis at 0. Draw a heavy line to the left of 0. Draw an arrow at the left of the line.

13. The real numbers greater than –1 are to the right of –1 on the number line. Draw a parenthesis at –1. Draw a heavy line to the right of –1. Draw an arrow at the right of the line.

15. The real numbers less than –5 are to the left of –5 on the number line. Draw a parenthesis at –5. Draw a heavy line to the left of –5. Draw an arrow at the left of the line.

17. Draw a parenthesis at 2 and a parenthesis at 5. Draw a heavy line between 2 and 5.

19. Draw a parenthesis at –4 and a parenthesis at 0. Draw a heavy line between –4 and 0.

21. Draw a parenthesis at –2 and a parenthesis at 6. Draw a heavy line between –2 and 6.

23. Draw a parenthesis at –6 and a parenthesis at 1. Draw a heavy line between –6 and 1.

98 Chapter 4: Decimals and Real Numbers

Objective B Exercises

25. a. $x > 9$
    $-3.8 > 9$ False

    b. $x > 9$
    $0 > 9$ False

    c. $x > 9$
    $9 > 9$ False

    d. $x > 9$
    $\sqrt{101} > 9$ True
    The number $\sqrt{101}$ makes the inequality true.

27. a. $x \geq -2$
    $-6 \geq 2$ False

    b. $x \geq -2$
    $-2 \geq -2$ True

    c. $x \geq -2$
    $0.4 \geq -2$ True

    d. $x \geq -2$
    $\sqrt{17} \geq -2$ True
    The numbers -2, 0.4, and $\sqrt{17}$ make the inequality true.

29. All real numbers less than 3 make the inequality true.

31. All real numbers greater than or equal to -1 make the inequality true.

33. Draw a parenthesis at -2. Draw a heavy line to the left of -2. Draw an arrow at the left of the line.

    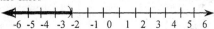

35. Draw a bracket at 0. Draw a heavy line to the right of 0. Draw an arrow at the right of the line.

    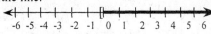

37. Draw a parenthesis at -5. Draw a heavy line to the right of -5. Draw an arrow at the right of the line.

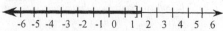

39. Draw a bracket at 2. Draw a heavy line to the left of 2. Draw an arrow at the left of the line.

Objective C Exercises

41. Strategy →To write the inequality, let s represent the number of sales. Since it is a minimum goal, the sales must be greater or equal to 50,000.
    →To determine if the sales goal has been reached, replace s in the inequality by 49,000. If the inequality is true, the sales quota has been reached. If the inequality is false, the sales quota has not been reached.

    Solution $s \geq 50{,}000$
    $49{,}000 \geq 50{,}000$ False
    No, the sales representative has not met the minimum quota for sales.

43. Strategy →To write the inequality, let h represent the number of credit hours allows per semester. Since h is a maximum, a part-time student carries credit hours that are less than or equal to 9.
    →To determine if a student taking 8.5 credit hours is a part-time student, replace h in the inequality by 8.5. If the inequality is true the student is part-time. If the inequality is false, the student is full-time student.

    Solution $h \leq 9$
    $8.5 \leq 9$ True
    Yes, the student is a part-time student.

45. Strategy →To write the inequality, let b represent the monthly budget. Since the budget is

a maximum, b is less than or equal to 2,400.
→To determine if the budget has been exceeded, replace b in the inequality by 2,380.50. If the inequality is true, the budget has not been exceeded. If the inequality is false, the monthly budget has been exceeded.

Solution $b \leq 2,400$
$2380.50 \leq 2,400$ True
Yes, the monthly budget has not been exceeded.

47. Strategy →To write the inequality, let T represent the temperature. Since the temperature is a minimum, T is greater than 85.
→To determine if the eggs are not safely stored, replace T in the inequality by 86.5. If the inequality is true, the temperature is not in the safe range. If the inequality is false, the temperature is suitable for storing eggs.

Solution $T > 85$
$86.5 > 85$ True
No, the eggs are not safely stored at 86.5°F.

Critical Thinking 4.5

49. a. −2 is an integer, a negative integer, a rational number, and a real number.

b. 18 is a whole number, an integer, a positive integer, a rational number, and a real number.

c. $-\frac{9}{37}$ is a rational number and a real number.

d. −6.606 is a rational number and a real number.

e. $4.5\overline{6}$ is a rational number and a real number.

f. 3.050050005 ... is an irrational number and a real number.

51. a.

$\|x\| < 9$	$\|x\| < 9$	$\|x\| < 9$	$\|x\| < 9$
$\|-2.5\| < 9$	$\|0\| < 9$	$\|9\| < 9$	$\|15.8\| < 9$
$2.5 < 9$	$0 < 9$	$9 < 9$	$15.8 < 9$
True	True	False	False

The numbers –2.5 and 0 make the inequality $\|x\| < 9$ true.

b.

$\|x\| > -3$	$\|x\| > -3$	$\|x\| > -3$	$\|x\| > -3$
$\|-6.3\| > -3$	$\|-3\| > -3$	$\|0\| > -3$	$\|6.7\| > -3$
$6.3 > -3$	$3 > -3$	$0 > -3$	$6.7 > -3$
True	True	True	True

The numbers –6.3, –3, 0, and 6.7 make the inequality $\|x\| > -3$ true.

c.

$\|x\| \geq 4$	$\|x\| \geq 4$	$\|x\| \geq 4$	$\|x\| \geq 4$
$\|-1.5\| \geq 4$	$\|0\| \geq 4$	$\|4\| \geq 4$	$\|13.6\| \geq 4$
$1.5 \geq 4$	$0 \geq 4$	$4 \geq 4$	$13.6 \geq 4$
False	False	True	True

The numbers 4 and 13.6 make the inequality $\|x\| \geq 4$ true.

d.

$\|x\| \leq 5$	$\|x\| \leq 5$	$\|x\| \leq 5$	$\|x\| \leq 5$
$\|-4.9\| \leq 5$	$\|0\| \leq 5$	$\|2.1\| \leq 5$	$\|5\| \leq 5$
$4.9 \leq 5$	$0 \leq 5$	$2.1 \leq 5$	$5 \leq 5$
True	True	True	True

The numbers –4.9, 0, 2.1 and 5 make the inequality $\|x\| \leq 5$ true.

53. a. $a > 0$ means a is a positive number.
$b > 0$ means b is a positive number.
ab is the product of a and b.
The product of two positive numbers is a positive number.
$ab > 0$ means the product of a and b is a positive number.
The statement is always true.

b. $a < 0$ means a is a negative number.
$a^2 > 0$ means that the square of the negative number a is a positive number.
The square of any negative number is a positive number.
The statement is always true.

c. $a > 0$ and $b > 0$ means that both a and b are positive numbers.
$a^2 > b$ means that the square of a is greater than b.
If $a = 5$ and $b = 2$, then $a^2 = 5^2 = 25$ and $a^2 > 2$.
If $a = 2$ and $b = 5$, then $a^2 = 2^2 = 4$ and a^2 is not greater than 5.
The statement is sometimes true.

55. In your students' definitions, you might look for the description of a rational number as a fraction, the concept of an irrational number as a nonterminating, nonrepeating decimal, and the idea that the real numbers are all the rationals and irrationals taken together.

Chapter Review Exercises

1. $3\sqrt{47} \approx 20.5670$

2. $0.918 \cdot 10^5 = 91,800$

3. $-\sqrt{121} = -11$

4. $-3.981 - 4.32 = -3.981 + (-4.32)$
   $= -8.301$

5. $a + b + c$
   $80.59 + (-3.647) + 12.3$
   $= 76.943 + 12.3$
   $= 89.243$

6. 5.034

7. $\sqrt{100} - 2\sqrt{49} = 10 - 2 \cdot 7$
   $= 10 - 14$
   $= 10 + (-14)$
   $= -4$

8. $14.2 \div 10^3 = 0.0142$

9. $4.2z = -1.428$
   $\dfrac{4.2z}{4.2} = \dfrac{-1.428}{4.2}$
   $z = -0.34$
   The solution is -0.34.

10. $8.31 = 8.310$
    $8.039 < 8.310$
    $8.039 < 8.31$

11. $\dfrac{x}{y}$
    $\dfrac{0.396}{3.6} = 0.11$

12. $\begin{array}{r} 9.47 \\ \times\, 0.26 \\ \hline 5682 \\ 1\,894 \\ \hline 2.4622 \end{array}$

13. a. $x \geq -1$
       $-6 \geq -1$
       False

    b. $x \geq -1$
       $-1 \geq -1$
       True

    c. $x \geq -1$
       $-0.5 \geq -1$
       True

    d. $x \geq -1$
       $\sqrt{10} \geq -1$
       True

    The numbers -1, -0.5, and $\sqrt{10}$ make the inequality true.

14. $\dfrac{3}{7} \approx 0.4286$ $0.429 = 0.4290$
    $0.4286 < 0.4290$
    $\dfrac{3}{7} < 0.429$

15. $0.28 = \dfrac{28}{100} = \dfrac{7}{25}$

16. $-6.8 \div 47.92 \approx -0.1$

17. $(4.5 + 0.9) - (3.4 + 0.8) = 5.4 - 4.2 = 1.2$
    The projected increase is 1.2 million workers over the age of 65.

18. Draw a parenthesis at -6 and a parenthesis at -2. Draw a heavy line between -6 and -2.

    ←(+++)+++++++++→
    -6 -5 -4 -3 -2 -1 0 1 2 3 4 5 6

19. Draw a bracket at –3. Draw a heavy line to the right of –3. Draw an arrow at the right end of the line.
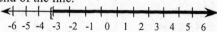

20. $-247.8 + (-193.4) = -441.2$

21. $614.3 \div 100 = 6.143$

22. $a - b$
    $80.32 - 29.577 = 80.32 + (-29.577)$
    $= 50.743$

23. $\sqrt{90} = \sqrt{9 \cdot 10}$
    $= \sqrt{9} \cdot \sqrt{10}$
    $= 3\sqrt{10}$

24. $60st$
    $60(5)(-3.7) = 300(-3.7)$
    $= -1,110$

25. $506.81 \rightarrow 500$
    $64.1 \rightarrow \underline{60}$
    440

26. Strategy
    →To write the inequality, let G represent the grade point average. Since it is a minimum qualification to qualify for a scholarship, the grade point average must be greater than or equal to 3.5
    →To determine if the grade point average qualifies the student for a scholarship, replace G by 3.48. If the inequality is true, the student qualifies. If the inequality is false, the student does not qualify for the scholarship.

    Solution
    $G \geq 3.5$
    $3.48 \geq 3.5$ False
    No, the student does not qualify for the scholarship.

27. Strategy   To find the difference, subtract the melting point of bromine (–7.2) from the boiling point of bromine (58.8).

    Solution $58.8 - (-7.2)$
    $= 58.8 + 7.2$
    $= 66$
    The difference in temperature between the melting and boiling point of bromine is 66°C.

28. Strategy
    a. To find the difference, subtract the cost of World War I (0.38) from the cost of World War II (3.1).
    b. To find how many times greater the monetary cost of the Vietnam War was, divide the cost of the Vietnam War by the cost of World War I.

    Solution
    a. $3.1 - 0.38 = 2.72$
    The difference between the monetary costs of the two World Wars was $2.72 trillion.

    b. $0.57 \div 0.38 = 1.5$
    The cost of the Vietnam War was 1.5 times the cost of World War I.

29. Strategy   To find the cost per ounce, divide 11.78 by 7.

    Solution $\dfrac{11.78}{7} \approx 1.683$
    To the nearest cent, the cost of instant coffee is $1.68 per ounce.

30. Strategy — To find the monthly lease payment, write and solve an equation using x to represent the monthly lease payment.

Solution

| the total of the monthly payments | is | the product of the number of months of the lease and the monthly lease payment |

$9977.76 = 24x$

$\dfrac{9977.76}{24} = \dfrac{24x}{24}$

$415.74 = x$

The monthly lease payment is $415.74.

31. Strategy — To find the price of the treadmill, substitute 369.99 for C and 129.50 for M in the given formula and solve for P.

Solution
$P = C + M$
$P = 369.99 + 129.50$
$P = 499.49$
The price of the treadmill is $499.49.

32. Strategy — To find the velocity of the falling object, substitute 25 for d in the given formula and solve for v.

Solution
$v = \sqrt{64d}$
$v = \sqrt{64 \cdot 25}$
$v = \sqrt{1,600}$
$v = 40$
The velocity of the falling object is 40 feet per second.

Chapter Test

1. 9.033

2. $4.003 < 4.009$

3. 6.05$\underline{1}$367 Given place value
   $3 < 5$
   6.051367 rounded to the nearest thousandth is 6.051.

4. $-30 - (-7.249) = -30 + 7.247$
   $= -22.753$

5. $x - y$
   $6.379 - (-8.28) = 6.379 + 8.28$
   $= 14.659$

6. $\begin{array}{r} 92.34 \\ -17.95 \\ \hline 74.39 \end{array} \rightarrow \begin{array}{r} 90 \\ -20 \\ \hline 70 \end{array}$

7. $4.58 - 3.9 + 6.017$
   $= 4.58 + (-3.9) + 6.017$
   $= 0.68 + 6.017$
   $= 6.697$

8. $-2.5(7.36) = -18.4$

9. $-20cd$
   $-20 \cdot 0.5 \cdot (-6.4) = (-10) \cdot (-6.4)$
   $= 64$

10. $5.488 = -3.92p$
    $\dfrac{5.488}{-3.92} = \dfrac{-3.92p}{-3.92}$
    $p = -1.4$

11. $\sqrt{256} - 2\sqrt{121} = 16 - 2 \cdot 11$
    $= 16 - 22$
    $= 16 + (-22)$
    $= -6$

12. $84.96 \div 100 = 0.8496$

13. $\dfrac{x}{y}$
    $\dfrac{52.7}{-6.2} = -8.5$

14. $0.22 = 0.2200 \qquad \dfrac{2}{9} \approx 0.2222$
    $0.2200 < 0.2222$
    $0.22 < \dfrac{2}{9}$

15. $2\sqrt{46} \approx 13.5647$

16. $\sqrt{68} = \sqrt{4 \cdot 17}$
    $= \sqrt{4} \cdot \sqrt{17}$
    $= 2\sqrt{17}$

17. $63.6 - 22.8 = 40.8$
    The gross from *Thunderball* was $40.8 million more than from *On Her Majesty's Secret Service*.

18. $8.4 = 5.9 + a$
    $\overline{8.4 \mid 5.9 + (-2.5)}$
    $8.4 \neq 3.4$
    No, -2.5 is not a solution of the equation.

19. $8.973 \cdot 10^4 = 89,730$

20. Draw a parenthesis at -2 and a parenthesis at 2. Draw a heavy line between -2 and 2.
    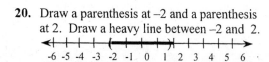

21. Draw a bracket at 3. Draw a heavy line to the right of 3. Draw an arrow at the right end of the line.
    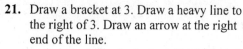

22. $x + y$
    $-233.81 + 71.3 = -162.51$

23. $-8v = 26$
    $\dfrac{-8v}{-8} = \dfrac{26}{-8}$
    $p = -3.25$

24. **Strategy** To find the difference, subtract the melting point of fluorine (-219.62) from the boiling point of fluorine (-188.14).

    Solution $-188.14 - (-219.62)$
    $= -188.14 + 219.62 = 31.48$
    The difference in temperature between the melting and boiling point of fluorine is $31.48°C$.

25. **Strategy** To find the velocity of the falling object, substitute 16 for d in the given formula and solve for v.

    Solution $v = \sqrt{64d}$
    $v = \sqrt{64 \cdot 16}$
    $v = \sqrt{1,024}$
    $v = 32$
    The velocity of the falling object is 32 feet per second.

26. **Strategy** To find the stockholders' equity, substitute 48.2 for A and 27.6 for L in the given formula and solve for S.

    Solution $A = L + S$
    $48.2 = 27.6 + S$
    $48.2 - 27.6 = 27.6 - 27.6 + S$
    $20.6 = S$
    The stockholders' equity is $20.6 million.

27. **Strategy** To find the perimeter, substitute 8.75, 5.25, and 4.5 for a, b, and c in the formula below and solve for P.

    Solution $P = a + b + c$
    $P = 8.75 + 5.25 + 4.5 = 18.5$
    The perimeter is 18.5 m.

28. **Strategy** → To write the inequality, let x represent the units sold. Since it is a minimum to reach the goal, the sales must be greater or equal to 65,000 per year.
→ To determine if the number of units sold has reached the goal, replace x by 57,000. If the inequality is true, the goal has been reached. If the inequality is false, the representative has not reached the sales goal.

   Solution $x \geq 65{,}000$
   $57{,}000 \geq 65{,}000$ False
   No, the representative has not reached the sales goal.

29. **Strategy** To find the force on the falling object, substitute 5.75 for m and -9.80 for g in the given formula and solve for F.

   Solution $F = ma$
   $F = 5.75(-9.80)$
   $F = -56.35$
   The force on the object is -56.35 newtons.

30. **Strategy** To find the temperature fall, subtract the low temperature (-20.56) from the temperature 15 minutes later (2.78).

   Solution $2.78 - (-20.56) = 2.78 + 20.56$
   $= 23.34$
   The temperature rose 23.34°C in 15 minutes.

Cumulative Review Exercises

1. $387.9 \div 10^4 = 0.03879$

2. $(x+y)^2 - 2z$
   $(-3+2)^2 - 2(-5) = (-1)^2 - 2(-5)$
   $= 1 - 2(-5)$
   $= 1 - (-10)$
   $= 1 + 10 = 11$

3. $-9.8 = -0.49c$
   $\dfrac{-9.8}{-0.49} = \dfrac{-0.49c}{-0.49}$
   $20 = c$
   The solution is 20.

4. 8,072,092

5. Draw a parenthesis at -4 and a parenthesis at 1. Draw a heavy line between -4 and 1.

   ⟵| | (—|—|—|—|—|)| | | | |⟶
   -6 -5 -4 -3 -2 -1 0 1 2 3 4 5 6

6. Draw a bracket at -2. Draw a heavy line to the left of -2. Draw an arrow at the left end of the line.

   ⟵|—|—|—|—|—|—|]| | | | | |⟶
   -6 -5 -4 -3 -2 -1 0 1 2 3 4 5 6

7. $-23 - (-19) = -23 + 19$
   $= -4$

8. $\begin{array}{rcr} 372 & \to & 400 \\ 541 & \to & 500 \\ 608 & \to & 600 \\ 429 & \to & +\ 400 \\ \hline & & 1{,}900 \end{array}$

9. $\sqrt{192} = \sqrt{64 \cdot 3}$
   $= \sqrt{64} \cdot \sqrt{3}$
   $= 8\sqrt{3}$

10. $x \div y$
    $3\dfrac{2}{3} \div 2\dfrac{4}{9} = \dfrac{11}{3} \div \dfrac{22}{9}$
    $= \dfrac{11}{3} \cdot \dfrac{9}{22}$
    $= \dfrac{11 \cdot 9}{3 \cdot 22}$
    $= \dfrac{11 \cdot 3 \cdot 3}{3 \cdot 2 \cdot 11}$
    $= \dfrac{3}{2} = 1\dfrac{1}{2}$

11. $-36.92 + 18.5 = -18.42$

12. $\left(\dfrac{5}{9}\right)\left(-\dfrac{3}{10}\right)\left(-\dfrac{6}{7}\right) = \left(\dfrac{5}{9} \cdot \dfrac{3}{10} \cdot \dfrac{6}{7}\right)$
    $= \dfrac{5 \cdot 3 \cdot 6}{9 \cdot 10 \cdot 7}$
    $= \dfrac{5 \cdot 3 \cdot 2 \cdot 3}{3 \cdot 3 \cdot 2 \cdot 5 \cdot 7}$
    $= \dfrac{1}{7}$

13. $x^4 y^2$
    $2^4 \cdot 10^2 = (2 \cdot 2 \cdot 2 \cdot 2) \cdot (10 \cdot 10)$
    $= 16 \cdot 100$
    $= 1{,}600$

14. $\begin{array}{r} 13 \\ 5\overline{)65} \\ 2\overline{)130} \end{array}$

$2\overline{)260}$

$260 = 2 \cdot 2 \cdot 5 \cdot 13 = 2^2 \cdot 5 \cdot 13$

15. $25\overline{)19.00}^{\,0.76}$

    $\frac{19}{25} = 0.76$

16. $10\sqrt{19} \approx 95.3939$

17. a. $32 > 28$
    Sweden mandates more vacation days than Ireland.

    b. $30 \div 20 = 1.5$
    Austria mandates 1.5 times more vacation days than Switzerland.

18. $\frac{-8}{0}$
    Division by zero is undefined.

19. $-\frac{5}{7} + \frac{4}{21} = \frac{-5}{7} + \frac{4}{21}$
    $= \frac{-15}{21} + \frac{4}{21}$
    $= \frac{-15 + 4}{21}$
    $= \frac{-11}{21} = -\frac{11}{21}$

20. $4\sqrt{25} - \sqrt{81} = 4 \cdot 5 - 9$
    $= 20 - 9$
    $= 11$

21. $\begin{array}{r} 62.8 \to 60 \\ 0.47 \to \times 0.5 \\ \hline 30 \end{array}$

22. $5(3 - 7) \div (-4) + 6(2)$
    $= 5(-4) \div (-4) + 6(2)$
    $= -20 \div (-4) + 6(2)$
    $= 5 + 6(2)$
    $= 5 + 12$
    $= 17$

23. $\dfrac{a}{b+c}$

    $\dfrac{\frac{3}{8}}{\frac{1}{2} + \frac{3}{4}} = \dfrac{\frac{3}{8}}{\frac{5}{4}}$

    $= \frac{3}{8} \div \frac{5}{4}$

    $= \frac{3}{8} \cdot \frac{4}{5}$

    $= \frac{3 \cdot 4}{8 \cdot 5}$

    $= \frac{3 \cdot 2 \cdot 2}{2 \cdot 2 \cdot 2 \cdot 5} = \frac{3}{10}$

24. $x - y + z$

    $\frac{5}{12} - \left(-\frac{3}{8}\right) + \left(-\frac{3}{4}\right) = \frac{5}{12} + \frac{3}{8} + \frac{-3}{4}$

    $= \frac{10}{24} + \frac{9}{24} + \frac{-18}{24}$

    $= \frac{10 + 9 + (-18)}{24}$

    $= \frac{1}{24}$

25. $2.617 \div 0.93 \approx 2.8$

26. Strategy — To calculate the cellular phone service bill:
    →Calculate the number of minutes the service was used after the first 50 min.
    →Add the monthly charge (39.99) to the product of 0.75 and the number of minutes the service was used after the first 50 min.

    Solution — $87 - 50 = 37$
    The service was used for 37 min after the first 50 min.
    $0.75(37) + 39.99$
    $= 27.75 + 39.99$
    $= 67.74$
    The cellular phone service bill is $67.74.

27. Strategy — To find the temperature fall, subtract the temperature at midnight (−29.4) from the temperature at noon (17.22).

    Solution — $17.22 - (-29.4) = 17.22 + 29.4$
    $= 46.62$
    The temperature fell 46.62°C in the 12-hour period.

28. Strategy — To find the cost per visit, substitute 390 for M and 125

for N in the given formula and solve for C.

Solution
$$C = \frac{M}{N}$$
$$C = \frac{390}{125}$$
$$C = 3.12$$
The cost per visit is $3.12.

29. Strategy
- **a.** To find the number of hours worked per week, add the number of hours spent in all five categories.

- **b.** To determine which takes more time, compare the amount of time spent on face-to-face selling (13.9) with the sum of the time spent on administrative work (7.0) and placing service calls (5.6).

Solution
- **a.** $13.9 + 11.5 + 8.5 + 7.0 + 5.6 = 46.5$
  On average, a salesperson works 46.5 h per week.

- **b.** $7.0 + 5.6 = 12.6$
  $13.9 > 12.6$
  The average salesperson spends more time on face-to-face selling.

30. Strategy To find the velocity, substitute 45 for d in the given equation and solve for v.

Solution
$$v = \sqrt{20d}$$
$$v = \sqrt{20 \cdot 45}$$
$$v = \sqrt{900}$$
$$v = 30$$
The velocity of the car is 30 mph.

Chapter 5: Variable Expressions

Prep Test

1. $54 > 45$
2. $-19 + 8 = -11$
3. $26 - 38 = -12$
4. $-2(44) = -88$
5. $-\dfrac{3}{4}(-8) = \dfrac{24}{4} = 6$
6. $3.97 \cdot 10^4 = 3.97 \cdot 10{,}000 = 39{,}700$
7. $(-3)^2 = (-3)(-3) = 9$
8. $(8-6)^2 + 12 \div 4 \cdot 3^2 = (2)^2 + 12 \div 4 \cdot 9$
   $= 4 + 3 \cdot 9$
   $= 4 + 27$
   $= 31$

Go Figure

From the third statement, we know that the order of three of the four men is either Luis, Kim, and Reggie or Reggie, Kim and Luis. From the fourth statement, Luis is standing between Dave and Kim. So building from what we already know, then either the order is Dave, Luis, Kim, and Reggie or Reggie, Kim, Luis, and Dave. From the second statement, we know that Dave is not first. Therefore, the order is Reggie, Kim, Luis, and Dave.

Section 5.1

Objective A Exercises

1. The Associative Property of Multiplication
3. The Commutative Property of Addition
5. The Inverse Property of Addition
7. The Inverse Property of Multiplication
9. The Commutative Property of Multiplication
11. a. The Associative Property of Multiplication
    b. The Inverse Property of Multiplication
    c. The Multiplication Property of One

13. $x + (4 + y) = (x + 4) + y$
15. $5 \cdot \dfrac{1}{5} = 1$
17. $a \cdot 0 = 0$
19. $-7y + 7y = 0$
21. $-\dfrac{3}{2}$
23. $6(2x) = (6 \cdot 2)x$
    $= 12x$
25. $-5(3x) = (-5 \cdot 3)x$
    $= -15x$
27. $(3t) \cdot 7 = 7 \cdot (3t)$
    $= (7 \cdot 3)t$
    $= 21t$
29. $(-3p) \cdot 7 = 7 \cdot (-3p)$
    $= [7 \cdot (-3)]p$
    $= -21p$
31. $(-2)(-6q) = [(-2)(-6)]q$
    $= 12q$
33. $\dfrac{1}{2}(4x) = \left(\dfrac{1}{2} \cdot 4\right)x$
    $= 2x$
35. $-\dfrac{5}{3}(9w) = \left(-\dfrac{5}{3} \cdot 9\right)w$
    $= -15w$
37. $-\dfrac{1}{2}(-2x) = \left[\left(-\dfrac{1}{2}\right)(-2)\right]x$
    $= 1 \cdot x = x$
39. $(2x)(3x) = (2 \cdot 3)(x \cdot x)$
    $= 6x^2$
41. $(-3x)(9x) = (-3 \cdot 9)(x \cdot x)$
    $= -27x^2$
43. $\left(\dfrac{1}{2}x\right)(2x) = \left(\dfrac{1}{2} \cdot 2\right)(x \cdot x)$
    $= 1 \cdot x^2 = x^2$

45. $\left(-\dfrac{2}{3}\right)(x)\left(-\dfrac{3}{2}\right) = \left[\left(-\dfrac{2}{3}\right)\left(-\dfrac{3}{2}\right)\right]x$
$= 1 \cdot x = x$

47. $6\left(\dfrac{1}{6}c\right) = \left(6 \cdot \dfrac{1}{6}\right)c$
$= 1 \cdot c = c$

49. $-5\left(-\dfrac{1}{5}a\right) = \left(-5 \cdot -\dfrac{1}{5}\right)a$
$= 1 \cdot a = a$

51. $\dfrac{4}{5}w \cdot 15 = \left(\dfrac{4}{5} \cdot 15\right)w$
$= 12w$

53. $2v \cdot 8w = (2 \cdot 8)(v \cdot w)$
$= 16vw$

55. $(-4b)(7c) = (-4 \cdot 7)(b \cdot c)$
$= -28bc$

57. $3x + (-3x) = 0$

59. $-12h + 12h = 0$

61. $9 + 2m + (-2m)$
$= 9 + 0$
$= 9$

63. $8x + 7 + (-8x)$
$= 8x + (-8x) + 7$
$= 0 + 7$
$= 7$

65. $6t - 15 + (-6t)$
$= -15 + 6t + (-6t)$
$= -15 + 0$
$= -15$

67. $8 + (-8) - 5y$
$= 0 - 5y$
$= -5y$

69. $(-4) + 4 + 13b$
$= 0 + 13b$
$= 13b$

Objective B Exercises

71. $2(5z + 2)$
$= 2(5z) + 2(2)$
$= 10z + 4$

73. $6(2y + 5z)$
$= 6(2y) + 6(5z)$
$= 12y + 30z$

75. $3(7x - 9)$
$= 3(7x) - 3(9)$
$= 21x - 27$

77. $-(2x - 7) = -2x + 7$

79. $-(-4x - 9) = 4x + 9$

81. $-5(y + 3)$
$= -5(y) + (-5)(3)$
$= -5y - 15$

83. $-6(2x - 3)$
$= -6(2x) - (-6)(3)$
$= -12x + 18$

85. $-5(4n - 8)$
$= -5(4n) - (-5)(8)$
$= -20n + 40$

87. $-8(-6z + 3)$
$= -8(-6z) + (-8)(3)$
$= 48z - 24$

89. $-6(-4p - 7)$
$= -6(-4p) - (-6)(7)$
$= 24p + 42$

91. $5(2a + 3b + 1)$
$= 5(2a) + 5(3b) + 5(1)$
$= 10a + 15b + 5$

93. $4(3x - y - 1)$
$= 4(3x) - 4(y) - 4(1)$
$= 12x - 4y - 4$

95. $9(4m - n + 2)$
$= 9(4m) - 9(n) + 9(2)$
$= 36m - 9n + 18$

97. $-6(-2v + 3w + 7)$
$= -6(-2v) + (-6)(3w) + (-6)(7)$
$= 12v - 18w - 42$

99. $-4(-5x - 1)$
$= -4(-5x) - (-4)(1)$
$= 20x + 4$

101. $5(4a - 5b + c)$
$= 5(4a) - 5(5b) + 5(c)$
$= 20a - 25b + 5c$

103. $-6(3p - 2r - 9)$
$= -6(3p) - (-6)(2r) - (-6)(9)$
$= -18p + 12r + 54$

105. $-(5a - 9b + 7) = -5a + 9b - 7$

107. $-(11p - 2q - r) = -11p + 2q + r$

Critical Thinking 5.1

109. No. The statement is not true for the number zero.

111. No. The number zero does not have a multiplicative inverse.

113. Students may provide different explanations for why division by zero is not allowed. They might describe the fact that it is not possible to divide a number into 0 equal parts whose sum is the number. Or they might use the operation of multiplication in their explanation, paraphrasing the discussion presented in the text.

Section 5.2

Objective A Exercises

1. $3x^2, 4x, \underline{-9}$

3. $b, \underline{5}$

5. $9\underline{a}^2, -12\underline{a}, 4\underline{b}^2$

7. $3\underline{x}^2$

9. coefficient of x^2: 1
   coefficient of $-6x$: -6

11. coefficient of $12a^2$: 12
    coefficient of $4ab$: 4

13. $7a + 9a = 16a$

15. $12x + 15x = 27x$

17. $9z - 6z = 3z$

19. $9x - x = 8x$

21. $8z - 15z = -7z$

23. $w - 7w = -6w$

25. $12v - 12v = 0$

27. $9s - 8s = s$

29. $\dfrac{n}{5} + \dfrac{3n}{5} = \dfrac{n+3n}{5}$
$= \dfrac{4n}{5}$

31. $\dfrac{x}{4} + \dfrac{x}{4} = \dfrac{x+x}{4}$
$= \dfrac{2x}{4}$
$= \dfrac{x}{2}$

33. $\dfrac{8y}{7} - \dfrac{4y}{7} = \dfrac{8y-4y}{7}$
$= \dfrac{4y}{7}$

35. $\dfrac{5c}{6} - \dfrac{c}{6} = \dfrac{5c-c}{6}$
$= \dfrac{4c}{6}$
$= \dfrac{2c}{3}$

37. $4x - 3y + 2x$
$= 4x + 2x - 3y$
$= 6x - 3y$

39. $4r + 8p - 2r + 5p$
$= 4r - 2r + 8p + 5p$
$= 2r + 13p$

41. $9w - 5v - 12w + 7v$
$= 9w - 12w - 5v + 7v$
$= -3w + 2v$

43. $-4p + 9 - 5p + 2$
$= -4p - 5p + 9 + 2$
$= -9p + 11$

45. $8p + 7 - 6p - 7$
$= 8p - 6p + 7 - 7$
$= 2p + 0$
$= 2p$

47. $7h + 15 - 7h - 9$
$= 7h - 7h + 15 - 9$
$= 0 + 6$
$= 6$

49. $9y^2 - 8 + 4y^2 + 9 = 9y^2 + 4y^2 - 8 + 9$
$= 13y^2 + 1$

51. $3w^2 - 7 - 9 + 9w^2 = 3w^2 + 9w^2 - 7 - 9$
    $= 12w^2 - 16$

53. $9w^2 - 15w + w - 9w^2$
    $= 9w^2 - 9w^2 - 15w + w$
    $= 0 - 14w$
    $= -14w$

55. $7a^2b + 5ab^2 - 2a^2b + 3ab^2$
    $= 7a^2b - 2a^2b + 5ab^2 + 3ab^2$
    $= 5a^2b + 8ab^2$

57. $8a - 9b + 2 - 8a + 9b + 3$
    $= 8a - 8a - 9b + 9b + 2 + 3$
    $= 0 + 0 + 5$
    $= 5$

59. $6x^2 - 7x + 1 + 5x^2 + 5x - 1$
    $= 6x^2 + 5x^2 - 7x + 5x + 1 - 1$
    $= 11x^2 - 2x + 0$
    $= 11x^2 - 2x$

61. $-3b^2 + 6b + 1 + 11b^2 - 8b - 1$
    $= -3b^2 + 11b^2 + 6b - 8b + 1 - 1$
    $= 8b^2 - 2b + 0$
    $= 8b^2 - 2b$

Objective B Exercises

63. $5x + 2(x + 1)$
    $= 5x + 2x + 2$
    $= 7x + 2$

65. $9n - 3(2n - 1)$
    $= 9n - 6n + 3$
    $= 3n + 3$

67. $7a - (3a - 4)$
    $= 7a - 3a + 4$
    $= 4a + 4$

69. $7 + 2(2a - 3)$
    $= 7 + 4a - 6$
    $= 4a + 1$

71. $6 + 4(2x + 9)$
    $= 6 + 8x + 36$
    $= 8x + 42$

73. $8 - 4(3x - 5)$
    $= 8 - 12x + 20$
    $= -12x + 28$

75. $2 - 9(2m + 6)$
    $= 2 - 18m - 54$
    $= -18m - 52$

77. $3(6c + 5) + 2(c + 4)$
    $= 18c + 15 + 2c + 8$
    $= 20c + 23$

79. $2(a - 2b) + 3(2a + 3b)$
    $= 2a - 4b + 6a + 9b$
    $= 8a + 5b$

81. $6(7z - 5) - 3(9z - 6)$
    $= 42z - 30 - 27z + 18$
    $= 15z - 12$

83. $-2(6y + 2) + 3(4y - 5)$
    $= -12y - 4 + 12y - 15$
    $= -19$

85. $-5(x - 2y) - 4(2x + 3y)$
    $= -5x + 10y - 8x - 12y$
    $= -13x - 2y$

87. $2 - 3(2v - 1) + 2(2v + 4)$
    $= 2 - 6v + 3 + 4v + 8$
    $= -2v + 13$

89. $2c - 3(c + 4) - 2(2c - 3)$
    $= 2c - 3c - 12 - 4c + 6$
    $= -5c - 6$

91. $8a + 3(2a - 1) + 6(4 - 2a)$
    $= 8a + 6a - 3 + 24 - 12a$
    $= 2a + 21$

93. $3n - 2[5 - 2(2n - 4)]$
    $= 3n - 2[5 - 4n + 8]$
    $= 3n - 2[13 - 4n]$
    $= 3n - 26 + 8n$
    $= 11n - 26$

95. $9x - 3[8 - 2(5 - 3x)]$
    $= 9x - 3[8 - 10 + 6x]$
    $= 9x - 3[-2 + 6x]$
    $= 9x + 6 - 18x$
    $= -9x + 6$

97. $-3v - 6[2(3-2v) - 5(3v - 7)]$
$= -3v - 6[6 - 4v - 15v + 35]$
$= -3v - 6[-19v + 41]$
$= -3v + 114v - 246$
$= 111v - 246$

99. $21r - 4[3(4 - 5r) - 3(2 - 7r)]$
$= 21r - 4[12 - 15r - 6 + 21r]$
$= 21r - 4[6 + 6r]$
$= 21r - 24 - 24r$
$= -3r - 24$

101. $9z^2 - 3\left[4(2z+3) - 3(2z^2 - 6)\right]$
$= 9z^2 - 3\left[8z + 12 - 6z^2 + 18\right]$
$= 9z^2 - 3\left[-6z^2 + 8z + 30\right]$
$= 9z^2 + 18z^2 - 24z - 90$
$= 27z^2 - 24z - 90$

Critical Thinking 5.2

103. $6 \cdot 527 = 6(500 + 20 + 7)$
$= 6 \cdot 500 + 6 \cdot 20 + 6 \cdot 7$
$= 3{,}000 + 120 + 42$
$= 3{,}162$

105. Any units of measure for distance and area can be used to show that units and square units cannot be combined. For example, inches are used to measure distance and square inches are used to measure area. We can describe the length of a rectangle as 8 inches, or the area of a rectangle as 24 square inches, but these two measures (8 in. and 24 in^2) cannot be combined.

Section 5.3

Objective A Exercises

1. Yes

3. No, variable in denominator

5. Yes

7. No, variable under square root

9. Yes

11. No, variable under square root

13. 3 terms

15. 1 term

17. Binomial

19. Monomial

21. $3x^3 + 8x^2 - 2x - 6$

23. $5a^3 - 3a^2 + 2a + 1$

25. $-b^2 + 4$

27. $(5y^2 + 3y - 7) + (6y^2 - 7y + 9)$
$= (5y^2 + 6y^2) + (3y - 7y) + (-7 + 9)$
$= 11y^2 - 4y + 2$

29. $(-4b^2 + 9b + 11) + (7b^2 - 12b - 13)$
$= (-4b^2 + 4x^2) + (9b - 12b) + (11 - 13)$
$= 3b^2 - 3b - 2$

31. $(3w^3 + 8w^2 - 2w) + (5w^2 - 6w - 5)$
$= 3w^3 + (8w^2 + 5w^2) + (-2w - 6w) - 5$
$= 3w^3 + 13w^2 - 8w - 5$

33. $(-9a^3 + 3a^2 + 2a - 7)$
$+ (7a^3 - 12a^2 - 10a + 8)$
$= (-9a^3 + 7a^3) + (3a^2 - 12a^2)$
$+ (2a - 10a) + (-7 + 8)$
$= -2a^3 - 9a^2 - 8a + 1$

35. $(7t^3 - 8t - 15) + (7t^2 + 8t - 20)$
$= 7t^3 + 7t^2 + (-8t + 8t) + (-15 - 20)$
$= 7t^3 + 7t^2 - 35$

37. $(6t^2 - 8t - 15) + (7t^2 + 8t - 20)$
$= (6t^2 + 7t^2) + (-8t + 8t) + (-15 - 20)$
$= 13t^2 - 35$

39. $5k^2 - 7k - 8$
$6k^2 + 9k - 10$
$\overline{11k^2 + 2k - 18}$

41. $8x^3 - 9x + 2$
$9x^3 + 9x - 7$
$\overline{17x^3 - 5}$

43. $12b^3 + 9b^2 + 5b - 10$
$4b^3 + 5b^2 - 5b + 11$
$\overline{16b^3 + 14b^2 + 1}$

45. $\dfrac{\begin{array}{r}8p^3 -7p\\ 9p^2+p-7\end{array}}{8p^3+9p^2-6p-7}$

47. $\dfrac{\begin{array}{r}7a^2-6a-7\\ -6a^3-7a^2+6a-10\end{array}}{-6a^3-17}$

49. $\dfrac{\begin{array}{r}9d^4-7d^2+5\\ -6d^4-3d^2-8\end{array}}{3d^4-10d^2-3}$

Objective B Exercises

51. $-8x^3-5x^2+3x+6$

53. $9a^3-a^2+2a-9$

55. $(7y^2-8y-10)-(3y^2+2y-9)$
$=(7y^2-8y-10)+(-3y^2-2y+9)$
$=4y^2-10y-1$

57. $(13w^3+3w^2-9)-(7w^3-9w+10)$
$=(13w^3+3w^2-9)+(-7w^3+9w-10)$
$=6w^3+3w^2+9w-19$

59. $(15t^3-9t^2+8t+11)-(17t^3-9t^2-8t+6)=(15t^3-9t^2+8t+11)+(-17t^3+9t^2+8t-6)$
$=-2t^3+16t+5$

61. $(8p^3+14p)-(9p^2-12)=(8p^3+14p)+(-9p^2+12)$
$=8p^3-9p^2+14p+12$

63. $(-4v^2+8v-2)-(6v^3-13v^2+7v+1)=(-4v^2+8v-2)+(-6v^3+13v^2-7v-1)$
$=-6v^3+9v^2+v-3$

65. $(7m^2-3m-6)-(2m^2-m+5)=(7m^2-3m-6)+(-2m^2+m-5)$
$=5m^2-2m-11$

67. $\dfrac{\begin{array}{r}8b^2-7b-6\\ -5b^2-8b-12\end{array}}{3b^2-15b-18}$

69. $\dfrac{\begin{array}{r}10y^3-8y-13\\ -6y^2-2y-7\end{array}}{10y^3-6y^2-10y-20}$

71. $\dfrac{\begin{array}{r}4a^2+8a+12\\ -3a^3-4a^2-7a+12\end{array}}{-3a^3+a+24}$

114 Chapter 5: *Variable Expressions*

73.
$$\begin{array}{r}7m-6\\ -2m^3+m^2\\ \hline -2m^3+m^2+7m-6\end{array}$$

75.
$$\begin{array}{r}4q^3+7q^2+8q-9\\ -14q^3-7q^2+8q+9\\ \hline -10q^3+16q\end{array}$$

77.
$$\begin{array}{r}7x^4+3x^2-11\\ 5x^4+8x^2-6\\ \hline 12x^4+11x^2-17\end{array}$$

Objective C Exercises

79. **Strategy**   To find the distance from Haley to Bedford, add the distance from Haley to Lincoln to the distance from Lincoln to Bedford.

   Solution
   $$\begin{array}{r}2y^2+y-4\\ 5y^2-y+3\\ \hline 7y^2-1\end{array}$$

   The distance from Haley to Bedford is ($7y^2-1$) km.

81. **Strategy**   To find the perimeter, substitute (n^2+3) for a and (n^2-2) for b and (n^2+5) for c in the formula below and solve for P.

   Solution $P=a+b+c$
   $P=(n^2+3)+(n^2-2)+(n^2+5)$
   $=3n^2+6$
   The perimeter is ($3n^2+6$) m.

83. **Strategy**   To find the monthly profit, substitute ($50n+4000$) for C and ($-0.6n^2+250n$) for R in the formula and solve for P.

   Solution $P=R-C$
   $P=-0.6n^2+250n-(50n+4000)$
   $=-0.6n^2+250n-50n-4000$
   $=-0.6n^2+200n-4000$
   The monthly profit is ($-0.6n^2+200n-4000$) dollars.

85. **Strategy**   To find the monthly profit, substitute ($100n+1500$) for C and ($-n^2+800n$) for R in the formula and solve for P.

   Solution $P=R-C$
   $P=-n^2+800n-(100n+1500)$
   $=-n^2+800n-100n-1500$
   $=-n^2+700n-1500$
   The monthly profit is ($-n^2+700n-1500$) dollars.

Critical Thinking 5.3

87. Strategy To find the polynomial, subtract $(5x^2 + 3x + 1)$ from $(2x^2 - x - 2)$.

    Solution $(2x^2 - x - 2) - (5x^2 + 3x + 1)$
    $= 2x^2 - x - 2 + (-5x^2 - 3x - 1)$
    $= -3x^2 - 4x - 3$

Section 5.4

Objective A Exercises

1. $a^4 \cdot a^5 = a^9$

3. $x^9 \cdot x^7 = x^{16}$

5. $n^4 \cdot n^2 = n^6$

7. $z^3 \cdot z \cdot z^4 = z^8$

9. $(a^3 b^2)(a^5 b) = (a^3 \cdot a^5)(b^2 \cdot b)$
    $= a^8 b^3$

11. $(-m^3 n)(m^6 n^2) = -(m^3 \cdot m^6)(n \cdot n^2)$
    $= -m^9 n^3$

13. $(2x^3)(5x^4) = (2 \cdot 5)(x^3 \cdot x^4)$
    $= 10x^7$

15. $(8x^2 y)(xy^5) = 8(x^2 \cdot x)(y \cdot y^5)$
    $= 8x^3 y^6$

17. $(-4m^3)(3m^4) = (-4 \cdot 3)(m^3 \cdot m^4)$
    $= -12m^7$

19. $(7v^3)(-2w) = [7 \cdot (-2)](v^3) \cdot w$
    $= -14v^3 w$

21. $(ab^2 c^3)(-2b^3 c^2) = -2 \cdot a(b^2 \cdot b^3)(c^3 \cdot c^2)$
    $= -2ab^5 c^5$

23. $(4b^4 c^2)(6a^3 b) = (4 \cdot 6)(a^3)(b^4 \cdot b)(c^2)$
    $= 24a^3 b^5 c^2$

25. $(-8r^2 t^3)(-5rt^4 v)$
    $= [(-8)(-5)](r^2 \cdot r)(t^3 \cdot t^4)(v)$
    $= 40r^3 t^7 v$

27. $(9mn^4 p)(-3mp^2)$
    $= [9 \cdot (-3)](m \cdot m)(n^4)(p \cdot p^2)$
    $= -27m^2 n^4 p^3$

29. $(2x)(3x^2)(4x^4)$
    $= (2 \cdot 3 \cdot 4)(x \cdot x^2 \cdot x^4)$
    $= 24x^7$

31. $(3ab)(2a^2 b^3)(a^3 b)$
    $= (3 \cdot 2)(a \cdot a^2 \cdot a^3)(b \cdot b^3 \cdot b)$
    $= 6a^6 b^5$

33. $(-xy^5)(3x^2)(5y^3)$
    $= -(3 \cdot 5)(x \cdot x^2)(y^5 \cdot y^3)$
    $= -15x^3 y^8$

35. $(8rt^3)(-2r^3 v^2)(-3t^5 v^2)$
    $= [8 \cdot (-2)(-3)](r \cdot r^3)(t^3 \cdot t^5)(v^2 \cdot v^2)$
    $= 48r^4 t^8 v^4$

37. $(-5ac^3)(-4b^3 c)(-3a^2 b^2)$
    $= [(-5)(-4)(-3)](a \cdot a^2)(b^3 \cdot b^2)(c^3 \cdot c)$
    $= -60a^3 b^5 c^4$

39. $(2ab^6)(-4a^5 b^4)$
    $= [2(-4)](a \cdot a^5)(b^6 \cdot b^4)$
    $= -8a^6 b^{10}$

Objective B Exercises

41. $(b^2)^4 = b^{2 \cdot 4} = b^8$

43. $(p^4)^7 = p^{4 \cdot 7} = p^{28}$

45. $(c^7)^4 = c^{7 \cdot 4} = c^{28}$

47. $(3x)^2 = 3^{1 \cdot 2} x^{1 \cdot 2}$
$= 3^2 x^2 = 9x^2$

49. $(x^2 y^3)^6 = x^{2 \cdot 6} y^{3 \cdot 6} = x^{12} y^{18}$

51. $(r^3 t)^4 = r^{3 \cdot 4} t^{1 \cdot 4} = r^{12} t^4$

53. $(-y^2)^2 = (-1)^{1 \cdot 2} y^{2 \cdot 2} = y^4$

55. $(2x^4)^3 = 2^{1 \cdot 3} x^{4 \cdot 3}$
$= 2^3 x^{12}$
$= 8x^{12}$

57. $(-2a^2)^3 = (-2)^{1 \cdot 3} a^{2 \cdot 3}$
$= (-2)^3 a^6$
$= -8a^6$

59. $(3x^2 y)^2 = 3^{1 \cdot 2} x^{2 \cdot 2} y^{1 \cdot 2}$
$= 3^2 x^4 y^2$
$= 9x^4 y^2$

61. $(2a^3 bc^2)^3 = 2^{1 \cdot 3} a^{3 \cdot 3} b^{1 \cdot 3} c^{2 \cdot 3}$
$= 2^3 a^9 b^3 c^6$
$= 8a^9 b^3 c^6$

63. $(-mn^5 p^3)^4$
$= (-1)^{1 \cdot 4} m^{1 \cdot 4} n^{5 \cdot 4} p^{3 \cdot 4}$
$= m^4 n^{20} p^{12}$

Critical Thinking 5.4

65. Strategy   To find the area, substitute $(3a^2 b^5)$ for L and $(a^4 b)$ for W in the formula below and solve for A.

     Solution $A = LW$
$A = 3a^2 b^5 \cdot a^4 b$
$= 3a^6 b^6$
The area is $(3a^6 b^6)$ ft².

67. $(2^3)^2 = 2^6 = 64$

$2^{(3^2)} = 2^9 = 512$

The results are not the same. $2^{(3^2)}$ is the larger number.

Section 5.5

Objective A Exercises

1. $x(x^2 - 3x - 4)$
$= x(x^2) - x(3x) - x(4)$
$= x^3 - 3x^2 - 4x$

3. $4a(2a^2 + 3a - 6)$
$= 4a(2a^2) + 4a(3a) - 4a(6)$
$= 8a^3 + 12a^2 - 24a$

5. $-2a(3a^2 + 9a - 7)$
$= -2a(3a^2) + (-2a)(9a) - (-2a)(7)$
$= -6a^3 - 18a^2 + 14a$

7. $m^3(4m - 9)$
$= m^3(4m) - m^3(9)$
$= 4m^4 - 9m^3$

9. $2x^3(5x^2 - 6xy + 2y^2)$
$= 2x^3(5x^2) - 2x^3(6xy) + 2x^3(2y^2)$
$= 10x^5 - 12x^4 y + 4x^3 y^2$

11. $-6r^5(r^2 - 2r - 6)$
$= -6r^5(r^2) - (-6r^5)(2r) - (-6r^5)(6)$
$= -6r^7 + 12r^6 + 36r^5$

13. $4a^2(3a^2 + 6a - 7)$
$= 4a^2(3a^2) + 4a^2(6a) - 4a^2(7)$
$= 12a^4 + 24a^3 - 28a^2$

15. $-2n^2(3 - 4n^3 - 5n^5)$
$= -2n^2(3) - (-2n^2)(4n^3) - (-2n^2)(5n^5)$
$= -6n^2 + 8n^5 + 10n^7$

17. $ab^2(3a^2 - 4ab + b^2)$
$= ab^2(3a^2) - ab^2(4ab) + (ab^2)(b^2)$
$= 3a^3 b^2 - 4a^2 b^3 + ab^4$

19. $-x^2 y^3(4x^5 y^2 - 5x^3 y - 7x)$
$= -x^2 y^3(4x^5 y^2) - (-x^2 y^3)(5x^3 y) - (-x^2 y^3)(7x)$
$= -4x^7 y^5 + 5x^5 y^4 + 7x^3 y^3$

21. $6r^2 t^3(1 - rt - r^3 t^3)$
$= 6r^2 t^3(1) - 6r^2 t^3(rt) - 6r^2 t^3(r^3 t^3)$
$= 6r^2 t^3 - 6r^3 t^4 - 6r^5 t^6$

23. $-4q(-9q+7)$
$= -4q(-9q) - 4q(7)$
$= 36q^2 - 28q$

Objective B Exercises

25. $(y+9)(y+3)$
$= (y)(y) + (y)(3) + 9(y) + 9(3)$
$= y^2 + 3y + 9y + 27$
$= y^2 + 12y + 27$

27. $(x+6)(x+5)$
$= (x)(x) + (x)(5) + 6(x) + 6(5)$
$= x^2 + 5x + 6x + 30$
$= x^2 + 11x + 30$

29. $(a-3)(a-8)$
$= (a)(a) + (a)(-8) + (-3)(a) + (-3)(-8)$
$= a^2 - 8a - 3a + 24$
$= a^2 - 11a + 24$

31. $(5z+2)(2z+1)$
$= (5z)(2z) + (5z)(1) + (2)(2z) + 2(1)$
$= 10z^2 + 5z + 4z + 2$
$= 10z^2 + 9z + 2$

33. $(8c-7)(5c+3)$
$= (8c)(5c) + 8c(3) + (-7)(5c) + (-7)(3)$
$= 40c^2 + 24c - 35c - 21$
$= 40c^2 - 11c - 21$

35. $(5v-3)(2v-1)$
$= (5v)(2v) + (5v)(-1) + (-3)(2v) + (-3)(-1)$
$= 10v^2 - 5v - 6v + 3$
$= 10v^2 - 11v + 3$

37. $(7t-2)(5t+4)$
$= (7t)(5t) + (7t)(4) + (-2)(5t) + (-2)(4)$
$= 35t^2 + 28t - 10t - 8$
$= 35t^2 + 18t - 8$

39. $(8x+5)(3x-2)$
$= (8x)(3x) + (8x)(-2) + 5(3x) + 5(-2)$
$= 24x^2 - 16x + 15x - 10$
$= 24x^2 - x - 10$

41. $(5r+2)(5r-2)$
$= (5r)(5r) + (5r)(-2) + 2(5r) + 2(-2)$
$= 25r^2 - 10r + 10r - 4$
$= 25r^2 - 4$

43. $(7y+5)(3y-8)$
$= (7y)(3y) + (7y)(-8) + 5(3y) + 5(-8)$
$= 21y^2 - 56y + 15y - 40$
$= 21y^2 - 41y - 40$

Critical Thinking 5.5

45. Strategy To find the area, substitute $(2x+3)$ for L and $(x-6)$ for W in the formula below and solve for A.

    Solution $A = LW$
    $A = (2x+3) \cdot (x-6)$
    $= (2x)(x) + (2x)(-6) + 3(x) + 3(-6)$
    $= 2x^2 - 12x + 3x - 18$
    $= 2x^2 - 9x - 18$
    The area is $(2x^2 - 9x - 18)$ mi^2.

47. a. $(5+x)^2 = 25 + x^2$ False
    $(5+x)^2$
    $= (5+x)(5+x)$
    $= 25 + 5x + 5x + x^2$
    $= 25 + 10x + x^2$

    b. $(5x)^2 = 25x^2$ True

    c. $(a-4)^2 = a^2 - 16$ False
    $(a-4)^2 = (a-4)(a-4)$
    $= a^2 - 8a + 16$

49. In explaining the similarities between multiplying two binomials and multiplying a binomial by a monomial, students should include the fact that in each case, each term of one factor must be multiplied by each term of the other factor. In explaining the differences, students might explain that in multiplying a monomial by a binomial, the Distributive Property is used only once, while in multiplying two binomials, the Distributive Property is actually being used twice. Some students may describe the FOIL method in their explanation of multiplying two binomials.

Section 5.6

Objective A Exercises

1. $27^0 = 1$

3. $-(17)^0 = -1$

5. $3^{-2} = \dfrac{1}{3^2} = \dfrac{1}{9}$

7. $2^{-3} = \dfrac{1}{2^3} = \dfrac{1}{8}$

9. $x^{-5} = \dfrac{1}{x^5}$

11. $w^{-8} = \dfrac{1}{w^8}$

13. $y^{-1} = \dfrac{1}{y}$

15. $\dfrac{1}{a^{-5}} = a^5$

17. $\dfrac{1}{b^{-3}} = b^3$

19. $\dfrac{a^8}{a^2} = a^{8-2} = a^6$

21. $\dfrac{q^5}{q} = q^{5-1} = q^4$

23. $\dfrac{m^4 n^7}{m^3 n^5} = m^{4-3} n^{7-5} = mn^2$

25. $\dfrac{t^4 u^8}{t^2 u^5} = t^{4-2} u^{8-5} = t^2 u^3$

27. $\dfrac{x^4}{x^9} = x^{4-9} = x^{-5} = \dfrac{1}{x^5}$

29. $\dfrac{b}{b^5} = b^{1-5} = b^{-4} = \dfrac{1}{b^4}$

Objective B Exercises

31. $2{,}370{,}000 = 2.37 \times 10^6$

33. $0.00045 = 4.5 \times 10^{-4}$

35. $309{,}000 = 3.09 \times 10^5$

37. $0.000000601 = 6.01 \times 10^{-7}$

39. $57{,}000{,}000{,}000 = 5.7 \times 10^{10}$

41. $0.000000017 = 1.7 \times 10^{-8}$

43. $7.1 \times 10^5 = 710{,}000$

45. $4.3 \times 10^{-5} = 0.000043$

47. $6.71 \times 10^8 = 671{,}000{,}000$

49. $7.13 \times 10^{-6} = 0.00000713$

51. $5 \times 10^{12} = 5{,}000{,}000{,}000{,}000$

53. $8.01 \times 10^{-3} = 0.00801$

55. $16{,}000{,}000{,}000$ mi $= 1.6 \times 10^{10}$ mi

57. 0.00000000000000000016 coulombs $= 1.6 \times 10^{-19}$ coulombs

59. $0.000000000001 = 1 \times 10^{-12}$

Critical Thinking 5.6

61. a. $3.45 \times 10^{-14} > 6.45 \times 10^{-15}$

    b. $5.23 \times 10^{18} > 5.23 \times 10^{17}$

    c. $3.12 \times 10^{12} > 4.23 \times 10^{11}$

    d. $-6.81 \times 10^{-24} < -9.37 \times 10^{-25}$

63. A student's explanation of how to divide exponential expressions should include a description of what "like bases" means and the concept that exponents of like bases are subtracted. Students should provide a description of how to simplify an expression after applying the Rule for Dividing Exponential Expressions and how to rewrite an expression with positive exponents. A thorough discussion will also include the idea of dividing an exponential expression by itself, which leads to a zero exponent, which simplifies to 1.

Section 5.7

Objective A Exercises

1. Three more than t
   $t + 3$

3. five less than the product of six and m
   $6m - 5$

5. the difference between three times b and seven
   $3b - 7$

7. the product of n and seven
   $7n$

9. twice the sum of three and w
   $2(3 + w)$

11. four times the difference between twice r and five
    $4(2r - 5)$

13. the quotient of v and the difference between v and 4
    $\dfrac{v}{v - 4}$

15. four times the square of t
    $4t^2$

17. The sum of the square of m and the cube of m
    $m^2 + m^3$

19. smaller numbers: s
    larger number: $31 - s$
    five more than the larger number
    $(31 - s) + 5$

Objective B Exercises

21. Let the number be x.
    a number decreased by the total of the number and twelve
    $x - (x + 12)$
    $x - x - 12$
    -12

23. Let the number be x.
    the difference between two thirds of a number and three eighths of the number
    $\dfrac{2}{3}x - \dfrac{3}{8}x$
    $\dfrac{16}{24}x - \dfrac{9}{24}x$
    $\dfrac{7}{24}x$

25. Let the number be x.
    twice the sum of seven times a number and six
    $2(7x + 6)$
    $14x + 12$

27. Let the number be x.
    the sum of eleven times a number and the product of three and the number
    $11x + 3x$
    $14x$

29. Let the number be x.
    nine times the sum of a number and seven
    $9(x + 7)$
    $9x + 63$

31. Let the number be x.
    seven more than the sum of a number and five
    $(x + 5) + 7$
    $x + 12$

33. Let the number be x.
    the product of seven and the difference between a number and four
    $7(x - 4)$
    $7x - 28$

35. Let the number be x.
    the difference between ten times a number and the product of three and the number
    $10x - 3x$
    $7x$

37. Let the number be x.
    the sum of a number and twice the difference between the number and four
    $x + 2(x - 4)$
    $x + 2x - 8$
    $3x - 8$

39. Let the number be x.
seven <u>times</u> the <u>difference</u> between a number and fourteen
$7(x - 14)$
$7x - 98$

41. Let the number be x.
the <u>product</u> of eight and the <u>sum</u> of a number and ten
$8(x + 10)$
$8x + 80$

43. Let the number be x.
a number <u>increased</u> by the <u>difference</u> between seven <u>times</u> the number and eight
$x + (7x - 8)$
$x + 7x - 8$
$8x - 8$

45. Let the number be x.
five <u>increased</u> by <u>twice</u> the <u>sum</u> of a number and fifteen
$5 + 2(x + 15)$
$5 + 2x + 30$
$2x + 35$

47. Let the number be x.
fourteen <u>decreased</u> by the <u>sum</u> of a number and thirteen
$14 - (x + 13)$
$14 - x - 13$
$-x + 1$

49. Let the number be x.
the <u>product</u> of eight <u>times</u> a number and two
$8x \cdot 2$
$16x$

51. Let the number be x.
a number <u>plus</u> nine <u>added</u> to the <u>difference</u> between four times the number and three
$(x + 9) + (4x - 3)$
$x + 9 + 4x - 3$
$5x + 6$

53. Let the smaller number be y.
The larger number is $9 - y$.
five <u>times</u> the larger number
$5(9 - y)$
$-5y + 45$

55. Let the larger number be m.
The smaller number is $17 - m$.
nine <u>less than</u> three <u>times</u> the smaller number
$3(17 - m) - 9$
$51 - 3m - 9$
$-3m + 42$

Objective C Exercises

57. the distance from Earth to the moon: d
the distance from Earth to the sun: $390d$

59. the number of genes in the round worm genome: G
the number of genes in the human genome: $G + 11,000$

61. the amount of cashews: A
the amount of peanuts: $3A$

63. the regular price of a suit: c
the sale price of the suit: $\frac{3}{4}c$

65. the longer piece of the string: L
the shorter piece of the string: $3 - L$

67. the length of the shorter piece: L
the length of the longer piece: $12 - L$

Critical Thinking 5.7

69. There are twice as many hydrogen atoms as oxygen atoms in the water. Thus if there are x oxygen atoms, there will be $2x$ hydrogen atoms.

71. For each 7 turns of the smaller wheel, the larger wheel turns 4 times. Thus the larger wheel will make $\frac{4}{7}$ as many turns as the smaller wheel.

73. "The difference between x and 5" translates to $x - 5$. "5 less than x" also translates to $x - 5$. Both the "difference between" and "less than" indicate the operation of subtraction. However, in the first instance, x and 5 are written in the order in which they occur in the expression, whereas in the second instance, the order of 5 and x must be reversed when writing the mathematical expression.

Chapter Review Exercises

1. $4z^2 + 3z - 9z + 2z^2$
    $= (4z^2 + 2z^2) + (3z - 9z)$
    $= 6z^2 - 6z$

2. $-2(9z + 1) = -18z - 2$

3. $(3z^2 + 4z - 7) + (7z^2 - 5z - 8)$
    $= (3z^2 + 7z^2) + (4z - 5z) + (-7 - 8)$
    $= 10z^2 - z - 15$

4. $(2m^3n)(-4m^2n)$
    $= [2(-4)](m^3 \cdot m^2)(n \cdot n)$
    $= -8m^5n^2$

5. $3^{-5} = \dfrac{1}{3^5} = \dfrac{1}{243}$

6. The additive inverse of $\dfrac{3}{7}$ is $-\dfrac{3}{7}$.

7. $\dfrac{2}{3}\left(\dfrac{3}{2}x\right) = x$

8. $-5(2s - 5t) + 6(3t + s)$
    $= -10s + 25t + 18t + 6s$
    $= -4s + 43t$

9. $(-5xy^4)(-3x^2y^3)$
    $= [(-5)(-3)](x \cdot x^2)(y^4 \cdot y^3)$
    $= 15x^3y^7$

10. $(7a + 6)(3a - 4)$
    $= 21a^2 - 28a + 18a - 24$
    $= 21a^2 - 10a - 24$

11. $\begin{array}{r} 6b^3 - 7b^2 + 5b - 9 \\ -9b^3 + 7b^2 - b - 9 \\ \hline -3b^3 \qquad\quad + 4b - 18 \end{array}$

12. $(2z^4)^5 = 2^{1\cdot 5}z^{4\cdot 5} = 2^5 z^{20} = 32z^{20}$

13. $-\dfrac{3}{4}(-8w) = 6w$

14. $5xyz^2(-3x^2z + 6yz^2 - x^3y^4)$
    $= 5xyz^2(-3x^2z) + 5xyz^2(6yz^2)$
    $\quad - 5xyz^2(x^3y^4)$
    $= -15x^3yz^3 + 30xy^2z^4 - 5x^4y^5z^2$

15. The multiplicative inverse of $-\dfrac{9}{4}$ is $-\dfrac{4}{9}$.

16. $-4(3c - 8) = -12c + 32$

17. $2m - 6n + 7 - 4m + 6n + 9$
    $= (2m - 4m) + (-6n + 6n) + (7 + 9)$
    $= -2m + 16$

18. $(4a^3b^8)(-3a^2b^7)$
    $= [4(-3)](a^3 \cdot a^2)(b^8 \cdot b^7)$
    $= -12a^5b^{15}$

19. The Distributive Property

20. $(p^2q^3)^3 = p^{2\cdot 3}q^{3\cdot 3}$
    $= p^6q^9$

21. $\dfrac{a^4}{a^{11}} = a^{4-11}$
    $= a^{-7}$
    $= \dfrac{1}{a^7}$

22. $0.0000397 = 3.97 \times 10^{-5}$

23. The Commutative Property of Addition

24. $(9y^3 + 8y^2 - 10) + (-6y^3 + 8y - 9)$
    $= (9y^3 - 6y^3) + 8y^2 + 8y + (-10 - 9)$
    $= 3y^3 + 8y^2 + 8y - 19$

25. $8(2c - 3d) - 4(c - 5d)$
    $= 16c - 24d - 4c + 20d$
    $= 12c - 4d$

26. $7(2m - 6) = 14m - 42$

27. $\dfrac{x^3y^5}{xy} = x^{3-1}y^{5-1} = x^2y^4$

28. $7a^2 + 9 - 12a^2 + 3a$
    $= (7a^2 - 12a^2) + 3a - 9$
    $= -5a^2 + 3a + 9$

29. $(3p-9)(4p+7)$
    $= 12p^2 + 21p - 36p - 63$
    $= 12p^2 - 15p - 63$

30. $-2a^2b(4a^3 - 5ab^2 + 3b^4)$
    $= -2a^2b(4a^3) - (-2a^2b)(5ab^2)$
    $\quad + (-2a^2b) + (3b^4)$
    $= -8a^5b + 10a^3b^3 - 6a^2b^5$

31. $-12x + 7y + 15x - 11y$
    $= (-12x + 15x) + (7y - 11y)$
    $= 3x - 4y$

32. $-7(3a - 4b) - 5(3b - 4a)$
    $= -21a + 28b - 15b + 20a$
    $= -a + 13b$

33. $c^{-5} = \dfrac{1}{c^5}$

34. $\quad 12x^3 + 9x^2 - 5x - 1$
    $\quad \underline{-6x^3 - 9x^2 - 5x + 1}$
    $\quad 6x^3 -10x$

35. $2.4 \times 10^5 = 240,000$

36. Strategy To find the perimeter, substitute $(b^2 - 4)$ for a, $(b^2 + 2)$ for b and $(b^2 + 5)$ for c in the formula below and solve for P.

    Solution $P = a + b + c$
    $= (b^2 - 4) + (b^2 + 2) + (b^2 + 5)$
    $= 3b^2 + 3$
    The perimeter is $(3b^2 + 3)$ ft.

37. nine less than the quotient of four times a number and seven
    $\dfrac{4x}{7} - 9$

38. the sum of three times a number and twice the difference between the number and seven
    $3x + 2(x - 7)$
    $3x + 2x - 14$
    $5x - 14$

39. $602,300,000,000,000,000,000,000$
    $= 6.023 \times 10^{23}$

40. the number of pounds of mocha java beans:
    p

    the number of pounds of espresso beans:
    $30 - p$

Chapter Test

1. $\dfrac{2}{3}\left(-\dfrac{3}{2}r\right) = -r$

2. $-3(5y - 7) = -15y + 21$

3. $7y - 3 - 4y + 6$
   $= (7y - 4y) + (-3 + 6)$
   $= 3y + 3$

4. $4x^2 - 2z + 7z - 8x^2$
   $= (4x^2 - 8x^2) + (-2z + 7z)$
   $= -4x^2 + 5z$

5. $2a - 4b + 12 - 5a - 2b + 6$
   $= (2a - 5a) + (-4b - 2b) + (12 + 6)$
   $= -3a - 6b + 18$

6. The multiplicative inverse of $\dfrac{5}{4}$ is $\dfrac{4}{5}$.

7. $-2(3x - 4y) + 5(2x + y)$
   $= -6x + 8y + 10x + 5y$
   $= 4x + 13y$

8. $9 - 2(4b - a) + 3(3b - 4a)$
   $= 9 - 8b + 2a + 9b - 12a$
   $= -10a + b + 9$

9. $0.00000079 = 7.9 \times 10^{-7}$

10. $4.9 \times 10^6 = 4,900,000$

11. $(4x^2 - 2x - 2) + (2x^2 - 3x + 7)$
    $= (4x^2 + 2x^2) + (-2x - 3x) + (-2 + 7)$
    $= 6x^2 - 5x + 5$

12. $\left(v^2 w^5\right)^4 = v^{2 \cdot 4} w^{5 \cdot 4} = v^8 w^{20}$

13. $(3m^2n^3)^3 = 3^{1\cdot 3} m^{2\cdot 3} n^{3\cdot 3}$
    $= 3^3 m^6 n^9 = 27m^6 n^9$

14. $(-5v^2z)(2v^3z^2)$
    $= (-5)(2)(v^2 \cdot v^3)(z \cdot z^2)$
    $= -10v^5z^3$

15. $(3p-8)(2p+5)$
    $= 6p^2 + 15p - 16p - 40$
    $= 6p^2 - p - 40$

16. $(2m^2n^2)(-4mn^3 + 2m^3 - 3n^4)$
    $= 2m^2n^2(-4mn^3) + 2m^2n^2(2m^3) + 2m^2n^2(-3n^4)$
    $= -8m^3n^5 + 4m^5n^2 - 6m^2n^6$

17. $3z + 4w = 4w + 3z$

18. $\dfrac{x^2 y^5}{xy^2} = x^{2-1} y^{5-2} = xy^3$

19. $a^{-5} = \dfrac{1}{a^5}$

20. The Associative Property of Multiplication

21. $\begin{array}{r} 5a^3 - 6a^2 + 4a - 8 \\ -8a^3 + 7a^2 - 4a - 2 \\ \hline -3a^3 + a^2 \quad\quad -10 \end{array}$

22. $\dfrac{1}{c^{-6}} = c^6$

23. The Distributive Property

24. $6w \cdot 0 = 0$

25. $(3x - 7y)(3x + 7y)$
    $= 9x^2 + 21xy - 21xy - 49y^2$
    $= 9x^2 - 49y^2$

26. The additive inverse of $-\dfrac{4}{7}$ is $\dfrac{4}{7}$.

27. $720{,}000{,}000 = 7.2 \times 10^8$

28. $(3a - 6)(4a + 2)$
    $= 12a^2 + 6a - 24a - 12$
    $= 12a^2 - 18a - 12$

29. $2(4a - 3b) + 3(5a - 2b)$
    $= 8a - 6b + 15a - 6b$
    $= 23a - 12b$

30. $\dfrac{m^4 n^2}{m^2 n^5} = m^{4-2} n^{2-5} = \dfrac{m^2}{n^3}$

31. five <u>more than</u> three <u>times</u> a number
    $3x + 5$

32. the <u>sum</u> of a number and four <u>times</u> the <u>difference</u> between the number and seven
    $x + 4(x - 7)$
    $x + 4x - 28$
    $5x - 28$

33. the number of cups of sugar in the batter: s
    the number of cups of flour in the batter: $s + 3$

Cumulative Review Exercises

1. $\dfrac{4.712}{-0.38} = -12.4$

2. $9v - 10 + 5v + 8$
   $= (9v + 5v) + (-10 + 8)$
   $= 14v - 2$

3. $(3x - 5)(2x + 4)$
   $= 6x^2 + 12x - 10x - 20$
   $= 6x^2 + 2x - 20$

4. $-a - b$, $a = \dfrac{11}{24}$ and $b = -\dfrac{5}{6}$
   $-\dfrac{11}{24} - \left(-\dfrac{5}{6}\right) = \dfrac{-11}{24} - \dfrac{-5}{6}$
   $= \dfrac{-11}{24} - \dfrac{-20}{24}$
   $= \dfrac{-11 - (-20)}{24} = \dfrac{-11 + 20}{24}$
   $= \dfrac{9}{24} = \dfrac{3 \cdot 3}{3 \cdot 8} = \dfrac{3}{8}$

5. $\sqrt{81} + 3\sqrt{25} = 9 + 3 \cdot 5$
   $= 9 + 15$
   $= 24$

6. Draw a parenthesis at –3. Draw a line to the right of –3. Draw an arrow at the right end of the line.

7. $\dfrac{1}{x^{-7}} = x^7$

8. $-4t = 36$
   $\dfrac{-4t}{-4} = \dfrac{36}{-4}$
   $t = -9$
   The solution is -9.

9. $0.00000084 = 8.4 \times 10^{-7}$

10. $(5x^2 - 3x + 2) + (4x^2 + x - 6)$
    $= (5x^2 + 4x^2) + (-3x + x) + (2 - 6)$
    $= 9x^2 - 2x - 4$

11. $-5\sqrt{x+y},\ x = 18$ and $y = 31$
    $-5\sqrt{18+31} = -5\sqrt{49}$
    $= -5 \cdot 7 = -35$

12. $\dfrac{\frac{5}{8}+\frac{3}{4}}{3-\frac{1}{2}} = \dfrac{\frac{11}{8}}{\frac{5}{2}}$
    $= \dfrac{11}{8} \div \dfrac{5}{2}$
    $= \dfrac{11}{8} \cdot \dfrac{2}{5}$
    $= \dfrac{11 \cdot 2}{8 \cdot 5} = \dfrac{11 \cdot 2}{2 \cdot 2 \cdot 2 \cdot 5} = \dfrac{11}{20}$

13. $(-3a^2b)(4a^5b^8)$
    $= (-3 \cdot 4)(a^2 \cdot a^5)(b \cdot b^8)$
    $= -12a^7b^9$

14. $\dfrac{x^3}{x^5} = x^{3-5}$
    $= x^{-2}$
    $= \dfrac{1}{x^2}$

15. $x^3y^2,\ x = \dfrac{2}{5}$ and $y = 2\dfrac{1}{2}$
    $\left(\dfrac{2}{5}\right)^3 \left(2\dfrac{1}{2}\right)^2 = \left(\dfrac{2}{5} \cdot \dfrac{2}{5} \cdot \dfrac{2}{5}\right) \cdot \left(\dfrac{5}{2} \cdot \dfrac{5}{2}\right)$
    $= \dfrac{2 \cdot 2 \cdot 2 \cdot 5 \cdot 5}{5 \cdot 5 \cdot 5 \cdot 2 \cdot 2} = \dfrac{2}{5}$

16. $-8p(6) = (6)(-8p) = [(6)(-8)]p = -48p$

17. $\begin{array}{r} 829.43 \to 800 \\ 567.109 \to -600 \\ \hline 200 \end{array}$

18. $-3ab^2(4a^2b + 5ab - 2ab^2)$
    $= -3ab^2(4a^2b) + (-3ab^2)(5ab) - (-3ab^2)(-2ab^2)$
    $= -12a^3b^3 - 15a^2b^3 + 6a^2b^4$

19. $6(5x - 4y) - 12(x - 2y)$
    $= 30x - 24y - 12x + 24y$
    $= 18x$

20. $\dfrac{a}{-b},\ a = -56$ and $b = -8$
    $\dfrac{-56}{-(-8)} = \dfrac{-56}{8} = -7$

21. $0.5625 = \dfrac{5625}{10{,}000} = \dfrac{625 \cdot 9}{625 \cdot 16} = \dfrac{9}{16}$

22. $6 \cdot (-2)^3 \div 12 - (-8) = 6 \cdot (-8) \div 12 - (-8)$
    $= -48 \div 12 - (-8)$
    $= -4 - (-8)$
    $= -4 + 8 = 4$

23. $\sqrt{300} = \sqrt{100 \cdot 3}$
    $= \sqrt{100} \cdot \sqrt{3}$
    $= 10\sqrt{3}$

24. $(8y^2 - 7y + 4) - (3y^2 - 5y + 9)$
    $= (8y^2 - 7y + 4) + (-3y^2 + 5y - 9)$
    $= 5y^2 - 2y - 5$

25. $-6cd,\ c = -\dfrac{2}{9}$ and $d = \dfrac{3}{4}$
    $-6\left(-\dfrac{2}{9}\right)\left(\dfrac{3}{4}\right) = \dfrac{6}{1} \cdot \dfrac{2}{9} \cdot \dfrac{3}{4}$
    $= \dfrac{6 \cdot 2 \cdot 3}{1 \cdot 9 \cdot 4}$
    $= \dfrac{2 \cdot 3 \cdot 2 \cdot 3}{1 \cdot 3 \cdot 3 \cdot 2 \cdot 2} = 1$

26. $-(3a^2)^0 = -1$

27. $(2a^4b^3)^5 = 2^{1 \cdot 5} a^{4 \cdot 5} b^{3 \cdot 5}$
    $= 2^5 a^{20} b^{15}$
    $= 32a^{20}b^{15}$

28. $(a-b)^2 + 5c;\ a = -4,\ b = 6,\ c = -2$
    $(-4 - 6)^2 + 5(-2) = (-10)^2 + 5(-2)$
    $= 100 + 5(-2)$
    $= 100 + (-10)$
    $= 90$

29. $2\frac{4}{5} \cdot \frac{6}{7} = \frac{14}{5} \cdot \frac{6}{7}$

$= \frac{14 \cdot 6}{5 \cdot 7}$

$= \frac{2 \cdot 7 \cdot 2 \cdot 3}{5 \cdot 7}$

$= \frac{12}{5} = 2\frac{2}{5}$

30. $6.23 \times 10^{-5} = 0.0000623$

31. Let the unknown number be x.
the <u>quotient</u> of ten and the <u>difference</u> between a number and nine
$\frac{10}{x-9}$

32. Let the unknown number be x.
two <u>less than</u> twice the <u>sum</u> of a number and four
$2(x+4) - 2$
$2x + 8 - 2$
$2x + 6$

33. Strategy   To find the difference, subtract the average rainfall in El Paso (7.82) from the average rainfall in Seattle (38.6).

     Solution $38.6 - 7.82 = 30.78$
The difference between the average annual rainfall in Seattle and El Paso is 30.78 in.

34. Strategy   To find the difference, subtract the amount of trash thrown away by a person in 1960 (2.7) from the amount of trash thrown away by a person in the 2000 (4.5). Multiply the difference by the number of days in a year (365). Round your answer.

     Solution $4.5 - 2.7 = 1.8$
$1.8 \cdot 365 = 657$
The person in 2000 throws away 657 lb more trash than a person in 1960.

35. the distance from Earth to the sun: d
the distance from Neptune to the sun: $30d$

36. Strategy   To find the cost, substitute $15\frac{3}{8}$ for S and 200 for N in the given formula and solve for C.

     Solution $C = SN$
$C = 15\frac{3}{8} \cdot 200$
$C = 3,075$
The cost of the stock was $3,075.

Chapter 6: First-Degree Equations

Prep Test

1. $8 - 12 = -4$
2. $-\dfrac{3}{4}\left(-\dfrac{4}{3}\right) = 1$
3. $-\dfrac{5}{8}(16) = -5(2) = -10$
4. $\dfrac{-3}{-3} = 1$
5. $-16 + 7y + 16 = 7y$
6. $8x - 9 - 8x = -9$
7. $2x + 3$
   $2(-4) + 3 = -8 + 3 = -5$
8. $y = -4x + 5$
   $y = -4(-2) + 5$
   $\quad = 8 + 5$
   $\quad = 13$

Go Figure

With the donut on a table, slice the donut parallel to the table. Now cut the halved donut into quarters.

Section 6.1

Objective A Exercises

1. **a.** The equation $7 + p = -23$ is one of the form $x + a = b$ because, on the left side of the equation, a number is added to the variable p. The right side is a constant, -23.

   b. The equation $-16 = -2s$ is one of the form $ax = b$ because, on the right side of the equation, a number is multiplied times the variable s. The left side is a constant, -16.

   c. The equation $-\dfrac{7}{8}g = 49$ is one of the form $ax = b$ because, on the left side of the equation, a number is multiplied times the variable g. The right side is a constant, 49.

   d. The equation $2.8 = q - 9$ is one of the form $x + a = b$. Subtraction is addition of the opposite; therefore, the right side of the equation can be rewritten as $q + (-9)$. In the expression $q + (-9)$, a number is added to the variable q. The left side is a constant, 2.8.

3. $x + 3 = 9$
   $x + 3 - 3 = 9 - 3$
   $x = 6$
   The solution is 6.

5. $4 + x = 13$
   $4 - 4 + x = 13 - 4$
   $x = 9$
   The solution is 9.

7. $m - 12 = 5$
   $m - 12 + 12 = 5 + 12$
   $m = 17$
   The solution is 17.

9. $x - 3 = -2$
   $x - 3 + 3 = -2 + 3$
   $x = 1$
   The solution is 1.

11. $a + 5 = -2$
    $a + 5 - 5 = -2 - 5$
    $a = -7$
    The solution is -7.

13. $3 + m = -6$
    $3 - 3 + m = -6 - 3$
    $m = -9$
    The solution is -9.

15. $8 = x + 3$
    $8 - 3 = x + 3 - 3$
    $5 = x$
    The solution is 5.

17. $3 = w - 6$
    $3 + 6 = w - 6 + 6$
    $9 = w$
    The solution is 9.

19. $-7 = -7 + m$
    $-7 + 7 = -7 + 7 + m$
    $0 = m$
    The solution is 0.

21. $-3 = v + 5$
    $-3 - 5 + v + 5 - 5$

$-8 = v$
The solution is -8.

23. $-5 = 1 + x$
$-5 - 1 = 1 - 1 + x$
$-6 = x$
The solution is -6.

25. $3 = -9 + m$
$3 + 9 = -9 + 9 + m$
$12 = m$
The solution is 12.

27. $4 + x - 7 = 3$
$x - 3 = 3$
$x - 3 + 3 = 3 + 3$
$x = 6$
The solution is 6.

29. $8t + 6 - 7t = -6$
$t + 6 = -6$
$t + 6 - 6 = -6 - 6$
$t = -12$
The solution is -12.

31. $y + \frac{4}{7} = \frac{6}{7}$
$y + \frac{4}{7} - \frac{4}{7} = \frac{6}{7} - \frac{4}{7}$
$y = \frac{2}{7}$
The solution is $\frac{2}{7}$.

33. $x - \frac{3}{8} = \frac{1}{8}$
$x - \frac{3}{8} + \frac{3}{8} = \frac{1}{8} + \frac{3}{8}$
$x = \frac{4}{8}$
$x = \frac{1}{2}$
The solution is $\frac{1}{2}$.

35. $c + \frac{2}{3} = \frac{3}{4}$
$c + \frac{2}{3} - \frac{2}{3} = \frac{3}{4} - \frac{2}{3}$
$c = \frac{9}{12} - \frac{8}{12}$
$c = \frac{1}{12}$
The solution is $\frac{1}{12}$.

37. $w - \frac{1}{4} = \frac{3}{8}$
$w - \frac{1}{4} + \frac{1}{4} = \frac{3}{8} + \frac{1}{4}$
$w = \frac{3}{8} + \frac{2}{8}$
$w = \frac{5}{8}$
The solution is $\frac{5}{8}$.

Objective B Exercises

39. $3x = 9$
$\frac{3x}{3} = \frac{9}{3}$
$x = 3$
The solution is 3.

41. $4c = -12$
$\frac{4c}{4} = \frac{-12}{4}$
$c = -3$
The solution is -3.

43. $-2r = 16$
$\frac{-2r}{-2} = \frac{16}{-2}$
$r = -8$
The solution is -8.

45. $-4m = -28$
$\frac{-4m}{-4} = \frac{-28}{-4}$
$m = 7$
The solution is 7.

47. $-3y = 0$
$\frac{-3y}{-3} = \frac{0}{-3}$
$y = 0$
The solution is 0.

49. $12 = 2c$
$\frac{12}{2} = \frac{2c}{2}$
$6 = c$
The solution is 6.

51. $-72 = 18v$
$\frac{-72}{18} = \frac{18v}{18}$
$-4 = v$
The solution is -4.

53. $-68 = -17t$
$\dfrac{-68}{-17} = \dfrac{-17t}{-17}$
$4 = t$
The solution is 4.

55. $12x = 30$
$\dfrac{12x}{12} = \dfrac{30}{12}$
$x = \dfrac{5}{2}$
The solution is $\dfrac{5}{2}$.

57. $-6a = 21$
$\dfrac{-6a}{-6} = \dfrac{21}{-6}$
$a = -\dfrac{7}{2}$
The solution is $-\dfrac{7}{2}$.

59. $28 = -12y$
$\dfrac{28}{-12} = \dfrac{-12y}{-12}$
$-\dfrac{7}{3} = y$
The solutions is $-\dfrac{7}{3}$.

61. $-52 = -18a$
$\dfrac{-52}{-18} = \dfrac{-18a}{-18}$
$\dfrac{26}{9} = a$
The solution is $\dfrac{26}{9}$.

63. $\dfrac{2}{3}x = 4$
$\dfrac{3}{2} \cdot \dfrac{2}{3}x = \dfrac{3}{2}(4)$
$x = 6$
The solution is 6.

65. $\dfrac{1}{3}a = -12$
$\dfrac{3}{1} \cdot \dfrac{1}{3}a = \dfrac{3}{1}(-12)$
$a = -36$
The solution is -36.

67. $-\dfrac{4c}{7} = 16$
$-\dfrac{7}{4}\left(-\dfrac{4}{7}c\right) = -\dfrac{7}{4}(16)$
$c = -28$
The solution is -28.

69. $-\dfrac{z}{4} = -3$
$-\dfrac{4}{1}\left(-\dfrac{1}{4}z\right) = -\dfrac{4}{1}(-3)$
$z = 12$
The solution is 12.

71. $8 = \dfrac{4}{5}y$
$\dfrac{5}{4}(8) = \dfrac{5}{4} \cdot \dfrac{4}{5}y$
$10 = y$
The solution is 10.

73. $\dfrac{5y}{6} = \dfrac{7}{12}$
$\dfrac{6}{5}\left(\dfrac{5}{6}y\right) = \dfrac{6}{5}\left(\dfrac{7}{12}\right)$
$y = \dfrac{7}{10}$
The solution is $\dfrac{7}{10}$.

75. $7y - 9y = 10$
$-2y = 10$
$\dfrac{-2y}{-2} = \dfrac{10}{-2}$
$y = -5$
The solution is -5.

77. $m - 4m = 21$
$-3m = 21$
$\dfrac{-3m}{-3} = \dfrac{21}{-3}$
$m = -7$
The solution is -7.

Critical Thinking 6.1

79. **a.** $x + a = b$
$x + a - a = b - a$
$x = b - a$
The solution is valid for all real numbers a and b.

   b. $ax = b$
$\dfrac{ax}{a} = \dfrac{b}{a}$
$x = \dfrac{b}{a}$
No, the solution is not valid for $a = 0$.

81. Look for the following explanations in your students' work.

$\frac{2}{3}x = 6$

$\frac{3}{2} \cdot \frac{2}{3}x = \frac{3}{2} \cdot 6$ By the Multiplication Property of Equations, each side of an equation can be multiplied by the same nonzero number without changing the solution of the equation. Multiply each side of the equation by the reciprocal of the coefficient of x.

$\frac{3}{2} \cdot \frac{2}{3}x = 9$ Multiply $\frac{3}{2}$ by 6.

$\left(\frac{3}{2} \cdot \frac{2}{3}\right)x = 9$ Use the Associative Property of Multiplication to group $\frac{3}{2}$ and $\frac{2}{3}$.

$1x = 9$ By the Inverse Property of Multiplication, the product of a nonzero number and its reciprocal is 1.

$x = 9$ By the Multiplication Property of 1, the product of a number and 1 is the number.

Section 6.2

Objective A Exercises

1. Students should rephrase the Addition Property of Equations (The same number or variable expression can be added to each side of an equation without changing the solution of the equation.) They should explain that this property is used to remove a term from one side of an equation by adding the opposite of that term to each side of the equation.

3. $5y + 1 = 11$
$5y + 1 - 1 = 11 - 1$
$5y = 10$
$\frac{5y}{5} = \frac{10}{5}$
$y = 2$
The solution is 2.

5. $2z - 9 = 11$
$2z - 9 + 9 = 11 + 9$
$2z = 20$
$\frac{2z}{2} = \frac{20}{2}$
$z = 10$
The solution is 10.

7. $12 = 2 + 5a$
$12 - 2 = 2 - 2 + 5a$
$10 = 5a$
$\frac{10}{5} = \frac{5a}{5}$
$2 = a$
The solution is 2.

9. $-5y + 8 = 13$
$-5y + 8 - 8 = 13 - 8$
$-5y = 5$
$\frac{-5y}{-5} = \frac{5}{-5}$
$y = -1$
The solution is -1.

11. $-12a - 1 = 23$
$-12a - 1 + 1 = 23 + 1$
$-12a = 24$
$\frac{-12a}{-12} = \frac{24}{-12}$
$a = -2$
The solution is -2.

13. $10 - c = 14$
$10 - 10 - c = 14 - 10$
$-c = 4$
$\frac{-1c}{-1} = \frac{4}{-1}$
$c = -4$
The solution is -4.

15. $4 - 3x = -5$
    $4 - 4 - 3x = -5 - 4$
    $-3x = -9$
    $\dfrac{-3x}{-3} = \dfrac{-9}{-3}$
    $x = 3$
    The solution is 3.

17. $-33 = 3 - 4z$
    $-33 - 3 = 3 - 3 - 4z$
    $-36 = -4z$
    $\dfrac{-36}{-4} = \dfrac{-4z}{-4}$
    $9 = z$
    The solution is 9.

19. $-4t + 16 = 0$
    $-4t + 16 - 16 = 0 - 16$
    $-4t = -16$
    $\dfrac{-4t}{-4} = \dfrac{-16}{-4}$
    $t = 4$
    The solution is 4.

21. $5a + 9 = 12$
    $5a + 9 - 9 = 12 - 9$
    $5a = 3$
    $\dfrac{5a}{5} = \dfrac{3}{5}$
    $a = \dfrac{3}{5}$
    The solution is $\dfrac{3}{5}$.

23. $2t - 5 = 2$
    $2t - 5 + 5 = 2 + 5$
    $2t = 7$
    $\dfrac{2t}{2} = \dfrac{7}{2}$
    $t = \dfrac{7}{2}$
    The solution is $\dfrac{7}{2}$.

25. $8x + 1 = 7$
    $8x + 1 - 1 = 7 - 1$
    $8x = 6$
    $\dfrac{8x}{8} = \dfrac{6}{8}$
    $x = \dfrac{3}{4}$
    The solution is $\dfrac{3}{4}$.

27. $4z - 5 = 1$
    $4z - 5 + 5 = 1 + 5$
    $4z = 6$
    $\dfrac{4z}{4} = \dfrac{6}{4}$
    $z = \dfrac{3}{2}$
    The solution is $\dfrac{3}{2}$.

29. $25 = 11 + 8v$
    $25 - 11 = 11 - 11 + 8v$
    $14 = 8v$
    $\dfrac{14}{8} = \dfrac{8v}{8}$
    $\dfrac{7}{4} = v$
    The solution is $\dfrac{7}{4}$.

31. $-3 = 7 + 4y$
    $-3 - 7 = 7 - 7 + 4y$
    $-10 = 4y$
    $\dfrac{-10}{4} = \dfrac{4y}{4}$
    $-\dfrac{5}{2} = y$
    The solution is $-\dfrac{5}{2}$.

33. $8a - 5 = 31$
    $8a - 5 + 5 = 31 + 5$
    $8a = 36$
    $\dfrac{8a}{8} = \dfrac{36}{8}$
    $a = \dfrac{9}{2}$
    The solution is $\dfrac{9}{2}$.

35. $7 - 12y = 7$
    $7 - 7 - 12y = 7 - 7$
    $-12y = 0$
    $\dfrac{-12y}{-12} = \dfrac{0}{-12}$
    $y = 0$
    The solution is 0.

37. $-9 - 12y = 5$
    $-9 + 9 - 12y = 5 + 9$
    $-12y = 14$
    $\dfrac{-12y}{-12} = \dfrac{14}{-12}$
    $y = -\dfrac{7}{6}$.

39. $6z - \frac{1}{3} = \frac{5}{3}$

$6z - \frac{1}{3} + \frac{1}{3} = \frac{5}{3} + \frac{1}{3}$

$6z = \frac{6}{3}$

$6z = 2$

$\frac{6z}{6} = \frac{2}{6}$

$z = \frac{1}{3}$

The solution is $\frac{1}{3}$.

41. $3p - \frac{5}{8} = \frac{19}{8}$

$3p - \frac{5}{8} + \frac{5}{8} = \frac{19}{8} + \frac{5}{8}$

$3p = \frac{24}{8}$

$3p = 3$

$\frac{3p}{3} = \frac{3}{3}$

$p = 1$

The solution is 1.

43. $\frac{4}{5}y + 3 = 11$

$\frac{4}{5}y + 3 - 3 = 11 - 3$

$\frac{4}{5}y = 8$

$\frac{5}{4}\left(\frac{4}{5}y\right) = \frac{5}{4} \cdot 8$

$y = 10$

The solution is 10.

45. $\frac{3v}{7} - 2 = 10$

$\frac{3v}{7} - 2 + 2 = 10 + 2$

$\frac{3}{7}v = 12$

$\frac{7}{3}\left(\frac{3}{7}v\right) = \frac{7}{3} \cdot 12$

$v = 28$

The solution is 28.

47. $\frac{4z}{9} + 23 = 3$

$\frac{4z}{9} + 23 - 23 = 3 - 23$

$\frac{4z}{9} = -20$

$\frac{9}{4}\left(\frac{4}{9}z\right) = \frac{9}{4}(-20)$

$z = -45$

The solution is -45.

49. $\frac{y}{4} + 5 = 2$

$\frac{y}{4} + 5 - 5 = 2 - 5$

$\frac{y}{4} = -3$

$\frac{4}{1}\left(\frac{1}{4}y\right) = \frac{4}{1}(-3)$

$y = -12$

The solution is -12.

51. $\frac{2}{5}y - 3 = 1$

$\frac{2}{5}y - 3 + 3 = 1 + 3$

$\frac{2}{5}y = 4$

$\frac{5}{2}\left(\frac{2}{5}y\right) = \frac{5}{2}(4)$

$y = 10$

The solution is 10.

53. $5 - \frac{7}{8}y = 2$

$5 - 5 - \frac{7}{8}y = 2 - 5$

$-\frac{7}{8}y = -3$

$\left(-\frac{8}{7}\right)\left(-\frac{7}{8}y\right) = \left(-\frac{8}{7}\right)(-3)$

$y = \frac{24}{7}$

The solution is $\frac{24}{7}$.

55. $\frac{3}{5}y + \frac{1}{4} = \frac{3}{4}$

$\frac{3}{5}y + \frac{1}{4} - \frac{1}{4} = \frac{3}{4} - \frac{1}{4}$

$\frac{3}{5}y = \frac{1}{2}$

$\frac{5}{3}\left(\frac{3}{5}y\right) = \frac{5}{3}\left(\frac{1}{2}\right)$

$y = \frac{5}{6}$

The solution is $\frac{5}{6}$.

57. $\frac{3}{5} = \frac{2}{7}t + \frac{1}{5}$
$\frac{3}{5} - \frac{1}{5} = \frac{2}{7}t + \frac{1}{5} - \frac{1}{5}$
$\frac{2}{5} = \frac{2}{7}t$
$\frac{7}{2}\left(\frac{2}{5}\right) = \frac{7}{2}\left(\frac{2}{7}t\right)$
$\frac{7}{5} = t$
The solution is $\frac{7}{5}$.

59. $\frac{z}{3} - \frac{1}{2} = \frac{1}{4}$
$\frac{z}{3} - \frac{1}{2} + \frac{1}{2} = \frac{1}{4} + \frac{1}{2}$
$\frac{z}{3} = \frac{1}{4} + \frac{2}{4}$
$\frac{z}{3} = \frac{3}{4}$
$\frac{3}{1}\left(\frac{1}{3}z\right) = \frac{3}{1}\left(\frac{3}{4}\right)$
$z = \frac{9}{4}$
The solution is $\frac{9}{4}$.

61. $5.6t - 5.1 = 1.06$
$5.6t - 5.1 + 5.1 = 1.06 + 5.1$
$5.6t = 6.16$
$\frac{5.6t}{5.6} = \frac{6.16}{5.6}$
$t = 1.1$
The solution is 1.1.

63. $6.2 - 3.3t = -12.94$
$6.2 - 6.2 - 3.3t = -12.94 - 6.2$
$-3.3t = -19.14$
$\frac{-3.3t}{-3.3} = \frac{-19.14}{-3.3}$
$t = 5.8$
The solution is 5.8.

65. $6c - 2 - 3c = 10$
$3c - 2 = 10$
$3c - 2 + 2 = 10 + 2$
$3c = 12$
$\frac{3c}{3} = \frac{12}{3}$
$c = 4$
The solution is 4.

67. $4y + 5 - 12y = -3$
$-8y + 5 = -3$
$-8y + 5 - 5 = -3 - 5$
$-8y = -8$
$\frac{-8y}{-8} = \frac{-8}{-8}$
$y = 1$
The solution is 1.

69. $17 = 12p - 5 - 6p$
$17 = 6p - 5$
$17 + 5 = 6p - 5 + 5$
$22 = 6p$
$\frac{22}{6} = \frac{6p}{6}$
$\frac{11}{3} = p$
The solution is $\frac{11}{3}$.

71. $3 = 6n + 23 - 10n$
$3 = -4n + 23$
$3 - 23 = -4n + 23 - 23$
$-20 = -4n$
$\frac{-20}{-4} = \frac{-4n}{-4}$
$5 = n$
The solution is 5.

Objective B Exercises

73. Strategy To find the number of years, substitute 47,500 for V and 63,000 for C in the given equation and solve for t.

Solution
$V = C - 5{,}500t$
$47{,}500 = 63{,}000 - 5{,}500t$
$47{,}500 - 63{,}000 = 63{,}000 - 63{,}000 - 5{,}500t$
$-15{,}500 = -5{,}500t$
$\dfrac{-15{,}500}{-5{,}500} = \dfrac{-5{,}500t}{-5{,}500}$
$2.818 \approx t$
In approximately 2.8 years, the X-ray machine will have a value of $47,500.

75. Strategy To find the maximum loan amount you can afford, substitute 325 for P in the given formula and solve for L.

Solution
$P = 0.02076L$
$325 = 0.02076L$
$\dfrac{325}{0.02076} = \dfrac{0.02076L}{0.02076}$
$15{,}655.11 \approx L$
The maximum loan amount you can afford is $15,655.11.

77. Strategy To find the year, substitute 3.77 for t in the given equation and solve for y.

Solution
$t = 17.08 - 0.0067y$
$3.77 = 17.08 - 0.0067y$
$3.77 - 17.08 = 17.08 - 17.08 - 0.0067y$
$-13.31 = 0.0067y$
$\dfrac{-13.31}{-0.0067} = \dfrac{-0.0067y}{-0.0067}$
$1986.567 \approx y$
The 3.77-minute mile was predicted for 1987.

79. Strategy To find the distance, substitute -11 for C in the given equation and solve for D.

Solution
$C = \tfrac{1}{4}D - 45$
$-11 = \tfrac{1}{4}D - 45$
$-11 + 45 = \tfrac{1}{4}D - 45 + 45$
$34 = \tfrac{1}{4}D$
$\tfrac{4}{1}(34) = \tfrac{4}{1}\left(\tfrac{1}{4}D\right)$
$136 = D$
The car sill slide 136 ft.

Critical Thinking 6.2

81. Answers will vary.
$x = -3$. Let $a = 2$ and $b = 5$, and solve for c.
$ax + b = c$
$2(-3) + 5 = c$
$-6 + 5 = c$
$-1 = c$
Thus the equation $2x + 5 = -1$ has a solution of -3.

83. The expression "Solve $2x - 3(4x + 1)$" has no meaning because there is no equal sign. It is not an equation.

Section 6.3

Objective A Exercises

1. $4x + 3 = 2x + 9$
$4x - 2x + 3 = 2x - 2x + 9$
$2x + 3 = 9$
$2x + 3 - 3 = 9 - 3$
$2x = 6$
$\dfrac{2x}{2} = \dfrac{6}{2}$
$x = 3$
The solution is 3.

3. $7y - 6 = 3y + 6$
$7y - 3y - 6 = 3y - 3y + 6$
$4y - 6 = 6$
$4y - 6 + 6 = 6 + 6$
$4y = 12$
$\dfrac{4y}{4} = \dfrac{12}{4}$
$y = 3$
The solution is 3.

5. $12m + 11 = 5m + 4$
$12m - 5m + 11 = 5m - 5m + 4$
$7m + 11 = 4$
$7m + 11 - 11 = 4 - 11$
$7m = -7$
$\dfrac{7m}{7} = \dfrac{-7}{7}$
$m = -1$
The solution is -1.

7. $7c - 5 = 2c - 25$
$7c - 2c - 5 = 2c - 2c - 25$
$5c - 5 = -25$
$5c - 5 + 5 = -25 + 5$
$5c = -20$
$\dfrac{5c}{5} = \dfrac{-20}{5}$
$c = -4$
The solution is -4.

9. $2n - 3 = 5n - 18$
$2n - 5n - 3 = 5n - 5n - 18$
$-3n - 3 = -18$
$-3n - 3 + 3 = -18 + 3$
$-3n = -15$
$\dfrac{-3n}{-3} = \dfrac{-15}{-3}$
$n = 5$
The solution is 5.

11. $3z + 5 = 19 - 4z$
$3z + 4z + 5 = 19 - 4z + 4z$
$7z + 5 = 19$
$7z + 5 - 5 = 19 - 5$
$7z = 14$
$\dfrac{7z}{7} = \dfrac{14}{7}$
$z = 2$
The solution is 2.

13. $5v - 3 = 4 - 2v$
$5v + 2v - 3 = 4 - 2v + 2v$
$7v - 3 = 4$
$7v - 3 + 3 = 4 + 3$
$7v = 7$
$\dfrac{7v}{7} = \dfrac{7}{7}$
$v = 1$
The solution is 1.

15. $7 - 4a = 2a$
$7 - 4a + 4a = 2a + 4a$
$7 = 6a$
$\dfrac{7}{6} = \dfrac{6a}{6}$
$\dfrac{7}{6} = a$
The solution is $\dfrac{7}{6}$.

17. $12 - 5y = 3y - 12$
$12 - 5y - 3y = 3y - 3y - 12$
$12 - 8y = -12$
$12 - 12 - 8y = -12 - 12$
$-8y = -24$
$\dfrac{-8y}{-8} = \dfrac{-24}{-8}$
$y = 3$
The solution is 3.

19. $7r = 8 + 2r$
$7r - 2r = 8 + 2r - 2r$
$5r = 8$
$\dfrac{5r}{5} = \dfrac{8}{5}$
$r = \dfrac{8}{5}$
The solution is $\dfrac{8}{5}$.

21. $5a + 3 = 3a + 10$
$5a - 3a + 3 = 3a - 3a + 10$
$2a + 3 = 10$
$2a + 3 - 3 = 10 - 3$
$2a = 7$
$\dfrac{2a}{2} = \dfrac{7}{2}$
The solution is $\dfrac{7}{2}$.

23. $9w - 2 = 5w + 4$
$9w - 5w - 2 = 5w - 5w + 4$
$4w - 2 = 4$
$4w - 2 + 2 = 4 + 2$
$4w = 6$
$\dfrac{4w}{4} = \dfrac{6}{4}$
$w = \dfrac{3}{2}$
The solution is $\dfrac{3}{2}$.

25. $x - 7 = 5x - 21$
$x - 5x - 7 = 5x - 5x - 21$
$-4x - 7 = -21$
$-4x - 7 + 7 = -21 + 7$
$-4x = -14$
$\dfrac{-4x}{-4} = \dfrac{-14}{-4}$
$x = \dfrac{7}{2}$
The solution is $\dfrac{7}{2}$.

27. $5n - 1 + 2n = 4n + 8$
$7n - 1 = 4n + 8$
$7n - 4n - 1 = 4n - 4n + 8$
$3n - 1 = 8$
$3n - 1 + 1 = 8 + 1$
$3n = 9$
$\dfrac{3n}{3} = \dfrac{9}{3}$
$n = 3$
The solution is 3.

29. $3z - 2 - 7z = 4z + 6$
$-4z - 2 = 4z + 6$
$-4z - 4z - 2 = 4z - 4z + 6$
$-8z - 2 = 6$
$-8z - 2 + 2 = 6 + 2$
$-8z = 8$
$\dfrac{-8z}{-8} = \dfrac{8}{-8}$
$z = -1$
The solution is -1.

31. $4t - 8 + 12t = 3 - 4t - 11$
$16t - 8 = -8 - 4t$
$16t + 4t - 8 = -8 - 4t + 4t$
$20t - 8 = -8$
$20t - 8 + 8 = -8 + 8$
$20t = 0$
$\dfrac{20t}{20} = \dfrac{0}{20}$
$t = 0$
The solution is 0.

Objective B Exercises

33. $3(4y + 5) = 25$
$12y + 15 = 25$
$12y + 15 - 15 = 25 - 15$
$12y = 10$
$\dfrac{12y}{12} = \dfrac{10}{12}$
$y = \dfrac{5}{6}$
The solution is $\dfrac{5}{6}$.

35. $-2(4x + 1) = 22$
$-8x - 2 = 22$
$-8x - 2 + 2 = 22 + 2$
$-8x = 24$
$\dfrac{-8x}{-8} = \dfrac{24}{-8}$
$x = -3$
The solution is -3.

37. $5(2k + 1) - 7 = 28$
$10k + 5 - 7 = 28$
$10k - 2 = 28$
$10k - 2 + 2 = 28 + 2$
$10k = 30$
$\dfrac{10k}{10} = \dfrac{30}{10}$
$k = 3$
The solution is 3.

39. $3(3v - 4) + 2v = 10$
$9v - 12 + 2v = 10$
$11v - 12 = 10$
$11v - 12 + 12 = 10 + 12$
$11v = 22$
$\dfrac{11v}{11} = \dfrac{22}{11}$
$v = 2$
The solution is 2.

41. $3y + 2(y + 1) = 12$
$3y + 2y + 2 = 12$
$5y + 2 = 12$
$5y + 2 - 2 = 12 - 2$
$5y = 10$
$\dfrac{5y}{5} = \dfrac{10}{5}$
$y = 2$
The solution is 2.

43. $7v - 3(v - 4) = 20$
$7v - 3v + 12 = 20$
$4v + 12 = 20$
$4v + 12 - 12 = 20 - 12$
$4v = 8$
$\dfrac{4v}{4} = \dfrac{8}{4}$
$v = 2$
The solution is 2.

45. $6 + 3(3x - 3) = 24$
$6 + 9x - 9 = 24$
$9x - 3 = 24$
$9x - 3 + 3 = 24 + 3$
$9x = 27$
$\dfrac{9x}{9} = \dfrac{27}{9}$
$x = 3$
The solution is 3.

47. $9 - 3(4a - 2) = 9$
$9 - 12a + 6 = 9$
$-12a + 15 = 9$
$-12a + 15 - 15 = 9 - 15$
$-12a = -6$
$\dfrac{-12a}{-12} = \dfrac{-6}{-12}$
$a = \dfrac{1}{2}$
The solution is $\dfrac{1}{2}$.

49. $3(2z - 5) = 4z + 1$
$6z - 15 = 4z + 1$
$6z - 4z - 15 = 4z - 4z + 1$
$2z - 15 = 1$
$2z - 15 + 15 = 1 + 15$
$2z = 16$
$\dfrac{2z}{2} = \dfrac{16}{2}$
$z = 8$
The solution is 8.

51. $2 - 3(5x + 2) = 2(3 - 5x)$
$2 - 15x - 6 = 6 - 10x$
$-15x - 4 = 6 - 10x$
$-15x + 10x - 4 = 6 - 10x + 10x$
$-5x - 4 = 6$
$-5x - 4 + 4 = 6 + 4$
$-5x = 10$
$\dfrac{-5x}{-5} = \dfrac{10}{-5}$
$x = -2$
The solution is -2.

53. $4r + 11 = 5 - 2(3r + 3)$
$4r + 11 = 5 - 6r - 6$
$4r + 11 = -6r - 1$
$4r + 6r + 11 = -6r + 6r - 1$
$10r + 11 = -1$
$10r + 11 - 11 = -1 - 11$
$10r = -12$
$\dfrac{10r}{10} = \dfrac{-12}{10}$
$r = -\dfrac{6}{5}$
The solution is $-\dfrac{6}{5}$.

55. $7n - 2 = 5 - (9 - n)$
$7n - 2 = 5 - 9 + n$
$7n - 2 = -4 + n$
$7n - n - 2 = -4 + n - n$
$6n - 2 = -4$
$6n - 2 + 2 = -4 + 2$
$6n = -2$
$\frac{6n}{6} = \frac{-2}{6}$
$n = -\frac{1}{3}$
The solution is $-\frac{1}{3}$.

Objective C Exercises

57. Strategy — The distance of the fulcrum from the 190-pound person: x
The distance of the fulcrum from the 120-pound person: $15 - x$
To find the placement of the fulcrum, replace the variables F_1, F_2, and d by the given values and solve for x.

Solution $F_1 \cdot x = F_2 \cdot (d - x)$
$180 \cdot x = 120(15 - x)$
$180x = 1,800 - 120x$
$300x = 1,800$
$\frac{300x}{300} = \frac{1,800}{300}$
$x = 6$
The 180-pound person should be 6 ft from the fulcrum.

59. Strategy — To find the minimum force, substitute 150 for F_1, 8 for d, and 1.5 for x in the given equation and solve for F_2.

Solution $F_1 \cdot x = F_2 \cdot (d - x)$
$150 \cdot 1.5 = F_2(8 - 1.5)$
$225 = 6.5F_2$
$\frac{225}{6.5} = \frac{6.5F_2}{6.5}$
$34.6 \approx F_2$
The minimum force is approximately 34.6 lb.

61. Strategy — To find the break-even point, substitute 1,600 for P, 950 for C, and 211,250 for F in the given equation and solve for x.

Solution $Px = Cx + F$
$1,600x = 950x + 211,250$
$650x = 211,250$
$\frac{650x}{650} = \frac{211,250}{650}$
$x = 325$
325 units must be sold to break even.

63. Strategy — To find the break-even point, substitute 99 for P, 38 for C, and 24,400 for F in the given equation and solve for x.

Solution $Px = Cx + F$
$99x = 38x + 24,400$
$61x = 24,400$
$\frac{61x}{61} = \frac{24,400}{61}$
$x = 400$
400 units must be sold to break even.

Critical Thinking 6.3

65. Solve for a:
$5a - 4 = 3a + 2$
$5a - 3a = 2 + 4$
$2a = 6$
$a = 3$
Substitute 3 for a in $4a^3$:
$4a^3 = 4(3^3) = 4(27) = 108$

67. Students should explain that the solution of the original equation is $x = 0$. Therefore, the fourth line, where each side of the equation is divided by x, involves division by zero, which is not defined.

69. Students can use a lower-level physics textbook to find information concerning levers. A lever is a simple machine. Even though we cannot get more work out of a machine than we put into it, a machine can produce a larger force at the output than the force we apply at the input. However, the force at the input must be applied for a greater distance than the force at the output. The equation shown below, where F_O is the force at the output, D_O is the distance measured at the output, F_I is the force at the input, D_I is the distance measured at the input, applies to ideal machines (ones for which the effects of friction are ignored).

$$F_O \cdot D_O = F_I \cdot D_I$$

Suppose we use a lever to lift a 720-pound boulder. If we use a 10-foot lever and we are 6 ft from the fulcrum and the boulder is 4 ft from the fulcrum, then $720 \cdot 4 = F_I \cdot 6$, and the force we exert at the input is 480 lb. If we are 8 ft from the fulcrum and the boulder is 2 ft from the fulcrum, then $720 \cdot 2 = F_I \cdot 8$, and the force we exert at the input is 180 lb. If we are 9 ft from the fulcrum and the boulder is 1 ft from the fulcrum, then $720 \cdot 1 = F_I \cdot 9$, and the force we exert at the input is 80 lb. The greater the distance, the less the force. If we increase the distance by using a 21-foot lever to move the same boulder and we are 20 ft from the fulcrum and the boulder is 1 ft from the fulcrum, then $720 \cdot 1 = F_I \cdot 20$, and the force we exert at the input is 36 lb. If we use 51-ft lever and we are 50 ft from the fulcrum and the boulder is 1 ft from the fulcrum, then $720 \cdot 1 = F_I \cdot 50$, and the force we exert at the input is 14.4 lb. If we keep increasing the distance, we keep decreasing the amount of force we need to exert at the input. Taken to the extreme, given a long enough lever, we can move the world!

Section 6.4

Objective A Exercises

1. The unknown number: x

    | The sum of a number and twelve | is | twenty |

    $x + 12 = 20$
    $x + 12 - 12 = 20 - 12$
    $x = 8$
    The number is 8.

3. The unknown number: x

    | Three-fifths of a number | is | negative thirty |

    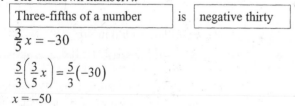

    $x = -50$
    The number is -50.

5. The unknown number: x

    | Four more than three times a number | is | thirteen |

    $3x + 4 = 13$
    $3x + 4 - 4 = 13 - 4$
    $3x = 9$
    $\dfrac{3x}{3} = \dfrac{9}{3}$
    $x = 3$
    The number is 3.

Section 6.4   **139**

7. The unknown number: x

   | The difference between nine times a number and six | is | twelve |

   $9x - 6 = 12$
   $9x - 6 + 6 = 12 + 6$
   $9x = 18$
   $\dfrac{9x}{9} = \dfrac{18}{9}$
   $x = 2$
   The number is 2.

9. The unknown number: x

   | a number and twice the number | is | nine |

   $x + 2x = 9$
   $3x = 9$
   $\dfrac{3x}{3} = \dfrac{9}{3}$
   $x = 3$
   The number is 3.

11. The unknown number: x

    | seventeen less than the product of five and a number | is | three |

    $5x - 17 = 3$
    $5x - 17 + 17 = 3 + 17$
    $5x = 20$
    $\dfrac{5x}{5} = \dfrac{20}{5}$
    $x = 4$
    The number is 4.

13. The unknown number: x

    | seven more than the product of six and a number | is | eight less than the product of three and the number |

    $6x + 7 = 3x - 8$
    $6x - 3x + 7 = 3x - 3x - 8$
    $3x + 7 = -8$
    $3x + 7 - 7 = -8 - 7$
    $3x = -15$
    $\dfrac{3x}{3} = \dfrac{-15}{3}$
    $x = -5$
    The number is -5.

15. The unknown number: x

    | forty | is | nine less than the product of seven and a number |

    $40 = 7x - 9$
    $40 + 9 = 7x - 9 + 9$
    $49 = 7x$
    $\dfrac{49}{7} = \dfrac{7x}{7}$
    $x = 7$
    The number is 7.

© Houghton Mifflin Company. All rights reserved.

17. The unknown number: x

| twice the difference between a number and twenty-five | is | three times the number |

$2(x - 25) = 3x$
$2x - 50 = 3x$
$2x - 2x - 50 = 3x - 2x$
$-50 = x$
The number is -50.

19. The unknown number: x

| The product of four and the Number minus three | is | eight less than the product of six and the number |

$4(x - 3) = 6x - 8$
$4x - 12 = 6x - 8$
$4x - 6x - 12 = 6x - 6x - 8$
$-2x - 12 + 12 = -8 + 12$
$-2x = 4$
$\dfrac{-2x}{-2} = \dfrac{4}{-2}$
$x = -2$
The number is -2.

21. The unknown number: x

| Six more than twice the sum of three times a number and eight | is | negative two |

$2(3x + 8) + 6 = -2$
$6x + 16 + 6 = -2$
$6x + 22 = -2$
$6x + 22 - 22 = -2 - 22$
$6x = -24$
$\dfrac{6x}{6} = \dfrac{-24}{6}$
$x = -4$
The number is -4.

23. The smaller number: x
The larger number: $20 - x$

| three times the smaller number | is | two times the larger number |

$3x = 2(20 - x)$
$3x = 40 - 2x$
$3x + 2x = 40 - 2x + 2x$
$5x = 40$
$\dfrac{5x}{5} = \dfrac{40}{5}$
$x = 8$
$20 - x = 20 - 8 = 12$
The smaller number is 8.
The larger number is 12.

Section 6.4   141

25. The smaller number: x
    The larger number: $21 - x$

    | twice the smaller number | is | three more than the larger number |

    $2x = (21 - x) + 3$
    $2x = 24 - x$
    $2x + x = 24 - x + x$
    $3x = 24$
    $\dfrac{3x}{3} = \dfrac{24}{3}$
    $x = 8$
    $21 - x = 21 - 8 = 13$
    The smaller number is 8.
    The larger number is 13.

27. The smaller number: x
    The larger number: $23 - x$

    | the larger number | is | five more than twice the smaller number |

    $23 - x = 2x + 5$
    $23 - x - 2x = 2x - 2x + 5$
    $23 - 3x = 5$
    $23 - 23 - 3x = 5 - 23$
    $-3x = -18$
    $\dfrac{-3x}{-3} = \dfrac{-18}{-3}$
    $x = 6$
    $23 - x = 23 - 6 = 17$
    The smaller number is 6.
    The larger number is 17.

Objective B Exercises

29. **Strategy** To find the original value of the car, write and solve an equation using v to represent the original value.

    Solution | $19,900 | is | $\dfrac{4}{5}$ of its original value |

    $19,900 = \dfrac{4}{5}v$

    $\dfrac{5}{4}(19,900) = \dfrac{5}{4}\left(\dfrac{4}{5}v\right)$

    $24,875 = v$
    The original value of the car was $24,875.

© Houghton Mifflin Company. All rights reserved.

Chapter 6: First-Degree Equations

31. Strategy   To find the number of people in Times Square, write and solve an equation using p to represent the number of people in Times Square in New York City.

Solution

| 5,000,000 | is | $2\frac{1}{2}$ times as in Times Square |

$$5,000,000 = \frac{5}{2}p$$
$$\frac{2}{5}(5,000,000) = \frac{2}{5}\left(\frac{5}{2}p\right)$$
$$2,000,000 = p$$

There were 2,000,000 people in Times Square in New York City.

33. Strategy   To find the amount spent on advertising and administrative costs, write and solve an equation using a to represent the amount spent on advertising and administrative costs.

Solution

| Costs plus $0.44 | is | $0.50 |

$$a + 0.44 = 0.50$$
$$a - 0.44 = 0.50 - 0.44$$
$$a = 0.06$$

The amount spent on advertising and administrative costs is 6 cents.

35. Strategy   To find the number of tons of paper that is thrown away by American office workers each year, write and solve an equation using n to represent the number of tons of paper.

Solution

| the amount recycled | is | two million more than four times the amount thrown away |

$$18,000,000 = 2,000,000 + 4n$$
$$18,000,000 - 2,000,000 = 2,000,000 - 2,000,000 + 4n$$
$$16,000,000 = 4n$$
$$\frac{16,000,000}{4} = \frac{4n}{4}$$
$$4,000,000 = n$$

Each year Americans throw away 4,000,000 tons of paper.

37. Strategy   To find the number of hours, write and solve an equation using h to represent the number of hours to paint the inside of the house.

Solution

| $250 for materials and $66 per hour for labor | is | $2,692 |

$$250 + 66 \cdot h = 2,692$$
$$250 - 250 + 66h = 2,692 - 250$$
$$66h = 2,442$$
$$\frac{66h}{66} = \frac{2,442}{66}$$
$$h = 37$$

37 h of labor was required to paint the house.

39. Strategy To find the length of each piece, write and solve an equation using x to represent the shorter piece and $12 - x$ to represent the longer piece.

Solution

| Twice the shorter piece | is | three feet less than the longer piece |

$2x = (12 - x) - 3$
$2x = 9 - x$
$2x + x = 9 - x + x$
$3x = 9$
$\frac{3x}{3} = \frac{9}{3}$
$x = 3$
$12 - x = 12 - 3 = 9$
The shorter piece is 3 ft.
The longer piece is 9 ft.

41. Strategy To find the amount of each scholarship, write and solve an equation using x to represent the smaller scholarship and $7{,}000 - x$ to represent the larger scholarship.

Solution

| twice the smaller scholarship | is | 1,000 less than the larger scholarship |

$2x = (7{,}000 - x) - 1{,}000$
$2x = 7{,}000 - x - 1{,}000$
$2x + x = 6{,}000 - x + x$
$3x = 6{,}000$
$\frac{3x}{3} = \frac{6{,}000}{3}$
$x = 2{,}000$
$7{,}000 - x = 7{,}000 - 2{,}000 = 5{,}000$
The larger scholarship is $5,000.

43. Strategy To find the number of pounds of each coffee in the mixture, write and solve an equation using x to represent the amount of Colombian, $x + 1$ to represent the amount of French Roast, and $(x + 1) + 2 = x + 3$ to represent the amount of Java.

Solution

| the total amount of coffee | is | 10 lb |

$x + (x + 1) + (x + 3) = 10$
$3x + 4 = 10$
$3x = 6$
$\frac{3x}{3} = \frac{6}{3}$
$x = 2$
$x + 1 = 2 + 1 = 3$
$x + 3 = 2 + 3 = 5$
The coffee mixture contains 2 lb of Colombian, 3 lb of French Roast, and 5 lb of Java.

Critical Thinking 6.4

45. $6x + 2 = 5 + 3(2x - 1)$
$6x + 2 = 5 + 6x - 3$
$6x + 2 = 6x + 2$
Identity

144 Chapter 6: *First-Degree Equations*

47. $6 + 4(2y + 1) = 5 - 8y$
$6 + 8y + 4 = 5 - 8y$
$8y + 10 = 5 - 8y$
$8y + 8y + 10 = 5 - 8y + 8y$
$16y + 10 = 5$
$16y + 10 - 10 = 5 - 10$
$16y = -5$
$\dfrac{16y}{16} = \dfrac{-5}{16}$
$y = -\dfrac{-5}{16}$
Conditional equation
The solution is $-\dfrac{5}{16}$.

49. $3v - 2 = 5v - 2(2 + v)$
$3v - 2 = 5v - 4 - 2v$
$3v - 2 = 3v - 4$
$3v - 3v - 2 = 3v - 3v - 4$
$-2 = -4$
Contradiction

51. The problem states that a 4-quart mixture of fruit juice is made from apple juice and cranberry juice. There are 6 more quarts of apple juice than of cranberry juice. If we let x = the number of quarts of cranberry juice, then $x + 6$ = the number of quarts of apple juice. The total number of quarts is 4. Therefore, we can write the equation $x + (x + 6) = 4$.
$x + (x + 6) = 4$
$2x + 6 = 4$
$2x = -2$
$x = -1$
Since x = the number of quarts of cranberry juice, there are -1 quarts of cranberry juice in the mixture. We cannot add -1 qt to a mixture. The solution is not reasonable. We can see from the original problem that the answer will not be reasonable. If the total number of quarts in the mixture is 4, we cannot have more than 6 qt of apple juice in the mixture.

Section 6.5

Objective A Exercises

1. Student explanations should include the idea of starting at the origin, then moving 4 units to the left and then moving 3 units up.

3. (5, 4) is in quadrant I.

5. (−8, 1) is in quadrant II.

7. a. The abscissa is positive, and the ordinate is positive.

b. The abscissa is negative, and the ordinate is negative.

9.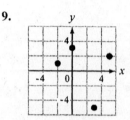

11.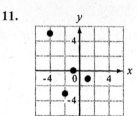

13.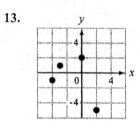

15.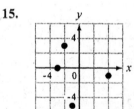

17.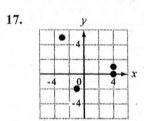

19. $A(0, 2)$
$B(-4, -1)$
$C(2, 0)$
$D(1, -3)$

21. $A(0, 4)$
$B(-4, 3)$
$C(-2, 0)$
$D(2, -3)$

23. A(3, 5)
    B(1, −4)
    C(−3, −5)
    D(−5, 0)

25. A(1, −4)
    B(−3, −6)
    C(−2, 0)
    D(3, 5)

27. a. Abscissa of point A: 2
       Abscissa of point C: −4

    b. Ordinate of point B: 1
       Ordinate of point D: −3

29. a. Abscissa of point A: 4
       Abscissa of point C: −3

    b. Ordinate of point B: −2
       Ordinate of point D: 2

Objective B Exercises

31.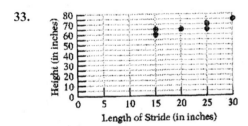

33.

35. The record time for the 800-meter race was 200 s.

37. a. For the car that has 30 mpg in the city has 34 mpg on the highway.

    b. For the car that has 36 mpg on the highway has 28 mpg in the city.

Critical Thinking 6.5

39. Longitude indicates the distance east or west of the zero meridian (or the prime meridian at Greenwich, England) and is expressed as a number of degrees between 0° and 180°. Latitude indicates the distance north or south of the equator and is expressed as a number of degrees between 0° and 180°. Students might observe that the degree measurement is further broken down into minutes and seconds. Most information almanacs provide the latitude and longitude of the major cities in the world.

Section 6.6

Objective A Exercises

1. No, this equation is not linear.

3. Yes

5. No, this equation is not in two variables.

7. $y = -x + 7$

   $$\begin{array}{c|c} 4 & -(3) + 7 \\ & -3 + 7 \\ & 4 \end{array}$$

   $4 = 4$
   Yes, (3, 4) is a solution of $y − x + 7$.

9. $y = \frac{1}{2}x - 1$

   $$\begin{array}{c|c} 2 & \frac{1}{2}(-1) - 1 \\ & -\frac{1}{2} - 1 \\ & -\frac{3}{2} \end{array}$$

   $2 \ne -\frac{3}{2}$
   No, (−1, 2) is not a solution of $y = \frac{1}{2}x - 1$.

11. $y = \frac{1}{4}x + 1$

    $$\begin{array}{c|c} 1 & \frac{1}{4}(4) + 1 \\ & 1 + 1 \\ & 2 \end{array}$$

    $1 \ne 2$
    No, (4, 1) is not a solution of $y = \frac{1}{4}x + 1$.

146 Chapter 6: *First-Degree Equations*

13. $y = \frac{3}{4}x + 4$

$$\begin{array}{c|c} 4 & \frac{3}{4}(0) + 4 \\ & 0 + 4 \\ & 4 \end{array}$$

$4 = 4$

Yes, (0, 4) is a solution of $y = \frac{3}{4}x + 4$.

15. $y = 3x + 2$

$$\begin{array}{c|c} 0 & 3(0) + 2 \\ & 0 + 2 \\ & 2 \end{array}$$

$0 \neq 2$

No, (0, 0) is not a solution of $y = 3x + 2$.

17. $y = 3x - 2$
   $= 3(3) - 2$
   $= 9 - 2$
   $= 7$

The ordered-pair solution is (3, 7).

19. $y = \frac{2}{3}x - 1$
   $= \frac{2}{3}(6) - 1$
   $= 4 - 1$
   $= 3$

The ordered-pair solution is (6, 3).

21. $y = -3x + 1$
   $= -3(0) + 1$
   $= 0 + 1$
   $= 1$

The ordered-pair solution is (0, 1).

23. $y = \frac{2}{5}x + 2$
   $= \frac{2}{5}(-5) + 2$
   $= -2 + 2$
   $= 0$

The ordered-pair solution is (-5, 0).

Objective B Exercises

25. Yes

27. No, the graph is not linear.

29.
x	y
0	-4
2	0
4	4

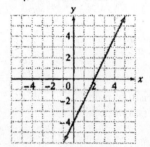

31.
x	y
-2	4
2	0
4	-2

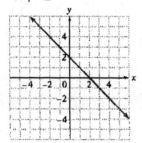

33.
x	y
-1	-4
2	-1
4	1

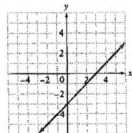

35.
x	y
0	3
1	1
3	-3

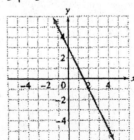

37.
x	y
0	4
1	1
2	-2

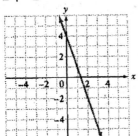

39.
x	y
-1	-3
0	-1
2	3

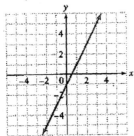

41.
x	y
-1	-3
0	0
1	3

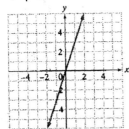

43.
x	y
3	1
0	0
-3	-1

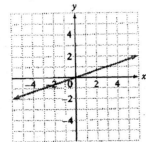

45.
x	y
-3	4
0	0
3	-4

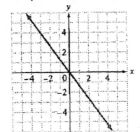

47.
x	y
-2	-4
0	-1
2	2

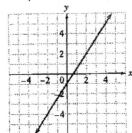

49.
x	y
-5	-3
0	-1
5	1

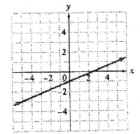

51.
x	y
-3	3
0	1
3	-1

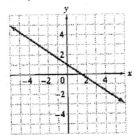

148 Chapter 6: *First-Degree Equations*

53.

x	y
−3	3
0	−2
1	−11/3

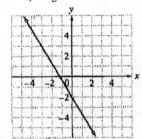

55.

x	y
2	4
0	−1
−2	−6

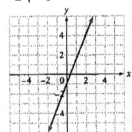

57.

x	y
1	1
0	0
−1	−1

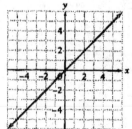

59. When $x = 3$, $y = 4$.

61. When $x = 4$, $y = −1$.

63. When $x = 3$, $y = 2$.

65. When $y = −2$, $x = 2$.

67. When $y = −2$, $x = 1$.

69. When $y = −1$, $x = −3$.

Critical Thinking 6.6

71. **a.** To find the coordinates of the point at which the graph of $y = 2x + 1$ crosses the y-axis, substitute 0 for x and solve for y.
$y = 2x + 1$
$y = 2(0) + 1$
$y = 1$
The graph of $y = 2x + 1$ will cross the y-axis at the point (0,1).

   b. To find the coordinates of the point at which the graph of $y = 3x − 6$ crosses the x–axis, substitute 0 for y and solve for x.
$y = 3x − 6$
$0 = 3x − 6$
$6 = 3x$
$2 = x$
The graph of $y = 3x − 6$ will cross the x-axis at (2, 0).

73.

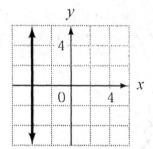

75.

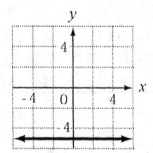

77. Since the results are not equal, the ordered pair is not a solution of the equation.

79. Students should note that as the value of m in $y = mx + 1$ increases, the graph of the line rotates counterclockwise.

Chapter Review Exercises

1. $z + 5 = 2$
$z + 5 - 5 = 2 - 5$
$z = -3$
The solution is -3.

2. $-8x + 4x = -12$
$-4x = -12$
$\dfrac{-4x}{-4} = \dfrac{-12}{-4}$
$x = 3$
The solution is 3.

3. $7 = 8a - 5$
$7 + 5 = 8a - 5 + 5$
$12 = 8a$
$\dfrac{12}{8} = \dfrac{8a}{8}$
$\dfrac{3}{2} = a$
The solution is $\dfrac{3}{2}$.

4. $7 + a = 0$
$7 - 7 + a = 0 - 7$
$a = -7$
The solution is -7.

5. $40 = -\dfrac{5}{3}y$
$-\dfrac{3}{5}(40) = \left(-\dfrac{3}{5}\right)\left(-\dfrac{5}{3}y\right)$
$-24 = y$
The solution is -24.

6. $-\dfrac{3}{8} = \dfrac{4}{5}z$
$\dfrac{5}{4}\left(-\dfrac{3}{8}\right) = \dfrac{5}{4}\left(\dfrac{4}{5}z\right)$
$-\dfrac{15}{32} = z$
The solution is $-\dfrac{15}{32}$.

7. $9 - 5y = -1$
$9 - 9 - 5y = -1 - 9$
$-5y = -10$
$\dfrac{-5y}{-5} = \dfrac{-10}{-5}$
$y = 2$
The solution is 2.

8. $-4(2 - x) = x + 9$
$-8 + 4x = x + 9$
$-8 + 4x - x = x - x + 9$
$-8 + 3x = 9$
$-8 + 8 + 3x = 9 + 8$
$3x = 17$
$\dfrac{3x}{3} = \dfrac{17}{3}$
$x = \dfrac{17}{3}$
The solution is $\dfrac{17}{3}$.

9. $3a + 8 = 12 - 5a$
$3a + 5a + 8 = 12 - 5a + 5a$
$8a + 8 = 12$
$8a + 8 - 8 = 12 - 8$
$8a = 4$
$\dfrac{8a}{8} = \dfrac{4}{8}$
$a = \dfrac{1}{2}$
The solution is $\dfrac{1}{2}$.

10. $12p - 7 = 5p - 21$
$12p - 5p - 7 = 5p - 5p - 21$
$7p - 7 = -21$
$7p - 7 + 7 = -21 + 7$
$7p = -14$
$\dfrac{7p}{7} = \dfrac{-14}{7}$
$p = -2$
The solution is -2.

11. $3(2n - 3) = 2n + 3$
$6n - 9 = 2n + 3$
$6n - 2n - 9 = 2n - 2n + 3$
$4n - 9 = 3$
$4n - 9 + 9 = 3 + 9$
$4n = 12$
$\dfrac{4n}{4} = \dfrac{12}{4}$
$n = 3$
The solution is 3.

12. $3m = -12$
$\dfrac{3m}{3} = \dfrac{-12}{3}$
$m = -4$
The solution is -4.

150 Chapter 6: First-Degree Equations

13. $4 - 3(2p + 1) = 3p + 11$
    $4 - 6p - 3 = 3p + 11$
    $-6p + 1 = 3p + 11$
    $-6p - 3p + 1 = 3p - 3p + 11$
    $-9p + 1 = 11$
    $-9p + 1 - 1 = 11 - 1$
    $-9p = 10$
    $\dfrac{-9p}{-9} = \dfrac{10}{-9}$
    $p = -\dfrac{10}{9}$
    The solution is $-\dfrac{10}{9}$.

14. $1 + 4(2c - 3) = 3(3c - 5)$
    $1 + 8c - 12 = 9c - 15$
    $8c - 11 = 9c - 15$
    $8c - 9c - 11 = 9c - 9c - 15$
    $-c - 11 = -15$
    $-c - 11 + 11 = -15 + 11$
    $-c = -4$
    $\dfrac{-1c}{-1} = \dfrac{-4}{-1}$
    $c = 4$
    The solution is 4.

15. $\dfrac{3x}{4} + 10 = 7$
    $\dfrac{3x}{4} + 10 - 10 = 7 - 10$
    $\dfrac{3x}{4} = -3$
    $\left(\dfrac{4}{3}\right)\left(\dfrac{3}{4}x\right) = \dfrac{4}{3}(-3)$
    $x = -4$
    The solution is −4.

16. $y = \dfrac{1}{5}x + 2$

0	$\dfrac{1}{5}(-10) + 2$
0	$-2 + 2$

    $0 = 0$
    Yes, (−10, 0) is a solution of the equation.

17.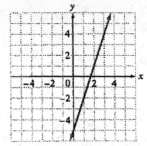

18. | x | y |
    |---|---|
    | 0 | −5 |
    | 1 | −2 |
    | 2 | 1 |

    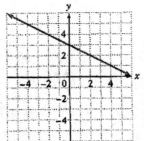

19. | x | y |
    |---|---|
    | −2 | 4 |
    | 0 | 3 |
    | 2 | 2 |

20. $y = 4x - 9$
    $y = 4(2) - 9$
    $y = 8 - 9$
    $y = -1$
    The ordered-pair solution is (2, −1).

21. The unknown number: x

    | the difference between seven and the product of five and a number | is | thirty-seven |

    $7 - 5x = 37$
    $7 - 7 - 5x = 37 - 7$
    $-5x = 30$
    $\dfrac{-5x}{-5} = \dfrac{30}{-5} = -6$
    $x = -6$
    The number is −6.

22. Strategy   To find the length of each piece, write and solve an equation using x to represent the shorter piece and $24 - x$ to represent the longer piece.

Solution

| Twice the length of the shorter piece | equals | the length of the longer piece |

$2x = 24 - x$
$2x + x = 24 - x + x$
$3x = 24$
$\dfrac{3x}{3} = \dfrac{24}{3}$
$x = 8$
$24 - x = 24 - 8 = 16$
The longer piece is 16 in.

23. Strategy   To find the number of hours of consultation, write and solve an equation using n to represent the number of hours.

Solution

| $250 plus $150 per hour | is | $1,300 |

$250 + 150 \cdot n = 1,300$
$250 - 250 + 150n = 1,300 - 250$
$150n = 1,050$
$\dfrac{150n}{150} = \dfrac{1,050}{150}$
$n = 7$
The consulting fee was for 7 h of consultation.

24. Strategy   To find the height of the leaning tower of Pisa, write and solve an equation using h to represent the height of the tower.

Solution

| 302 m | is | 28 m less than six times the height |

$302 = 6h - 28$
$302 + 28 = 6h - 28 + 28$
$330 = 6h$
$\dfrac{330}{6} = \dfrac{6h}{6}$
$55 = h$
The leaning tower of Pisa is 55 m high.

25.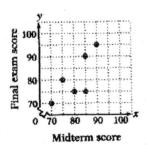

26. Strategy To find the force, substitute 18 for d, 6 for x, and 25 for F_1 in the given equation and solve for F_2.

Solution
$$F_1 x = F_2(d-x)$$
$$25(6) = F_2(18-6)$$
$$150 = 12 F_2$$
$$\frac{150}{12} = \frac{12 F_2}{12}$$
$$12.5 = F_2$$
The force needed to balance the system is 12.5 lb.

27. Strategy To find the number of amplifiers, substitute 38,669 for T, 127 for U, and 20,000 for F in the given equation and solve for N.

Solution
$$T = U \cdot N + F$$
$$38,669 = 127 \cdot N + 20,000$$
$$38,669 - 20,000 = 127N$$
$$18,669 = 127N$$
$$\frac{18,699}{127} = \frac{127N}{127}$$
$$147 = N$$
147 amplifiers were produced during the month.

Chapter Test

1. $7 + x = 2$
$7 - 7 + x = 2 - 7$
$a = -5$
The solution is -5.

2. $-\frac{3}{5} y = 6$
$\left(-\frac{5}{3}\right)\left(-\frac{3}{5} y\right) = 6\left(-\frac{5}{3}\right)$
$y = -10$
The solution is -10.

3. $2d - 7 = -13$
$2d - 7 + 7 = -13 + 7$
$2d = -6$
$\frac{2d}{2} = \frac{-6}{2}$
$d = -3$
The solution is -3.

4. $4 - 5c = -11$
$4 - 4 - 5c = -11 - 4$
$-5c = -15$
$\frac{-5c}{-5} = \frac{-15}{-5}$
$c = 3$
The solution is 3.

5. $3x + 4 = 24 - 2x$
$3x + 2x + 4 = 24 - 2x + 2x$
$5x + 4 = 24$
$5x + 4 - 4 = 24 - 4$
$5x = 20$
$\frac{5x}{5} = \frac{20}{5}$
$x = 4$
The solution is 4.

6. $7 - 5y = 6y - 26$
$7 - 5y + 5y = 6y + 5y - 26$
$7 = 11y - 26$
$7 + 26 = 11y - 26 + 26$
$33 = 11y$
$\frac{33}{11} = \frac{11y}{11}$
$y = 3$
The solution is 3.

7. $2t - 3(4 - t) = t - 8$
$2t - 12 + 3t = t - 8$
$5t - 12 = t - 8$
$5t - t - 12 = t - t - 8$
$4t - 12 = -8$
$4t - 12 + 12 = -8 + 12$
$4t = 4$
$\frac{4t}{4} = \frac{4}{4}$
$t = 1$
The solution is 1.

8. $12 - 3(n - 5) = 5n - 3$
$12 - 3n + 15 = 5n - 3$
$-3n + 27 = 5n - 3$
$-3n - 5n + 27 = 5n - 5n - 3$
$-8n + 27 = -3$
$-8n + 27 - 27 = -3 - 27$
$-8n = -30$
$\frac{-8n}{-8} = \frac{-30}{-8}$
$n = \frac{15}{4}$
The solution is $\frac{15}{4}$.

9. $\dfrac{3}{8} - n = \dfrac{2}{3}$

$\dfrac{3}{8} - \dfrac{3}{8} - n = \dfrac{2}{3} - \dfrac{3}{8}$

$-n = \dfrac{7}{24}$

$\dfrac{-n}{-1} = \left(\dfrac{7}{24}\right)\left(\dfrac{1}{-1}\right)$

$n = \dfrac{-7}{24}$

The solution is $-\dfrac{7}{24}$.

10. $3p - 2 + 5p = 2p + 12$
$8p - 2 = 2p + 12$
$8p - 2p - 2 = 2p - 2p + 12$
$6p - 2 = 12$
$6p - 2 + 2 = 12 + 2$
$6p = 14$
$\dfrac{6p}{6} = \dfrac{14}{6}$
$p = \dfrac{7}{3}$

The solution is $\dfrac{7}{3}$.

11. $A(-3, 1)$

12.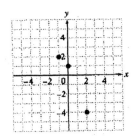

13.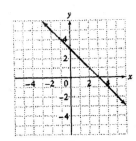

14.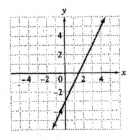

15.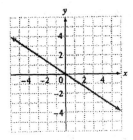

16.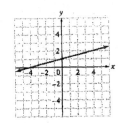

17. $2(4b - 14) = b - 7$
$8b - 28 = b - 7$
$8b - b - 28 = b - b - 7$
$7b - 28 = -7$
$7b - 28 + 28 = -7 + 28$
$7b = 21$
$\dfrac{7b}{7} = \dfrac{21}{7}$
$b = 3$
The solution is 3.

18. $\dfrac{5y}{3} + 12 = 2$

$\dfrac{5y}{3} + 12 - 12 = 2 - 12$

$\dfrac{5y}{3} = -10$

$\left(\dfrac{3}{5}\right)\left(\dfrac{5y}{3}\right) = -10\left(\dfrac{3}{5}\right)$

$y = -6$
The solution is -6.

19. $y = \dfrac{1}{3}x - 4$

    $y = \dfrac{1}{3}(6) - 4$

    $y = 2 - 4$
    $y = -2$
    The ordered-pair solution is (6, –2).

20. The unknown number: n

    | Four plus one third of a number | is | nine |

    $4 + \dfrac{1}{3}n = 9$

    $4 - 4 + \dfrac{1}{3}n = 9 - 4$

    $\dfrac{1}{3}n = 5$

    $\left(\dfrac{3}{1}\right)\left(\dfrac{1}{3}n\right) = 5\left(\dfrac{3}{1}\right)$

    $n = 15$
    The number is 15.

21. The unknown number: n

    | Sum of eight and the product of two and a number | is | negative 4 |

    $8 + 2n = -4$
    $8 - 8 + 2n = -4 - 8$
    $2n = -12$

    $\dfrac{2n}{2} = \dfrac{-12}{2}$

    $n = -6$
    The number is –6.

22. The smaller number: x
    The larger number: $17 - x$

    | Four times the smaller number and two times the larger number | is | Forty four |

    $4x + 2(17 - x) = 44$
    $4x + 34 - 2x = 44$
    $2x + 34 = 44$
    $2x + 34 - 34 = 44 - 34$
    $2x = 10$

    $\dfrac{2x}{2} = \dfrac{10}{2}$

    $x = 5$
    $17 - x = 17 - 5 = 12$
    The smaller number is 5.
    The larger number is 12.

23.

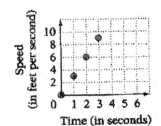

24. Strategy   To find the number of hours of labor, write and solve an equation using n to represent the number of hours.

Solution | $165 plus $58 per hour | is | $455 |

$165 + 58 \cdot n = 455$
$165 - 165 + 58n = 455 - 165$
$58n = 290$
$\dfrac{58n}{58} = \dfrac{290}{58}$
$n = 5$
The job required 5 h of labor.

25. Strategy   To find the depth, substitute 65 for P in the given equation and solve for D.

Solution $P = 15 + \dfrac{1}{2}D$

$65 = 15 + \dfrac{1}{2}D$

$65 - 15 = 15 - 15 + \dfrac{1}{2}D$

$50 = \dfrac{1}{2}D$

$\left(\dfrac{2}{1}\right)50 = \left(\dfrac{2}{1}\right)\left(\dfrac{1}{2}D\right)$

$100 = D$
The depth is 100 ft.

Cumulative Review Exercises

1. $-3ab$
$-3(-2)(3) = 6(3)$
$= 18$

2. $-3(4p - 7) = -12p + 21$

3. $\left(\dfrac{2}{3}\right)\left(-\dfrac{9}{8}\right) + \dfrac{3}{4} = -\dfrac{2 \cdot 9}{3 \cdot 8} + \dfrac{3}{4}$
$= -\dfrac{3}{4} + \dfrac{3}{4}$
$= 0$

4. $-\dfrac{2}{3}y = 12$

$\left(-\dfrac{3}{2}\right)\left(-\dfrac{2}{3}y\right) = -\dfrac{3}{2}(12)$

$y = -18$
The solution is -18.

5. $(-b)^3$
   $[-(-2)]^3 = 2^3$
   $= 8$

6. $4xy^2 - 2xy$
   $4(-2)(3^2) - 2(-2)(3)$
   $= 4(-2)(9) - 2(-2)(3)$
   $= -8(9) - 2(-2)(3)$
   $= -72 - 2(-2)(3)$
   $= -72 - (-4)(3)$
   $= -72 - (-12)$
   $= -72 + 12$
   $= -60$

7. $\sqrt{121} = 11$

8. $\sqrt{48} = \sqrt{16 \cdot 3}$
   $= \sqrt{16} \cdot \sqrt{3}$
   $= 4\sqrt{3}$

9. $4(3v - 2) - 5(2v - 3)$
   $= 12v - 8 - 10v + 15$
   $= (12v - 10v) + (-8 + 15)$
   $= 2v + 7$

10. $-4(-3m) = [(-4)(-3)]m$
    $= 12m$

11. $-5d = -45$
    $\overline{-5(-9) \mid -45}$
    $45 \ne -45$
    No, −9 is not a solution of the equation.

12. $5 - 7a = 3 - 5a$
    $5 - 7a + 5a = 3 - 5a + 5a$
    $5 - 2a = 3$
    $5 - 5 - 2a = 3 - 5$
    $-2a = -2$
    $\dfrac{-2a}{-2} = \dfrac{-2}{-2}$
    $a = 1$
    The solution is 1.

13. $6 - 2(7z - 3) + 4z = 6 - 14z + 6 + 4z$
    $= (-14z + 4z) + (6 + 6)$
    $= -10z + 12$

14. $\dfrac{a^2 + b^2}{2ab}$
    $\dfrac{(-2)^2 + (-1)^2}{2(-2)(-1)} = \dfrac{4+1}{4}$
    $= \dfrac{5}{4}$

15. $8z - 9 = 3$
    $8z - 9 + 9 = 3 + 9$
    $8z = 12$
    $\dfrac{8z}{8} = \dfrac{12}{8}$
    $z = \dfrac{3}{2}$
    The solution is $\dfrac{3}{2}$.

16. $(2m^2 n^5)^5 = 2^{1 \cdot 5} m^{2 \cdot 5} n^{5 \cdot 5}$
    $= 32 m^{10} n^{25}$

17. $-3a^3(2a^2 + 3ab - 4b^2)$
    $= -3a^3(2a^2) + (-3a^3)(3ab) - (-3a^3)(4b^2)$
    $= -6a^5 - 9a^4 b + 12a^3 b^2$

18. $(2x - 3)(3x + 1) = 6x^2 + 2x - 9x - 3$
    $= 6x^2 - 7x - 3$

19. $2^{-4} = \dfrac{1}{2^4} = \dfrac{1}{16}$

20. $\dfrac{x^8}{x^2} = x^{8-2} = x^6$

21. $(-5x^3 y)(-3x^5 y^2)$
    $= [(-5)(-3)](x^3 x^5)(y y^2)$
    $= 15x^8 y^3$

22. $5 - 3(2x - 8) = -2(1 - x)$
    $5 - 6x + 24 = -2 + 2x$
    $-6x + 29 = -2 + 2x$
    $-6x - 2x + 29 = -2 + 2x - 2x$
    $-8x + 29 = -2$
    $-8x + 29 - 29 = -2 - 29$
    $-8x = -31$
    $\dfrac{-8x}{-8} = \dfrac{-31}{-8}$
    $x = \dfrac{31}{8}$
    The solution is $\dfrac{31}{8}$.

23.
x	y
3	6
0	1
-3	-4

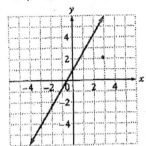

24.
x	y
5	-2
0	0
-5	2

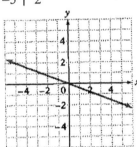

25. $3.5 \times 10^{-8} = 0.000000035$

26. The <u>product</u> of five and the <u>sum</u> of a number and two
$5(n + 2)$
$5n + 10$

27. **Strategy** To find the time, replace v by 98 and v_0 by 50 in the given formula and solve for t.

   Solution
   $v = v_0 + 32t$
   $98 = 50 + 32t$
   $98 - 50 = 50 - 50 + 32t$
   $48 = 32t$
   $\dfrac{48}{32} = \dfrac{32t}{32}$
   $1.5 = t$
   The time is 1.5 s.

28. The number of wolves in the world: w
    The number of dogs in the world: $1{,}000w$

29. **Strategy** To find the total box office gross, add the totals for the four films.

   Solution $461.0 + 290.2 + 260.0 + 181.3$
   $= 1{,}192.5$
   The total box office gross for all four films is $1,192.5 million.

30. **Strategy** To find the interest payment, write and solve an equation using I to represent the amount of interest and $I - 204$ to represent the amount of principal.

   Solution
   $I + \text{Principal} = 949$
   $I + I - 204 = 949$
   $2I = 1{,}153$
   $\dfrac{2I}{2} = \dfrac{1{,}153}{2}$
   $I = 576.50$
   The interest payment is $576.50.

31. **Strategy** To find the number of stories, divide the depth of the Aleutian Trench (8,100) by the height of one story (4.2).

   Solution $8{,}100 \div 4.2 \approx 1{,}929$
   A skyscraper with 1,929 stories is as tall as the Aleutian Trench.

32. **Strategy** To find the amount donated to each charity, write and solve an equation using x to represent the donation to one charity and $2x$ to represent the donation to the other charity.

   Solution
   $x + 2x = 12{,}000$
   $3x = 12{,}000$
   $\dfrac{3x}{3} = \dfrac{12{,}000}{3}$
   $x = 4{,}000$
   $2x = 2(4{,}000) = 8{,}000$
   One charity received $4,000 and the other charity received $8,000.

Chapter 7: Measurement and Proportion

Prep Test

1. $\dfrac{8}{10} = \dfrac{4}{5}$

2. $\dfrac{372}{15} = 24.8$

3. $36 \times \dfrac{1}{9} = \dfrac{\cancel{3} \cdot \cancel{3} \cdot 2 \cdot 2}{\cancel{3} \cdot \cancel{3}} = 4$

4. $\dfrac{5}{3} \times 6 = \dfrac{5 \cdot 2 \cdot \cancel{3}}{\cancel{3}} = 10$

5. $5\dfrac{3}{4} \times 8 = \dfrac{23}{4} \times 8$
   $= \dfrac{23 \cdot 2 \cdot \cancel{2} \cdot \cancel{2}}{\cancel{2} \cdot \cancel{2}} = 46$

6. $\begin{array}{r} 238 \\ 3\overline{)714} \\ \underline{6} \\ 11 \\ \underline{9} \\ 24 \\ \underline{24} \\ 0 \end{array}$

7. 37,320

8. 0.04107

9. 5.125

10. 5.96

11. 0.13

12. $35 \times \dfrac{1.61}{1} = 35 \times 1.61 = 56.35$

13. $1.67 \times \dfrac{1}{3.34} = 1.67 \div 3.34$

    $3.34\overline{)1.67.0}^{\,0.5}$

14. $315 \div 84 = 3.75$

Go Figure

Adding an even amount (6) of odd numbers results in an even number. Using that rule eliminates 15, 29, and 31 as possible solutions. Also, 4 cannot be a solution since the lowest possible score is 6. The highest possible score is 56, so 58 can also be eliminated as a solution. A possible score is 28, $28 = 9 + 7 + 5 + 5 + 3$.

Section 7.1

Objective A Exercises

1. In the metric system, the basic unit of length is the meter. The basic unit of liquid measure is the liter. The basic unit of weight is the gram.

3. **a.** giga-: 10^9
   mega-: 1,000,000
   kilo-: k, 10^4
   hecto-: 10^2
   deca-: 10
   deci-: 0.1
   centi-: c, 0.01
   milli-: m, $\dfrac{1}{10^3}$
   micro-: 0.000001
   nano-: 0.000000001
   pico-: $\dfrac{1}{10^{12}}$

   b. The exponent 10 indicates the number of places to move the decimal point. For prefixes tera-, giga-, mega-, kilo-, hecto-, and deca-, move the decimal point to the right. For other prefixes shown, move the decimal point to the left.

5. kilogram

7. milliliter

9. kiloliter

11. centimeter

13. kilogram

15. liter

17. gram

© Houghton Mifflin Company. All rights reserved.

Section 7.1

19. meter or centimeter

21. kilogram

23. milliliter

25. kilogram

27. kilometer

29. 91 cm = 910 mm

31. 1,856 g = 1.856 kg

33. 7,285 ml = 7.285 L

35. 8 m = 8,000 mm

37. 34 mg = 0.034 g

39. 0.0297 L = 29.7 ml

41. 7,530 m = 7.530 km

43. 9.2 kg = 9,200 g

45. 0.036 kl = 36 L

47. 2.35 km = 2,350 m

49. 0.083 g = 83 mg

51. 71.6 cm = 0.716 m

53. 6,302 L = 6.302 kl

55. 4.58 m = 458 cm

57. 92 mm = 9.2 cm

59. **Strategy**   To find the weight:
→Multiply the number of carats (10) by the number of milligrams in one carat (200).
→Convert milligrams to grams.

    Solution   10(200) = 2,000
2,000 mg = 2 g
A 10-carat precious stone weights 2 g.

61. **Strategy**   To find the length of the walk:
→Add the distances.
→Convert the sum to kilometers.

    Solution   1,400 + 1,200 + 1,800 = 4,400
4,400 m = 4.4 km
The walk was 4.4 km.

63. **Strategy**   To find the number of liters:
→Convert 800 ml to liters.
→Multiply 30 by the amount of chlorine used each day.

    Solution   800 ml = 0.8 L
30(0.8) = 24
The club uses 24 L of chlorine.

65. **Strategy**   To find the cost:
→Add the weights of the packages.
→Convert grams to kilograms.
→Multiply the sum in kilograms by the price per kilogram (9.89).

    Solution   540 + 670 + 890 = 2100
2100 g = 2.1 kg
2.1(9.89) ≈ 20.77
The cost of the meat is $20.77.

67. **Strategy**   To find the number of servings:
→Convert 230 ml to liters.
→Divide 3.78 by the amount of each serving.

    Solution   230 ml = 0.23 L
3.78 ÷ 0.23 ≈ 16.43
The container of milk will provide 16 servings.

69. **Strategy**   To find the number of grams:
→Convert 1.19 kg to grams.
→Divide the container size by the number of servings.

    Solution   1.19 kg = 1190 g
1190 ÷ 30 ≈ 39.67
One serving contains 40 g.

71. Strategy To find the number of liters:
→Convert 80 ml to liters.
→Multiply the total number of students by the amount of acid for each student.

Solution 80 ml = 0.08 L
30 · 3 · 0.08 = 7.2
The assistant should order 8 L.

73. Strategy
a. To find the amount of mix, convert serving size (31) from grams to kilograms and multiply by the servings per container (6).
b. To find the amount of sodium, convert the sodium (210) from milligrams to grams, and multiply by two.

Solution
a. 31 g = 0.031 kg
0.031 · 6 = 0.186
There are 0.186 kg of mix.

b. 210 mg = 0.21 g
0.210 · 2 = 0.42
There is 0.42 g of sodium.

75. Strategy To find the profit:
→Find the number of bottles needed by converting 5 L to milliliters, divide the amount by 250 ml.
→To find the total cost, multiply the number of bottles by $0.50 and add $190.
→To find revenue, multiply number of bottles by price per bottle.
→To find profit, subtract cost from revenue.

Solution 5 L = 5,000 ml
5,000 ÷ 250 = 20 bottles
20 · 0.50 + 190 = 200
20 · 15.78 = 315.60
315.60 − 200 = 115.60
The profit is $115.60.

77. Strategy To find the profit:
→Find the number of bottles needed by converting 32 kl to liters, divide the amount by 1.25 L.
→To find the total cost, multiply the number of bottles by $0.21 and add to $44,480.
→To find revenue, multiply number of bottles by price per bottle.
→To find profit, subtract cost from revenue.

Solution 32 kl = 32,000 L
32,000 ÷ 1.25 = 25,600 bottles
25,600 · 0.21 + 44,480 = 49,856
25,600 · 2.97 = 76,032
76,032 − 49,856 = 26,176
The profit is $26,176.

Critical Thinking 7.1

79. Strategy To find remaining amount:
→Convert 3 L to milliliters
→Subtract 280 ml serving from amount in bottle.
→Convert answer from milliliters to liters.

Solution 3 L = 3,000 ml
3,000 − 280 = 2,720
2,720 ml = 2.72 L
The water that remains is 2,720 ml or 2.72 L.

Section 7.2

Objective A Exercises

1. $\dfrac{16 \text{ in.}}{24 \text{ in.}} = \dfrac{16}{24} = \dfrac{2}{3}$
16 in.: 24 in. = 2:3
16 in. TO 24 in. = 2 TO 3

3. $\dfrac{9 \text{ h}}{24 \text{ h}} = \dfrac{9}{24} = \dfrac{3}{8}$
9 h: 24 h = 3:8
9 h TO 24 h = 3 TO 8

5. $\dfrac{9 \text{ ft}}{2 \text{ ft}} = \dfrac{9}{2}$
9 ft: 2 ft = 9:2
9 ft TO 2 ft = 9 TO 2

7. $\dfrac{30 \text{ ml}}{60 \text{ ml}} = \dfrac{30}{60} = \dfrac{1}{2}$
   30 ml : 60 ml = 1 : 2
   30 ml TO 60 ml = 1 TO 2

9. 42 − 3 = 39 plays in which no error was made.
   $\dfrac{39 \text{ plays}}{42 \text{ plays}} = \dfrac{39}{42} = \dfrac{13}{14}$

11. $\dfrac{24 \text{ teeth}}{36 \text{ teeth}} = \dfrac{24}{36} = \dfrac{2}{3}$

13. $\dfrac{150 \text{ mi}}{6 \text{ h}} = \dfrac{25 \text{ mi}}{1 \text{ h}}$

15. $\dfrac{\$6.56}{6 \text{ bars}} = \dfrac{\$3.28}{3 \text{ bars}}$

17. $\dfrac{9 \text{ children}}{4 \text{ families}}$

19. $\dfrac{\$38{,}700}{12} = \$3{,}225 \text{ / month}$

21. $\dfrac{364.8}{9.5} = 38.4 \text{ mi / gal}$

23. $\dfrac{\$20.16}{15 \text{ oz}} = \$1.344 \text{ / oz}$

25. **Strategy** To find the ratio of turns, divide the number of turns in primary coil (40) by the number of turns in the secondary coil (480).

    Solution $\dfrac{40}{480} = \dfrac{1}{12}$
    The ratio of turns in the primary coil to secondary coil is $\dfrac{1}{12}$.

27. **Strategy** To find the ratio of people per square mile, divide each country's population by the area.

    Solution $\dfrac{19{,}547{,}000}{2{,}968{,}000} \approx 6.586$

    $\dfrac{1{,}045{,}845{,}000}{1{,}269{,}000} \approx 824.149$

    $\dfrac{281{,}422{,}000}{3{,}718{,}000} \approx 75.692$

    The ratio for each country is Australia: 6.6 people/mi^2, India: 824.1 people/mi^2, and U.S. 75.7 people/mi^2.

Critical Thinking 7.2

29. For every 100 of Atkin's book sold, 7.3 of Grisham's book sold. Writing as a fraction
    $\dfrac{100}{7.3} = \dfrac{100}{7.3} \cdot \dfrac{10}{10} = \dfrac{1000}{73}$

Section 7.3

Objective A Exercises

1. $64 \text{ in} = \dfrac{64 \text{ in}}{1} \cdot \dfrac{1 \text{ ft}}{12 \text{ in}} = 5\dfrac{1}{3} \text{ ft}$

3. $42 \text{ oz} = \dfrac{42 \text{ oz}}{1} \cdot \dfrac{1 \text{ lb}}{16 \text{ oz}} = 2\dfrac{5}{8} \text{ lb}$

5. $7{,}920 \text{ ft} = \dfrac{7{,}920 \text{ ft}}{1} \cdot \dfrac{1 \text{ mi}}{5{,}280 \text{ ft}} = 1\dfrac{1}{2} \text{ mi}$

7. $500 \text{ lb} = \dfrac{500 \text{ lb}}{1} \cdot \dfrac{1 \text{ ton}}{2{,}000 \text{ lb}} = \dfrac{1}{4} \text{ ton}$

9. $10 \text{ qt} = \dfrac{10 \text{ qt}}{1} \cdot \dfrac{1 \text{ gal}}{4 \text{ qt}} = 2\dfrac{1}{2} \text{ gal}$

11. $2\dfrac{1}{2} \text{ c} = \dfrac{5 \text{ c}}{2} \cdot \dfrac{8 \text{ fl oz}}{1 \text{ c}} = 20 \text{ fl oz}$

13. $2\dfrac{1}{4} \text{ mi} = \dfrac{9 \text{ mi}}{4} \cdot \dfrac{5{,}280 \text{ ft}}{1 \text{ mi}} = 11{,}880 \text{ ft}$

15. $7\dfrac{1}{2} \text{ in} = \dfrac{15 \text{ in}}{2} \cdot \dfrac{1 \text{ ft}}{12 \text{ in}} = \dfrac{5}{8} \text{ ft}$

Chapter 7: Measurement and Proportion

17. $60 \text{ fl oz} = \dfrac{60 \text{ fl oz}}{1} \cdot \dfrac{1 \text{ c}}{8 \text{ fl oz}} = 7\dfrac{1}{2} \text{ c}$

19. $7\dfrac{1}{2} \text{ pt} = \dfrac{15 \text{ pt}}{2} \cdot \dfrac{2 \text{ c}}{1 \text{ pt}} \cdot \dfrac{1 \text{ qt}}{4 \text{ c}} = 3\dfrac{3}{4} \text{ qt}$

21. $1\dfrac{1}{2} \text{ mi} = \dfrac{3 \text{ mi}}{2} \cdot \dfrac{5{,}280 \text{ ft}}{1 \text{ mi}} \cdot \dfrac{1 \text{ yd}}{3 \text{ ft}} = 2{,}640 \text{ yd}$

Objective B Exercises

23. $35 \text{ yr} = \dfrac{35 \text{ yr}}{1} \cdot \dfrac{365 \text{ day}}{1 \text{ yr}} \cdot \dfrac{24 \text{ hr}}{1 \text{ day}} \cdot \dfrac{60 \text{ min}}{1 \text{ hr}} \cdot \dfrac{60 \text{ s}}{1 \text{ min}}$
    $= 1{,}103{,}760{,}000 \text{ s}$

25. **Strategy** To find the amount of punch:
    → Multiply the number of guests (200) by the amount of punch for each guest (1 c) to find the number of cups of punch to be prepared.
    → Convert cups to gallons.

    Solution $200 \cdot 1 = 200$ cups
    200 cups
    $= \dfrac{200 \text{ c}}{1} \cdot \dfrac{1 \text{ pt}}{2 \text{ c}} \cdot \dfrac{1 \text{ qt}}{2 \text{ pt}} \cdot \dfrac{1 \text{ gal}}{4 \text{ qt}} = 12\dfrac{1}{2} \text{ gal}$
    The guest list would require $12\dfrac{1}{2}$ gal of punch.

27. **Strategy** To find the number of gallons:
    → Convert from quarts to gal.
    → Multiply the amount by the number of days and students.

    Solution $2 \text{ qt} = \dfrac{2 \text{ qt}}{1} \cdot \dfrac{1 \text{ gal}}{4 \text{ qt}} = \dfrac{1}{2} \text{ gal}$
    $\dfrac{1}{2} \cdot 3 \cdot 5 = 7\dfrac{1}{2}$
    They need $7\dfrac{1}{2}$ gallons of water.

29. **Strategy** To find the amount of oil left:
    → Convert from gallons to qt.
    → Multiply the number of oil changes by the amount of oil required for each change.
    → Subtract the amount of oil used for oil changes from the amount in the container.

    Solution $50 \text{ gal} = \dfrac{50 \text{ gal}}{1} \cdot \dfrac{4 \text{ qt}}{1 \text{ gal}} = 200 \text{ qt}$
    $35 \cdot 5 = 175$
    $200 - 175 = 25$
    There is 25 qt of oil remaining.

31. **Strategy** To find the cost of the carpet:
    → Use the formula $A = LW$ to find the area.
    → Convert from square feet to square yards.
    → Multiply the square yards by \$33/yd² to find the cost.

    Solution $A = LW = 15 \text{ ft} \cdot 20 \text{ ft}$
    $A = 300 \text{ ft}^2$
    $300 \text{ ft}^2 = \dfrac{300 \text{ ft}^2}{1} \cdot \dfrac{1 \text{ yd}^2}{9 \text{ ft}^2} = 33\dfrac{1}{3} \text{ yd}^2$
    $\text{Cost} = \dfrac{100 \text{ yd}^2}{3} \cdot \dfrac{\$33}{1 \text{ yd}^2}$
    Cost = \$1100
    The cost of the carpet is \$1100.

33. **Strategy** To find the cost:
    → Convert from acres to ft.
    → Multiply lot size by price.

    Solution $\dfrac{1 \text{ acre}}{2} \cdot \dfrac{43{,}560 \text{ ft}^2}{1 \text{ acre}} = 21{,}780 \text{ ft}^2$
    $21{,}780 \cdot 3 = 65{,}340$
    The cost of the lot is \$65,340.

Objective C Exercises

35. $145 \text{ lb} = \dfrac{145 \text{ lb}}{1} \cdot \dfrac{1 \text{ kg}}{2.2 \text{ lb}} \approx 65.91 \text{ kg}$

37. $2 \text{ L} = \dfrac{2 \text{ L}}{1} \cdot \dfrac{1.06 \text{ qt}}{1 \text{ L}} \cdot \dfrac{4 \text{ c}}{1 \text{ qt}} = 8.48 \text{ c}$

39. $14.3 \text{ gal} = \dfrac{14.3 \text{ gal}}{1} \cdot \dfrac{3.97 \text{ L}}{1 \text{ gal}} \approx 54.20 \text{ L}$

41. $86 \text{ kg} = \dfrac{86 \text{ kg}}{1} \cdot \dfrac{2.2 \text{ lb}}{1 \text{ kg}} = 189.2 \text{ lb}$

43. $35 \text{ mm} = \dfrac{35 \text{ mm}}{1} \cdot \dfrac{1 \text{ cm}}{10 \text{ mm}} \cdot \dfrac{1 \text{ in}}{2.54 \text{ cm}} \approx 1.38 \text{ in}$

45. $24 \text{ L} = \dfrac{24 \text{ L}}{1} \cdot \dfrac{1 \text{ gal}}{3.79 \text{ L}} \approx 6.33 \text{ gal}$

47. $60 \text{ ft/s} = \dfrac{60 \text{ ft}}{1 \text{ s}} \cdot \dfrac{1 \text{ m}}{3.28 \text{ ft}} \approx 18.29 \text{ m/s}$

49. $\$0.99/\text{lb} = \dfrac{\$0.99}{1 \text{ lb}} \cdot \dfrac{2.2 \text{ lb}}{1 \text{ kg}} \approx \$2.18/\text{kg}$

51. $80 \text{ km/h} = \dfrac{80 \text{ km}}{1 \text{ h}} \cdot \dfrac{1 \text{ mi}}{1.61 \text{ km}} \approx 49.69 \text{ mph}$

53. $\$0.485/\text{L} = \dfrac{\$0.485}{1 \text{ L}} \cdot \dfrac{3.79 \text{ L}}{1 \text{ gal}} \approx \$1.84/\text{gal}$

55. **Strategy** To find the distance in kilometers, convert the distance from miles to km.

    Solution 24,887 mi
    $= \dfrac{24{,}887 \text{ mi}}{1} \cdot \dfrac{1.61 \text{ km}}{1 \text{ mi}} = 40{,}068.07 \text{ km}$
    The distance is 40,068.07 km.

Critical Thinking 7.3

57. a. Answers will vary.
    b. 2.54 cm/in; 2.2 lb/kg; 1.06 qt/L.

59. Answers will vary.

61. We can multiply an expression by the conversion rate $\dfrac{5{,}280 \text{ ft}}{1 \text{ mi}}$ because 5,280 ft = 1 mi. Therefore, we can substitute 1 mi for 5,280 ft: $\dfrac{5{,}280 \text{ ft}}{1 \text{ mi}} = \dfrac{1 \text{ mi}}{1 \text{ mi}}$. Since a nonzero expression divided by itself equals 1, $\dfrac{1 \text{ mi}}{1 \text{ mi}} = 1$. In multiplying an expression by $\dfrac{5{,}280 \text{ ft}}{1 \text{ mi}}$, we are multiplying by 1, and multiplying an expression by 1 does not change the value of the expression.

63. In reports stating the arguments for changing to the metric system in the United States, you might look for the following ideas: Changing to the metric system would ease international trade. Calculations in a decimal system are simpler than working with fractions, and the metric system is a decimal system. Conversion between units is far easier in the metric system.

In reports stating the arguments against changing to the metric system in the United States, you might look for the following ideas: U.S. industries oppose switching to the metric system because of the difficulty and expense of altering the present dimensions of machinery, tools, and products. Many citizens consider it difficult to learn a new system of measurement.

Section 7.4

Objective A Exercises

1. $\dfrac{27}{8} \times \dfrac{9}{4} \rightarrow \dfrac{72}{108}$
   The product of the extremes does not equal the product of the means.
   The proportion is not true.

3. $\dfrac{45}{135} \times \dfrac{3}{9} \rightarrow \dfrac{405}{405}$
   The product of the extremes equals the product of the means.
   The proportion is true.

5. $\dfrac{16}{3} \times \dfrac{48}{9} \rightarrow \dfrac{144}{144}$
   The product of the extremes equals the product of the means.
   The proportion is true.

7. $\dfrac{6 \text{ min}}{5 \text{ cents}} \times \dfrac{30 \text{ min}}{25 \text{ cents}} \rightarrow \dfrac{150}{150}$
   The product of the extremes equals the product of the means.
   The proportion is true.

9. $\dfrac{15 \text{ ft}}{3 \text{ yards}} \times \dfrac{90 \text{ ft}}{18 \text{ yards}} \rightarrow \dfrac{270}{270}$
   The product of the extremes equals the product of the means.
   The proportion is true.

11. $\dfrac{1 \text{ gal}}{4 \text{ qt}} \times \dfrac{7 \text{ gal}}{28 \text{ qt}} \rightarrow \dfrac{28}{28}$
    The product of the extremes equals the product of the means.
    The proportion is true.

13. $\dfrac{2}{3} = \dfrac{n}{15}$

$2 \cdot 15 = 3 \cdot n$

$30 = 3n$

$\dfrac{30}{3} = \dfrac{3n}{3}$

$10 = n$

15. $\dfrac{n}{5} = \dfrac{12}{25}$

$n \cdot 25 = 5 \cdot 12$

$25n = 60$

$\dfrac{25n}{25} = \dfrac{60}{25}$

$n = 2.4$

17. $\dfrac{3}{8} = \dfrac{n}{12}$

$3 \cdot 12 = 8 \cdot n$

$36 = 8n$

$\dfrac{36}{8} = \dfrac{8n}{8}$

$4.5 = n$

19. $\dfrac{3}{n} = \dfrac{7}{40}$

$3 \cdot 40 = n \cdot 7$

$120 = 7n$

$\dfrac{120}{7} = \dfrac{7n}{7}$

$17.14 \approx n$

21. $\dfrac{16}{n} = \dfrac{25}{40}$

$16 \cdot 40 = n \cdot 25$

$640 = 25n$

$\dfrac{640}{25} = \dfrac{25n}{25}$

$25.6 = n$

23. $\dfrac{120}{n} = \dfrac{144}{25}$

$120 \cdot 25 = n \cdot 144$

$3{,}000 = 144n$

$\dfrac{3{,}000}{144} = \dfrac{144n}{144}$

$20.83 \approx n$

25. $\dfrac{0.5}{2.3} = \dfrac{n}{20}$

$0.5 \cdot 20 = 2.3 \cdot n$

$10 = 23n$

$\dfrac{10}{2.3} = \dfrac{2.3n}{2.3}$

$n \approx 4.35$

27. $\dfrac{0.7}{1.2} = \dfrac{6.4}{n}$

$0.7 \cdot n = 1.2 \cdot 6.4$

$0.7n = 7.68$

$\dfrac{0.7n}{0.7} = \dfrac{7.68}{0.7}$

$n \approx 10.97$

29. $\dfrac{x}{6.25} = \dfrac{16}{87}$

$x \cdot 87 = 6.25 \cdot 16$

$87x = 100$

$\dfrac{87x}{87} = \dfrac{100}{87}$

$x \approx 1.15$

31. $\dfrac{1.2}{0.44} = \dfrac{y}{14.2}$

$1.2 \cdot 14.2 = 0.44 \cdot y$

$17.04 = 0.44y$

$\dfrac{17.04}{0.44} = \dfrac{0.44y}{0.44}$

$38.73 \approx y$

33. $\dfrac{n+2}{5} = \dfrac{1}{2}$

$(n+2)2 = 5 \cdot 1$

$2n + 4 = 5$

$2n = 1$

$\dfrac{2n}{2} = \dfrac{1}{2}$

$n = \dfrac{1}{2} = 0.5$

35. $\dfrac{4}{3} = \dfrac{n-2}{6}$

$4 \cdot 6 = 3(n-2)$

$24 = 3n - 6$

$30 = 3n$

$\dfrac{30}{3} = \dfrac{3n}{3}$

$10 = n$

37. $\dfrac{2}{n+3} = \dfrac{7}{12}$

$2 \cdot 12 = (n+3)(7)$

$24 = 7n + 21$

$3 = 7n$

$\dfrac{3}{7} = \dfrac{7n}{7}$

$0.43 \approx n$

39. $\dfrac{7}{10} = \dfrac{3+n}{2}$
$7 \cdot 2 = 10(3+n)$
$14 = 30 + 10n$
$-16 = 10n$
$\dfrac{-16}{10} = \dfrac{10n}{10}$
$-1.6 = n$

41. $\dfrac{x-4}{3} = \dfrac{3}{4}$
$(x-4)4 = 3 \cdot 3$
$4x - 16 = 9$
$4x = 25$
$\dfrac{4x}{4} = \dfrac{25}{4}$
$x = 6.25$

43. $\dfrac{6}{1} = \dfrac{x-2}{5}$
$6 \cdot 5 = 1(x-2)$
$30 = x - 2$
$32 = x$

45. $\dfrac{5}{8} = \dfrac{2}{x-3}$
$5(x-3) = 8 \cdot 2$
$5x - 15 = 16$
$5x = 31$
$\dfrac{5x}{5} = \dfrac{31}{5}$
$x = 6.2$

47. $\dfrac{3}{x-4} = \dfrac{5}{3}$
$3 \cdot 3 = (x-4)5$
$9 = 5x - 20$
$29 = 5x$
$\dfrac{29}{5} = \dfrac{5x}{5}$
$5.8 = x$

Objective B Exercises

49. Strategy — To find the length of the amoeba, write and solve a proportion using n to represent the length.

Solution $\dfrac{1 \text{ in.}}{0.002 \text{ in.}} = \dfrac{2.6 \text{ in.}}{n}$
$1 \cdot n = 0.002 \cdot 2.6$
$n = 0.0052$
The actual length of the amoeba is 0.0052 in.

51. Strategy — To find the number of robes, write and solve a proportion using n to represent the number of robes.

Solution $\dfrac{6 \text{ robes}}{6.5 \text{ yd}} = \dfrac{n}{26 \text{ yd}}$
$6 \cdot 26 = 6.5n$
$156 = 6.5n$
$\dfrac{156}{6.5} = \dfrac{6.5n}{6.5}$
$24 = n$
24 robes can be made.

53. Strategy — To find the property tax, write and solve a proportion using x to represent the amount of tax.

Solution $\dfrac{\$4{,}320}{\$180{,}000} = \dfrac{x}{\$280{,}000}$
$4{,}320 \cdot 280{,}000 = 180{,}000 \cdot x$
$1{,}209{,}600{,}000 = 180{,}000x$
$\dfrac{1{,}209{,}600{,}000}{180{,}000} = \dfrac{180{,}000x}{180{,}000}$
$6{,}720 = x$
The property tax is $6,720.

55. Strategy — To find the distance, write and solve a proportion using n to represent the miles driven.

Solution $\dfrac{84 \text{ mi}}{3 \text{ gal}} = \dfrac{n}{14.5 \text{ gal}}$
$84 \cdot 14.5 = 3 \cdot n$
$1218 = 3n$
$\dfrac{1218}{3} = \dfrac{3n}{3}$
$406 = n$
The car would travel 406 mi on 14.5 gal of gasoline.

57. Strategy — To find the cost of the grapefruit, write and solve a proportion using n to represent the cost of the grapefruit.

Solution $\dfrac{4 \text{ grapefruit}}{\$1.28} = \dfrac{14 \text{ grapefruit}}{n}$
$4 \cdot n = 1.28 \cdot 14$
$4n = 17.92$
$\dfrac{4n}{4} = \dfrac{17.92}{4}$
$n = 4.48$
The cost of 14 grapefruit is $4.48.

59. Strategy To find the number of light fixtures, write and solve a proportion using n to represent the required number of light fixtures.

Solution $\dfrac{5 \text{ lights}}{400 \text{ ft}^2} = \dfrac{n}{35{,}000 \text{ ft}^2}$
$5 \cdot 35{,}000 = 400 \cdot n$
$175{,}000 = 400n$
$\dfrac{175{,}000}{400} = \dfrac{400n}{400}$
$437.5 = n$
The office will require 438 lights.

61. Strategy To find the time necessary to lose 36 lb, write and solve a proportion using n to represent the time.

Solution $\dfrac{3 \text{ lb}}{5 \text{ weeks}} = \dfrac{36 \text{ lb}}{n}$
$3 \cdot n = 5 \cdot 36$
$3n = 180$
$\dfrac{3n}{3} = \dfrac{180}{3}$
$n = 60$
The dieter will lose 36 lb in 60 weeks.

63. Strategy To find the number of defects, write and solve a proportion using n to represent the number of defects.

Solution $\dfrac{22 \text{ defects}}{1{,}000 \text{ cars}} = \dfrac{n}{125{,}000 \text{ cars}}$
$22 \cdot 125{,}000 = 1{,}000 \cdot n$
$2{,}750{,}000 = 1{,}000n$
$2{,}750 = n$
There would be 2,750 defects found in the cars.

65. Strategy To find the number of miles, write and solve a proportion using n to represent the number of miles.

Solution 3 years is 36 months.
$\dfrac{22{,}000 \text{ mi}}{4 \text{ months}} = \dfrac{n}{36 \text{ months}}$
$22{,}000 \cdot 36 = 4 \cdot n$
$792{,}000 = 4n$
$\dfrac{792{,}000}{4} = \dfrac{4n}{4}$
$198{,}000 = n$
The account executive will drive 198,000 mi.

67. Strategy To find the additional amount of money, write and solve a proportion using x to represent the additional amount of money. Then $3{,}500 + x$ represents the total amount of money.

Solution $\dfrac{\$3{,}500}{\$280} = \dfrac{\$3{,}500 + x}{400}$
$3{,}500 \cdot 400 = 280(3{,}500 + x)$
$1{,}400{,}000 = 980{,}000 + 280x$
$420{,}000 = 280x$
$\dfrac{420{,}000}{280} = \dfrac{280x}{280}$
$1{,}500 = x$
To earn a dividend of $400, an additional $1,500 must be invested.

69. Strategy To find the number of inches the candle will burn, write and solve a proportion using x to represent the number of inches.

Solution $4 \text{ h} = 240 \text{ min}$
$\dfrac{1.5 \text{ in.}}{40 \text{ min}} = \dfrac{x}{240 \text{ min}}$
$1.5 \cdot 240 = 40 \cdot x$
$360 = 40x$
$\dfrac{360}{40} = \dfrac{40x}{40}$
$9 = x$
The candle will burn 9 in. in 4 h.

71. Strategy   To find the mid-management salary, write and solve a proportion using n to represent the salary.

Solution
$$\frac{7 \text{ mid-management}}{5 \text{ junior-management}} = \frac{n}{\$90,000}$$
$$7 \cdot 90,000 = 5 \cdot n$$
$$630,000 = 5n$$
$$\frac{630,000}{5} = \frac{5n}{5}$$
$$126,000 = n$$
The mid-management salary would be $126,000.

Critical Thinking 7.4

73. Yes, Assume that $\frac{a}{b} = \frac{c}{d}$ and that $\frac{a}{b}$ is in simplest form. The proportion $\frac{a}{b} = \frac{a+c}{b+d}$ is true. The numerator and denominator of $\frac{a+c}{b+d}$ have a common factor and the fraction reduces to $\frac{a}{b}$.

For example, $\frac{2}{3} = \frac{8}{12}$, thus
$$\frac{2}{3} = \frac{2+8}{3+12} = \frac{10}{15} = \frac{5 \cdot 2}{5 \cdot 3} = \frac{2}{3}.$$

75. $\frac{2}{5}$ cast a vote in favor of the amendment.

$\frac{3}{4}$ cast a vote against the amendment.

$\frac{2}{5} + \frac{3}{4} = \frac{8}{20} + \frac{15}{20} = \frac{23}{20}$

The number 1 represents the total population. The fraction $\frac{23}{20}$ indicates that there were more votes than voters. This is not possible.

77. Your students can learn about proportional representation in the United States House of Representatives by consulting an information almanac or a history or government text. From a historical point of view, your students might find it interesting to read The Constitution of the United States of America, Article 1, Section 2, and Article XIV, Section 2 of the Amendments to the Constitution, both of which deal with proportional representation in the House of Representatives.

Section 7.5

Objective A Exercises

1. a. The expression $y = kx$ is a direction variation because, on the right side of the equation, the constant of variation, k, is multiplied times x.

b. The expression $y = \frac{k}{x}$ is not a direct variation because, on the right side of the equation, the constant of variation, k, is divided by x, not multiplied times x.

c. The expression $y = k + x$ is not a direct variation because, on the right side of the equation, the constant of variation, k, is added to x, not multiplied times x.

d. The expression $y = \frac{k}{x^2}$ is not a direct variation because, on the right side of the equation, the constant of variation, k, is divided by x^2, not multiplied times x.

3. Strategy   To find the constant of variation, substitute 15 for y and 2 for x in the direct variation equation $y = kx$ and solve for k.

Solution
$$y = kx$$
$$15 = k \cdot 2$$
$$\frac{15}{2} = k$$
The constant of variation is $\frac{15}{2}$.

5. Strategy   To find the constant of variation, substitute 64 for n

and 2 for m in the direct variation equation $n = km^2$ and solve for k.

Solution $n = km^2$
$64 = k \cdot 2^2$
$64 = 4k$
$\frac{64}{4} = k$
$16 = k$
The constant of variation is 16.

7. Strategy To find P when $R = 6$:
→Write the basic direct variation equation, replace the variables by the given values, and solve for k.
→Write the direct variation equation, replacing k by its value. Substitute 6 for R and solve for P.

Solution $P = kR$
$20 = k \cdot 5$
$\frac{20}{5} = k$
$4 = k$
$P = 4R$
$P = 4 \cdot 6$
$P = 24$
The value of P is 24.

9. Strategy To find M when $P = 20$.
→Write the basic direct variation equation, replace the variables by the given values, and solve for k.
→Write the direct variation equation, replacing k by its value. Substitute 20 for P and solve for M.

Solution $M = kP$
$15 = k \cdot 30$
$\frac{15}{30} = k$
$\frac{1}{2} = k$
$M = \frac{1}{2} P$
$M = \frac{1}{2} \cdot 20$
$M = 10$
The value of M is 10.

11. Strategy To find y when $x = 0.5$:
→Write the basic direct variation equation, replace the variables by the given values, and solve for k.
→Write the direct variation equation, replacing k by its value. Substitute 0.5 for x and solve for y.

Solution $y = kx^2$
$10 = k \cdot 2^2$
$10 = 4k$
$\frac{5}{2} = k$
$y = \frac{5}{2} x^2$
$y = \frac{5}{2}(0.5)^2$
$y = \frac{5}{2}(0.25)$
$y = 0.625$
The value of y is 0.625.

13. Strategy To find the amount earned:
→Write the basic direct variation equation, replace the variables by the given values, and solve for k.
→Write the direct variation equation, replacing k by its value. Substitute 30 for h and solve for w.

Solution $w = kh$
$82 = k \cdot 8$
$\frac{82}{8} = k$
$10.25 = k$
$w = 10.25$
$w = 10.25(30)$
$w = 307.50$
The amount earned is $307.50.

15. Strategy To find the pressure:
→Write the basic direct variation equation, replace the variables by the given values, and solve for k.
→Write the direct variation equation, replacing k by its value. Substitute 12 for d and solve for P.

Solution
$P = kd$
$2.25 = k \cdot 5$
$\frac{2.25}{5} = k$
$0.45 = k$
$P = 0.45d$
$P = 0.45 \cdot 12$
$P = 5.4$
The pressure is 5.4 lb/in^2 at the depth of 12 ft.

17. Strategy To find the stopping distance:
→Write the basic direct variation equation, replace the variables by the given values, and solve for k.
→Write the direct variation equation, replacing k by its value. Substitute 65 for v and solve for s.

Solution
$s = kv^2$
$170 = k \cdot 50^2$
$170 = 2{,}500k$
$\frac{170}{2{,}500} = k$
$\frac{17}{250} = k$
$s = \frac{17}{250} v^2$
$s = \frac{17}{250}(65)^2$
$s = \frac{17}{250} \cdot 4{,}225$
$s = 287.3$
The stopping distance is 287.3 ft.

19. Strategy To find the current:
→Write the basic direct variation equation, replace the variables by the given values, and solve for k.
→Write the direct variation equation, replacing k by its value. Substitute 75 for V and solve for I.

Solution
$I = kV$
$4 = k \cdot 100$
$\frac{4}{100} = k$
$\frac{1}{25} = k$
$I = \frac{1}{25}V$
$I = \frac{1}{25} \cdot 75$
$I = 3$
The electric current is 3 amps.

21. Strategy To find the constant of variation, substitute 10 for y and 5 for x in the inverse variation equation $y = \frac{k}{x}$ and solve for k.

Solution
$y = \frac{k}{x}$
$10 = \frac{k}{5}$
$10 \cdot 5 = k$
$50 = k$
The constant of variation is 50.

23. Strategy To find the constant of variation, substitute 4 for p and 5 for q in the inverse variation equation $p = \frac{k}{q^2}$ and solve for k.

Solution
$p = \frac{k}{q^2}$
$4 = \frac{k}{5^2}$
$4 = \frac{k}{25}$
$4 \cdot 25 = k$
$100 = k$
The constant of variation is 100.

Chapter 7: *Measurement and Proportion*

25. Strategy   To find y when $x = 10$:
→ Write the basic inverse variation equation, replace the variables by the given values and solve for k.
→ Write the inverse variation equation, replacing k by its value. Substitute 10 for x and solve for y.

Solution
$y = \dfrac{k}{x}$
$500 = \dfrac{k}{4}$
$500 \cdot 4 = k$
$2{,}000 = k$
$y = \dfrac{2{,}000}{x}$
$y = \dfrac{2{,}000}{10}$
$y = 200$
The value of y is 200.

27. Strategy   To find y when $x = 10$:
→ Write the basic inverse variation equation, replace the variables by the given values, and solve for k.
→ Write the inverse variation equation, replacing k by its value. Substitute 10 for x and solve for y.

Solution
$y = \dfrac{k}{x^2}$
$40 = \dfrac{k}{4^2}$
$40 = \dfrac{k}{16}$
$40 \cdot 16 = k$
$640 = k$
$y = \dfrac{640}{x^2}$
$y = \dfrac{640}{10^2}$
$y = \dfrac{640}{100}$
$y = 6.4$
The value of y is 6.4 when $x = 10$.

29. Strategy   To find the length of the rectangle:
→ Write the basic inverse variation equation, replace the variables by the given values, and solve for k.
→ Write the inverse variation equation, replacing k by its value. Substitute 4 for the width and solve for the length.

Solution
$L = \dfrac{k}{W}$
$8 = \dfrac{k}{5}$
$8 \cdot 5 = k$
$40 = k$
$L = \dfrac{40}{W}$
$L = \dfrac{40}{4}$
$L = 10$
The length of the rectangle is 10 ft.

31. Strategy   To find the resistance in the circuit:
→ Write the basic inverse variation equation, replace the variables by the given values, and solve for k.
→ Write the inverse variation equation, replacing k by its value. Substitute 1.2 for the current and solve for the resistance.

Solution
$I = \dfrac{k}{R}$
$0.25 = \dfrac{k}{8}$
$0.25 \cdot 8 = k$
$2 = k$
$I = \dfrac{2}{R}$
$1.2 = \dfrac{2}{R}$
$1.2R = 2$
$R = \dfrac{2}{1.2}$
$R = 1.\overline{6}$
The resistance is $1.\overline{6}$ ohms.

33. Strategy To find the number of computers:
→Write the basic inverse variation equation, replace the variables by the given values, and solve for k.
→Write the inverse variation equation, replacing k by its value. Substitute 1,500 for the price and solve for the number of computers sold.

Solution
$S = \dfrac{k}{P}$
$1{,}800 = \dfrac{k}{1{,}800}$
$1{,}800 \cdot 1{,}800 = k$
$3{,}240{,}000 = k$
$S = \dfrac{3{,}240{,}000}{P}$
$S = \dfrac{3{,}240{,}000}{1{,}500}$
$S = 2{,}160$
At a price of $1,500, 2,160 computers can be sold.

35. Strategy To find the intensity of light:
→Write the basic inverse equation, replace the variables by the given values, and solve for k.
→Write the inverse variation equation, replacing k by its value. Substitute 5 for the distance and solve for the intensity.

Solution
$I = \dfrac{k}{d^2}$
$20 = \dfrac{k}{8^2}$
$20 = \dfrac{k}{64}$
$20 \cdot 64 = k$
$1{,}280 = k$
$I = \dfrac{1{,}280}{d^2}$
$I = \dfrac{1{,}280}{5^2}$
$I = \dfrac{1{,}280}{25} = 51.2$
The intensity of light is 51.2 lumens.

37. Strategy To find the pressure of the gas:
→Write the basic inverse variation equation, replace the variables by the given values, and solve for k.
→Write the inverse variation equation, replacing k by its value. Substitute 150 for the volume and solve for P.

Solution
$V = \dfrac{k}{P}$
$400 = \dfrac{k}{25}$
$25 \cdot 400 = k$
$10{,}000 = k$
$V = \dfrac{10{,}000}{P}$
$150 = \dfrac{10{,}000}{P}$
$P = \dfrac{10{,}000}{150} \approx 66.67 \text{ lb/in}^2$
The pressure is approximately 66.67 lb/in².

Critical Thinking 7.5

39. a. True. Because k is a constant, if x becomes larger, then f becomes larger.

b. False. Rewrite $x = \dfrac{k}{y}$ as the equivalent expression $xy = k$. If x increases, then y must decrease because the product is a constant.

c. False. If we double s in $T = ks^2$, we obtain $T = k(2s)^2 = k(4s^2) = 4ks^2$. Thus if we double s, we quadruple T.

41. a. If we double x in $y = kx^3$, we obtain $y = k(2x)^3 = k(8x^3) = 8kx^3$. Thus if we double x, y is 8 times larger.

b. If we double x in $y = \dfrac{k}{x^3}$, we obtain $y = \dfrac{k}{(2x)^3} = \dfrac{k}{8x^3} = \dfrac{1}{8} \cdot \dfrac{k}{x^3}$. Thus if we double x, y is $\dfrac{1}{8}$ as large.

43. You might encourage your students to provide an example of how a proportion can be used to price large quantities of a purchase as compared to smaller quantities. For example, if a box of 30 trash bags is priced at $4.69, the comparable price for 100 trash bags can be found by letting x equal the price of 100 trash bags and solving the proportion shown below for x.

$$\frac{4.69}{30} = \frac{x}{100}$$

The solution is that $x \approx 15.63$, or $15.63. Your students might find it interesting to visit a grocery store and find the unit prices of different sized containers of the same item, for example, peanut butter or condiments. They may be surprised to discover that the unit price of a large container is not always lower than the unit price of a smaller container of the same item.

45. In discussing the relationship between direct variation and proportion, students might note that the constant of variation is also referred to as the constant of *proportion*ality. It should be noted that we describe a direct variation by stating that one quantity is directly *proportional* to another. In the direct variation equation, y is a constant multiple of x. There $y \div x$ is a constant ratio, and the result is a proportion.

Chapter Review Exercises

1. $1.25 \text{ km} = 1,250 \text{ m}$

2. $0.450 \text{ g} = 450 \text{ mg}$

3. $\frac{100 \text{ lb}}{100 \text{ lb}} = \frac{1}{1}$, $1:1$, $1 \text{ TO } 1$

4. $\frac{18 \text{ roof supports}}{9 \text{ ft}} = \frac{2 \text{ roof supports}}{1 \text{ ft}}$

5. $\frac{\$628}{40 \text{ h}} = \$15.70/h$

6. $\frac{8 \text{ h}}{15 \text{ h}} = \frac{8}{15}$

7. $96 \text{ in} = \frac{96 \text{ in}}{1} \cdot \frac{1 \text{ ft}}{12 \text{ in}} \cdot \frac{1 \text{ yd}}{3 \text{ ft}} = 2\frac{2}{3} \text{ yd}$

8. $72 \text{ oz} = \frac{72 \text{ oz}}{1} \cdot \frac{1 \text{ lb}}{16 \text{ oz}} = 4\frac{1}{4} \text{ lb}$

9. $36 \text{ fl oz} = \frac{36 \text{ fl oz}}{1} \cdot \frac{1 \text{ c}}{8 \text{ fl oz}} = 4\frac{1}{2} \text{ c}$

10. $1\frac{1}{4} \text{ mi} = \frac{5 \text{ mi}}{4} \cdot \frac{5,280 \text{ ft}}{1 \text{ mi}} = 6,600 \text{ ft}$

11. $\frac{n}{3} = \frac{8}{15}$
    $n \cdot 15 = 3 \cdot 8$
    $15n = 24$
    $n = \frac{24}{15}$
    $n = 1.6$

12. $\frac{15 \text{ lb}}{12 \text{ trees}} = \frac{5 \text{ lb}}{4 \text{ trees}}$

13. $\frac{171 \text{ mi}}{3 \text{ h}} = 57 \text{ mph}$

14. $\frac{2}{3.5} = \frac{n}{12}$
    $2 \cdot 12 = 3.5 \cdot n$
    $24 = 3.5n$
    $\frac{24}{3.5} = n$
    $6.86 \approx n$

15. $1 \text{ qt} = \frac{1 \text{ qt}}{1} \cdot \frac{1 \text{ L}}{1.06 \text{ qt}} \cdot \frac{1000 \text{ ml}}{1 \text{ L}} \approx 943.40 \text{ ml}$

16. $29 \text{ ft} = \frac{29 \text{ ft}}{1} \cdot \frac{1 \text{ m}}{3.28 \text{ ft}} \approx 8.84 \text{ m}$

17. $100 \text{ m} = \frac{100 \text{ m}}{1} \cdot \frac{3.28 \text{ ft}}{1 \text{ m}} = 328 \text{ ft}$

18. $2.1 \text{ kg} = \frac{2.1 \text{ kg}}{1} \cdot \frac{2.2 \text{ lb}}{1 \text{ kg}} = 4.62 \text{ lb}$

19. $30 \text{ mph} = \frac{30 \text{ mi}}{1 \text{ h}} \cdot \frac{1.61 \text{ km}}{1 \text{ mi}} = 48.3 \text{ km/h}$

20. $75 \text{ km/h} = \frac{75 \text{ km}}{1 \text{ h}} \cdot \frac{1 \text{ mi}}{1.61 \text{ km}} \approx 46.58 \text{ mph}$

21. $\frac{18 \text{ c}}{24 \text{ pt}} = \frac{3 \text{ c}}{4 \text{ pt}}$

22. Strategy   To find the constant of variation, substitute 10 for y and 30 for x in the direct variation equation $y = kx$ and solve for k.

    Solution $y = kx$
$10 = k \cdot 30$
$\frac{10}{30} = k$
$\frac{1}{3} = k$

The constant of variation is $\frac{1}{3}$.

23. Strategy   To find T when $S = 120$:
→Write the basic direct variation equation, replace the variables by the given values, and solve for k.
→Write the direct variation equation, replacing k by its value. Substitute 120 for S and solve for T.

    Solution $T = kS^2$
$50 = k \cdot 5^2$
$50 = 25k$
$\frac{50}{25} = k$
$2 = k$
$T = 2S^2$
$T = 2 \cdot 120^2$
$T = 2 \cdot 14{,}400 = 28{,}800$
$T = 28{,}800$
The value of T is 28,800.

24. Strategy   To find y when $x = 25$:
→Write the inverse variation equation, replace the variables by the given values and solve for k.
→Write the inverse variation equation, replacing k by its value. Substitute 25 for x and solve for y.

    Solution $y = \frac{k}{x}$
$0.2 = \frac{k}{5}$
$0.2 \cdot 5 = k$
$1 = k$
$y = \frac{1}{x}$
$y = \frac{1}{25} = 0.04$
The value of y is 0.04.

25. Strategy   To find the number of pieces:
→Convert 1.21 m to cm.
→Add the distances.

    Solution   1.21 m = 121 cm
$42 + 18 + 121 = 181$
The total length of the shaft is 181 cm.

26. Strategy   To find the ratio:
→Subtract the present price (75) from the original price (125).
→Form the ratio of the decrease in price to the original price (125).

    Solution $125 - 75 = 50$
$\frac{\$50}{\$125} = \frac{2}{5}$
The ratio of the decrease in price to the original price is $\frac{2}{5}$.

27. Strategy   To find the number of ounces:
→Convert 12 lb to ounces.
→Divide the amount by 16.

    Solution $12 \text{ lb} = \frac{12 \text{ lb}}{1} \cdot \frac{16 \text{ oz}}{1 \text{ lb}} = 192 \text{ oz}$
$192 \div 16 = 12$
Each container has 12 oz of hamburger meat.

174 *Chapter 7: Measurement and Proportion*

28. Strategy To find the amount invested, write and solve for proportion using n to represent the amount of money invested.

Solution
$$\frac{8,000}{520} = \frac{n}{780}$$
$8,000 \cdot 780 = 520 \cdot n$
$6,240,000 = 520n$
$\frac{6,240,000}{520} = n$
$12,000 = n$
The amount of money invested is $12,000.

29. Strategy To find the distance:
→Write the basic direct variation equation, replace the variables by the given values, and solve for k.
→Write the direct variation equation, replacing k by its value. Substitute 28 for the weight and solve for the distance.

Solution
$d = kw$
$2 = k \cdot 5$
$\frac{2}{5} = k$
$d = \frac{2}{5}w$
$d = \frac{2}{5} \cdot 28$
$d = 11.2$
The weight will stretch the spring 11.2 in.

30. Strategy To find the amount of plant food, write and solve a proportion using n to represent the amount of plant food.

Solution
$$\frac{0.5}{50} = \frac{n}{275}$$
$0.5 \cdot 275 = 50 \cdot n$
$137.5 = 50n$
$\frac{137.5}{50} = n$
$2.75 = n$
2.75 lb of plant food should be used.

31. Strategy To convert mph to ft per sec, use the conversion factors $\frac{1 \text{ h}}{3,600 \text{ s}}$ and $\frac{5,280 \text{ ft}}{1 \text{ mi}}$.

Solution
87 mph
$= \frac{87 \text{ mi}}{1 \text{ h}} \cdot \frac{1 \text{ h}}{3,600 \text{ s}} \cdot \frac{5,280 \text{ ft}}{1 \text{ mi}}$
$= 127.6$ ft/s
The speed is 127.6 ft/s.

32. Strategy To find the volume:
→Write the basic inverse variation equation, replace the variables by the given values, and solve for k.
→Write the inverse variation equation, replacing k by its value. Substitute 12 for P and solve for V.

Solution
$V = \frac{k}{P}$
$2.5 = \frac{k}{6}$
$2.5 \cdot 6 = k$
$15 = k$
$V = \frac{15}{P}$
$V = \frac{15}{12}$
$V = 1.25$
The volume of the balloon is 1.25 ft³.

33. Strategy To find the amount the other attorney receives, write and solve a proportion using n to represent the amount.

Solution
$$\frac{3}{2} = \frac{96,000}{n}$$
$3 \cdot n = 2 \cdot 96,000$
$3n = 192,000$
$\frac{3n}{3} = \frac{192,000}{3}$
$n = 64,000$
The attorney receives $64,000.

Chapter Test

1. 4,650 cm = 46.5 m

2. 4.1 L = 4,100 ml

3. $\dfrac{3 \text{ yd}}{24 \text{ yd}} = \dfrac{1}{8}$, 1 : 8, 1 TO 8

4. $\dfrac{16 \text{ oz}}{64 \text{ cookies}} = \dfrac{1 \text{ oz}}{4 \text{ cookies}}$

5. $\dfrac{120 \text{ mi}}{200 \text{ min}} = 0.6 \text{ mi/min}$

6. $\dfrac{200 \text{ ft}}{100 \text{ ft}} = \dfrac{2}{1}$

7. $2\dfrac{3}{5} \text{ tons} = \dfrac{13 \text{ tons}}{5} \cdot \dfrac{2,000 \text{ lb}}{1 \text{ ton}} = 5,200 \text{ lb}$

8. $2\dfrac{1}{2} \text{ c} = \dfrac{5 \text{ c}}{2} \cdot \dfrac{8 \text{ fl oz}}{1 \text{ c}} = 20 \text{ fl oz}$

9. $3\dfrac{1}{4} \text{ lb} = \dfrac{13 \text{ lb}}{4} \cdot \dfrac{16 \text{ oz}}{1 \text{ lb}} = 52 \text{ oz}$

10. $8\dfrac{1}{2} \text{ ft} = \dfrac{17 \text{ ft}}{2} \cdot \dfrac{12 \text{ in.}}{1 \text{ ft}} = 102 \text{ in.}$

11. $\dfrac{n}{5} = \dfrac{3}{20}$
    $n \cdot 20 = 5 \cdot 3$
    $20n = 15$
    $n = \dfrac{15}{20}$
    $n = 0.75$

12. $\dfrac{8 \text{ ft}}{4 \text{ s}} = 2 \text{ ft/s}$

13. $4.3 \text{ c} = \dfrac{4.3 \text{ c}}{1} \cdot \dfrac{8 \text{ oz}}{1 \text{ c}} = 34.4 \text{ oz}$

14. $42 \text{ yd} = \dfrac{42 \text{ yd}}{1} \cdot \dfrac{3 \text{ ft}}{1 \text{ yd}} = 126 \text{ ft}$

15. $\dfrac{2,860 \text{ ft}^2}{6 \text{ h}} = 476.67 \text{ ft}^2/\text{h}$

16. $\dfrac{n}{4} = \dfrac{8}{9}$
    $n \cdot 9 = 4 \cdot 8$
    $9n = 32$
    $\dfrac{9n}{9} = \dfrac{32}{9}$
    $n \approx 3.56$

17. $12 \text{ oz} = \dfrac{12 \text{ oz}}{1} \cdot \dfrac{28.35 \text{ g}}{1 \text{ oz}} = 340.2 \text{ g}$

18. $547 \text{ ft} = \dfrac{547 \text{ ft}}{1} \cdot \dfrac{1 \text{ m}}{3.28 \text{ ft}} \approx 166.77 \text{ m}$

19. $1,000 \text{ m} = \dfrac{1,000 \text{ m}}{1} \cdot \dfrac{1.09 \text{ yd}}{1 \text{ m}} = 1,090 \text{ yd}$

20. $1.9 \text{ kg} = \dfrac{1.9 \text{ kg}}{1} \cdot \dfrac{2.2 \text{ lb}}{1 \text{ kg}} = 4.18 \text{ lb}$

21. $35 \text{ mph} = \dfrac{35 \text{ mi}}{1 \text{ h}} \cdot \dfrac{1.61 \text{ km}}{1 \text{ mi}} = 56.35 \text{ km/h}$

22. $60 \text{ km/h} = \dfrac{60 \text{ km}}{1 \text{ h}} \cdot \dfrac{1 \text{ mi}}{1.61 \text{ km}} \approx 37.27 \text{ mph}$

23. **Strategy**   To find the constant of proportionality, substitute 10 for y and 2 for x in the inverse variation equation $y = \dfrac{k}{x}$ and solve for k.

    Solution
    $y = \dfrac{k}{x}$
    $10 = \dfrac{k}{2}$
    $10 \cdot 2 = k$
    $20 = k$
    The constant of proportionality is 20.

176 *Chapter 7: Measurement and Proportion*

24. Strategy To find P when $R = 15$:
→ Write the basic direct variation equation, replace the variables by the given values, and solve for k.
→ Write the direct variation equation, replacing k by its value. Substitute 15 for R and solve for P.

Solution
$R = kP$
$4 = k \cdot 20$
$\frac{4}{20} = k$
$\frac{1}{5} = k$
$R = \frac{1}{5}P$
$15 = \frac{1}{5}P$
$5 \cdot 15 = P$
$75 = P$
The value of P is 75.

25. Strategy To find U when $V = 2$:
→ Write the inverse variation equation, replace the variables by the given values, and solve for k.
→ Write the inverse variation equation, replacing k by its value. Substitute 2 for V and solve for U.

Solution
$U = \frac{k}{V^2}$
$20 = \frac{k}{4^2}$
$20 = \frac{k}{16}$
$20 \cdot 16 = k$
$320 = k$
$U = \frac{320}{V^2}$
$U = \frac{320}{2^2}$
$U = \frac{320}{4} = 80$
The value of U is 80.

26. Strategy To find the ratio, form the ratio of the original weight (165) to the increased weight (190).

Solution $\frac{165}{190} = \frac{33}{38}$
The ratio of the original weight to the increased weight is $\frac{33}{38}$.

27. Strategy To find the sales tax, write and solve a proportion using x to represent the amount of tax.

Solution
$\frac{\$7.60}{\$95} = \frac{x}{\$39,200}$
$7.60 \cdot 39,200 = 95 \cdot x$
$297,920 = 95x$
$\frac{297,920}{95} = \frac{95x}{95}$
$3136 = x$
The sales tax is $3136.

28. Strategy To find the number of registered voters that would vote, write and solve a proportion using n to represent the number of people that would vote.

Solution
$\frac{3}{4} = \frac{n}{325,000}$
$3 \cdot 325,000 = 4n$
$975,000 = 4n$
$\frac{975,000}{4} = n$
$243,750 = n$
243,750 of the registered voters would vote.

29. Strategy To find the difference:
→ Convert from lb to ounces.
→ Divide the amount of cheese by the package size.
→ To find the selling price, multiply the number of packages by price (7.50).
→ Subtract the purchase price from the selling price.

Solution $24 \text{ lb} = \dfrac{24 \text{ lb}}{1} \cdot \dfrac{16 \text{ oz}}{1 \text{ lb}} = 384 \text{ oz}$

$\dfrac{384}{12} = 32$

$32 \cdot 7.50 = 240$

$240 - 126 = 114$

The difference is $114.

30. Strategy To find the length of the room, write and solve a proportion using n to represent the length.

Solution $\dfrac{4}{1} = \dfrac{n}{12\frac{1}{2}}$

$4 \cdot \dfrac{25}{2} = 1 \cdot n$

$50 = n$

The length of the room is 50 ft.

31. Strategy To convert mph to ft per sec, use the conversion factors $\dfrac{1 \text{ h}}{3{,}600 \text{ s}}$ and $\dfrac{5{,}280 \text{ ft}}{1 \text{ mi}}$.

Solution 52 mph
$= \dfrac{52 \text{ mi}}{1 \text{ h}} \cdot \dfrac{1 \text{ h}}{3{,}600 \text{ s}} \cdot \dfrac{5{,}280 \text{ ft}}{1 \text{ mi}}$
$= 76.27 \text{ ft/s}$
The speed is 76.27 ft/s.

32. Strategy To find the stopping distance:
→ Write the basic direct variation equation, replace the variables by the given values, and solve for k.
→ Write the direct variation equation, replacing k by its value. Substitute 60 for v and solve for d

Solution $d = kv^2$
$130 = k \cdot 40^2$
$130 = 1{,}600k$
$\dfrac{130}{1{,}600} = k$

$d = \dfrac{13}{160} v^2$

$d = \dfrac{13}{160} \cdot 60^2$

$d = \dfrac{13}{160} \cdot 3{,}600$

$d = 292.5$
The stopping distance of the car is 292.5 ft.

33. Strategy To find the number of revolutions per minute:
→ Write the basic inverse variation equation, replace the variables by the given values, and solve for k.
→ Write the inverse variation equation, replacing k by its value. Substitute 40 for the number of teeth and solve for the number of revolutions per minute.

Solution $s = \dfrac{k}{t}$

$160 = \dfrac{k}{25}$

$160 \cdot 25 = k$
$4{,}000 = k$

$s = \dfrac{4{,}000}{t}$

$s = \dfrac{4000}{40} = 100$

The gear will make 100 revolutions per minute.

Cumulative Review Exercises

1. $18 \div \dfrac{6-3}{9} - (-3)$
   $= 18 \div \dfrac{3}{9} - (-3)$
   $= 18 \div \dfrac{1}{3} - (-3)$
   $= 18 \cdot \dfrac{3}{1} - (-3)$
   $= 54 - (-3)$
   $= 54 + 3$
   $= 57$

2. $1.2 \text{ gal} = 1.2 \text{ gal} \cdot \dfrac{4 \text{ qt}}{1 \text{ gal}}$
   $= 4.8 \text{ qt}$

3. $7\dfrac{5}{12} - 3\dfrac{5}{9} = 7\dfrac{15}{36} - 3\dfrac{20}{36}$
   $= 6\dfrac{51}{36} - 3\dfrac{20}{36}$
   $= 3\dfrac{31}{36}$

4. $\dfrac{4}{5} \div \dfrac{4}{5} + \dfrac{2}{3} = \dfrac{4}{5} \cdot \dfrac{5}{4} + \dfrac{2}{3}$
   $= 1 + \dfrac{2}{3}$
   $= 1\dfrac{2}{3}$

5. $342 \div (-3) = -114$

6. $2a - 3ab$
   $2(2) - 3(2)(-3)$
   $= 4 - 3(2)(-3)$
   $= 4 - 6(-3)$
   $= 4 - (-18)$
   $= 4 + 18$
   $= 22$

7. $5x - 20 = 0$
   $5x - 20 + 20 = 0 + 20$
   $5x = 20$
   $\dfrac{5x}{5} = \dfrac{20}{5}$
   $x = 4$
   The solution is 4.

8. $3(x - 4) + 2x = 3$
   $3x - 12 + 2x = 3$
   $5x - 12 = 3$
   $5x - 12 + 12 = 3 + 12$
   $5x = 15$
   $\dfrac{5x}{5} = \dfrac{15}{5}$
   $x = 3$
   The solution is 3.

9. Draw a solid dot one half unit to the left of -3 on the number line.

10. Draw a parenthesis at -3. Draw a line to the left of -3. Draw an arrow at the left end of the line.

11. $(-5)^2 - (-8) \div (7 - 5)^2 \cdot 2 - 8$
    $= (-5)^2 - (-8) \div 2^2 \cdot 2 - 8$
    $= 25 - (-8) \div 4 \cdot 2 - 8$
    $= 25 - (-2) \cdot 2 - 8$
    $= 25 - (-4) - 8$
    $= 25 + 4 - 8$
    $= 29 - 8$
    $= 21$

12. $\left(-\dfrac{2}{3}\right)\left(-\dfrac{3}{4}\right)^2 = \left(-\dfrac{2}{3}\right)\left(\dfrac{9}{16}\right)$
    $= -\dfrac{2}{3} \cdot \dfrac{9}{16}$
    $= -\dfrac{2 \cdot 9}{3 \cdot 16}$
    $= -\dfrac{2 \cdot 3 \cdot 3}{3 \cdot 2 \cdot 2 \cdot 2 \cdot 2}$
    $= -\dfrac{3}{8}$

13. $\sqrt{169} = 13$

14. $5 - 2(1 - 3a) + 2(a - 3)$
    $= 5 - 2 + 6a + 2a - 6$
    $= 8a - 3$

15. $(4a^3b)(-5a^2b^3) = [4(-5)](a^3 a^2)(bb^3)$
    $= -20a^5 b^4$

16. $-3y^2 + 3y - y^2 - 6y$
    $= (-3y^2 - y^2) + (3y - 6y)$
    $= -4^2 - 3y$

17. $y = 3x - 2$
    $y = 3(-1) - 2$
    $y = -3 - 2$
    $y = -5$
    The ordered pair solution is $(-1, -5)$.

18. $\dfrac{30 \text{ cents}}{1 \text{ dollar}} = \dfrac{30 \text{ cents}}{100 \text{ cents}} = \dfrac{3}{10}$

19. $\dfrac{\$19{,}425}{5 \text{ months}} = \$3{,}885/\text{month}$

Cumulative Review

20. $\$1.97/gal = \dfrac{\$1.97}{1\ gal} \cdot \dfrac{1\ gal}{3.79\ L} \approx \$0.52/L$

21. $\dfrac{2}{3} = \dfrac{n}{48}$
    $2 \cdot 48 = 3 \cdot n$
    $96 = 3n$
    $\dfrac{96}{3} = n$
    $32 = n$

22. $\dfrac{\frac{1}{2} + \frac{3}{4}}{2 - \frac{5}{8}} = \dfrac{\frac{2}{4} + \frac{3}{4}}{\frac{16}{8} - \frac{5}{8}}$
    $= \dfrac{\frac{5}{4}}{\frac{11}{8}}$
    $= \dfrac{5}{4} \div \dfrac{11}{8}$
    $= \dfrac{5}{4} \cdot \dfrac{8}{11}$
    $= \dfrac{5 \cdot 8}{4 \cdot 11}$
    $= \dfrac{5 \cdot 2 \cdot 2 \cdot 2}{2 \cdot 2 \cdot 11} = \dfrac{10}{11}$

23. $-2\sqrt{x^2 - 3y}$
    $-2\sqrt{4^2 - 3(-3)}$
    $= -2\sqrt{16 - (-9)}$
    $= -2\sqrt{16 + 9}$
    $= -2\sqrt{25}$
    $= -2 \cdot 5 = -10$

24. $3x + 3(x + 4) = 4(x + 2)$
    $3x + 3x + 12 = 4x + 8$
    $6x + 12 = 4x + 8$
    $6x - 4x + 12 = 4x - 4x + 8$
    $2x + 12 = 8$
    $2x + 12 - 12 = 8 - 12$
    $2x = -4$
    $\dfrac{2x}{2} = \dfrac{-4}{2}$
    $x = -2$
    The solution is −2.

25. **Strategy** To find the monthly difference:
    →Find the difference in annual expenses between the northeast (587) and south (243).
    →Convert the difference from annual to monthly by dividing by 12.

    Solution $587 - 243 = 344$
    $344 \div 12 \approx 28.67$
    The monthly difference is $28.67.

26. The unknown number: x

    | five less than two thirds of a number | is | three |

    $\dfrac{2}{3}x - 5 = 3$
    $\dfrac{2}{3}x - 5 + 5 = 3 + 5$
    $\dfrac{2}{3}x = 8$
    $\dfrac{3}{2}\left(\dfrac{2}{3}x\right) = \dfrac{3}{2} \cdot 8$
    $x = 12$
    The number is 12.

27. Let the number be x.
    The <u>difference</u> between four <u>times</u> a number and three <u>times</u> the sum of the number and two.
    $4x - 3(x + 2)$
    $4x - 3x - 6$
    $x - 6$

28. **Strategy** To find the number of miles left to drive:
    →Find the number of miles already driven by subtracting the original odometer reading (18,325) from the present odometer reading (18,386).
    →Subtract the number of miles already drive from 125.

    Solution
    $18,386$
    $-18,325$
    $\overline{61}$

    125
    -61
    $\overline{64}$

    You have 64 mi left to drive.

Chapter 7: Measurement and Proportion

29. Strategy To find the new checking balance:
→ Add the deposit (122.35) to the checking account balance (422.89).
→ Subtract the check (279.76) from the new checking account balance.

Solution
$$\begin{array}{r} 422.89 \\ +122.35 \\ \hline 545.24 \end{array}$$

$$\begin{array}{r} 545.24 \\ -279.76 \\ \hline 265.48 \end{array}$$

The new checking account balance is $265.48.

30. Strategy To find the part that remains to be completed:
→ Add the amount already done $\left(\frac{2}{5}+\frac{1}{3}\right)$.
→ Subtract the amounts already done from the total job (1).

Solution
$\frac{2}{5}+\frac{1}{3}=\frac{6}{15}+\frac{5}{15}=\frac{11}{15}$

$1-\frac{11}{15}=\frac{15}{15}-\frac{11}{15}=\frac{4}{15}$

$\frac{4}{15}$ of the job remains to be done.

31. Strategy To find the number of votes cast, multiply the number of registered voters (31,281) by the fraction of those who voted $\left(\frac{2}{3}\right)$.

Solution
$31,281 \cdot \frac{2}{3} = \frac{31,281 \cdot 2}{3}$
$= 20,854$

In the city election 20,854 votes were cast.

32. Strategy To find the number of miles driven per gallon, divide the number of miles driven (402.5) by the number of gallons of gas used (11.5).

Solution $\frac{402.5}{11.5} = 35$

The car travels 35 mi on each gallon of gas.

33. Strategy To find the rpm in third gear, write and solve an equation using n to represent the rpm in third gear.

Solution

| $\frac{2}{3}$ of the rpm in third gear | is | 2,500 |

$\frac{2}{3}n = 2,500$

$\frac{3}{2}\left(\frac{2}{3}n\right) = \frac{3}{2}(2,500)$

$n = 3,750$

The rpm of the engine is 3,750 in third gear.

Chapter 8: Percent

Prep Test

1. $\dfrac{19}{100}$

2. 0.23

3. 47

4. 2850

5. $0.015\overline{)60.000}$ — quotient 4000.

6. $8 \div \dfrac{1}{4} = \dfrac{8}{1} \times \dfrac{4}{1} = 32$

7. $\dfrac{5}{8} \times \dfrac{100}{1} = \dfrac{5 \cdot 2 \cdot 2 \cdot 5 \cdot 5}{2 \cdot 2 \cdot 2} = \dfrac{125}{2} = 62\dfrac{1}{2} = 62.5$

8. $66\dfrac{2}{3}$

9. $16\overline{)28.00}$ — quotient 1.75

Go Figure

If my father's parents have 10 grandchildren, and I have 2 brothers and 1 sister, then my siblings and I account for 4 out of the 10 grandchildren. We have 6 first cousins on my father's side.
If my mother's parents have 11 grandchildren, and I have 2 brothers and 1 sister, then my siblings and I account for 4 out of the 11 grandchildren. We have 7 first cousins on my mother's side.
Adding the 6 first cousins on my father's side to the 7 first cousins on my mother's side, results in 13 first cousins.

Section 8.1

Objective A Exercises

1. **a.** Students should explain that to convert a percent to a fraction, drop the percent sign and multiply by $\dfrac{1}{100}$.

   b. Students should explain that to convert a percent to a decimal, drop the percent sign and multiply by 0.01.

3. $5\% = 5\left(\dfrac{1}{100}\right) = \dfrac{5}{100} = \dfrac{1}{20}$
   $5\% = 5(0.01) = 0.05$

5. $30\% = 30\left(\dfrac{1}{100}\right) = \dfrac{30}{100} = \dfrac{3}{10}$
   $30\% = 30(0.01) = 0.30$

7. $250\% = 250\left(\dfrac{1}{100}\right) = \dfrac{250}{100} = \dfrac{5}{2}$
   $250\% = 250(0.01) = 2.50$

9. $28\% = 28\left(\dfrac{1}{100}\right) = \dfrac{28}{100} = \dfrac{7}{25}$
   $28\% = 28(0.01) = 0.28$

11. $35\% = 35\left(\dfrac{1}{100}\right) = \dfrac{35}{100} = \dfrac{7}{20}$
    $35\% = 35(0.01) = 0.35$

13. $29\% = 29\left(\dfrac{1}{100}\right) = \dfrac{29}{100}$
    $29\% = 29(0.01) = 0.29$

15. $11\dfrac{1}{9}\% = 11\dfrac{1}{9}\left(\dfrac{1}{100}\right) = \dfrac{100}{9}\left(\dfrac{1}{100}\right) = \dfrac{1}{9}$

17. $37\dfrac{1}{2}\% = 37\dfrac{1}{2}\left(\dfrac{1}{100}\right) = \dfrac{75}{2}\left(\dfrac{1}{100}\right) = \dfrac{3}{8}$

19. $66\dfrac{2}{3}\% = 66\dfrac{2}{3}\left(\dfrac{1}{100}\right) = \dfrac{200}{3}\left(\dfrac{1}{100}\right) = \dfrac{2}{3}$

21. $6\dfrac{2}{3}\% = 6\dfrac{2}{3}\left(\dfrac{1}{100}\right) = \dfrac{20}{3}\left(\dfrac{1}{100}\right) = \dfrac{1}{15}$

23. $\dfrac{1}{2}\% = \dfrac{1}{2}\left(\dfrac{1}{100}\right) = \dfrac{1}{200}$

25. $6\dfrac{1}{4}\% = 6\dfrac{1}{4}\left(\dfrac{1}{100}\right) = \dfrac{25}{4}\left(\dfrac{1}{100}\right) = \dfrac{1}{16}$

27. $7.3\% = 7.3(0.01) = 0.073$

29. $15.8\% = 15.8(0.01) = 0.158$

31. $0.3\% = 0.3(0.01) = 0.003$

33. $121.2\% = 121.2(0.01) = 1.212$

35. $62.14\% = 62.14(0.01) = 0.6214$

37. $8.25\% = 8.25(0.01) = 0.0825$

39. $24\% = 24\left(\dfrac{1}{100}\right) = \dfrac{24}{100} = \dfrac{6}{25}$

$\dfrac{6}{25}$ of the owners would buy a house or car with their dog in mind.

Objective B Exercises

41. $0.37 = 0.37(100\%) = 37\%$

43. $0.02 = 0.02(100\%) = 2\%$

45. $0.125 = 0.125(100\%) = 12.5\%$

47. $1.36 = 1.36(100\%) = 136\%$

49. $0.96 = 0.96(100\%) = 96\%$

51. $0.07 = 0.07(100\%) = 7\%$

53. $\dfrac{83}{100} = \dfrac{83}{100}(100\%) = 83\%$

55. $\dfrac{1}{3} = \dfrac{1}{3}(100\%) \approx 33.3\%$

57. $\dfrac{4}{9} = \dfrac{4}{9}(100\%) \approx 44.4\%$

59. $\dfrac{9}{20} = \dfrac{9}{20}(100\%) = 45\%$

61. $2\dfrac{1}{2} = \dfrac{5}{2} = \dfrac{5}{2}(100\%) = 250\%$

63. $\dfrac{1}{6} = \dfrac{1}{6}(100\%) \approx 16.7\%$

65. $\dfrac{17}{25} = \dfrac{17}{25}(100\%) = 68\%$

67. $\dfrac{9}{16} = \dfrac{9}{16}(100\%) = 56\dfrac{1}{4}\%$

69. $2\dfrac{5}{8} = \dfrac{21}{8} = \dfrac{21}{8}(100\%) = 262\dfrac{1}{2}\%$

71. $2\dfrac{5}{6} = \dfrac{17}{6} = \dfrac{17}{6}(100\%) = 283\dfrac{1}{3}\%$

73. $\dfrac{7}{30} = \dfrac{7}{30}(100\%) = 23\dfrac{1}{3}\%$

75. $\dfrac{2}{9} = \dfrac{2}{9}(100\%) = 22\dfrac{2}{9}\%$

Critical Thinking 8.1

77. Employee B's salary is the highest after the raise. Explanations will vary. One possible explanation is: Because each employee receives a 5% raise, 5% of the largest original salary will be the largest raise. Adding the largest raise to the largest original salary ensures that that employee will have the largest resulting salary.

Section 8.2

Objective A Exercises

1. Strategy   To find the amount, solve the basic percent equation.
   Percent = 8% = 0.08, base = 100, amount = n

   Solution   Percent · base = amount
   $0.08 \cdot 100 = n$
   $8 = n$
   8% of 100 is 8.

3. Strategy   To find the amount, solve the basic percent equation.
   Percent = 0.05% = 0.0005, base = 150, amount = n.

   Solution   Percent · base = amount
   $0.0005 \cdot 150 = n$
   $0.075 = n$
   0.05% of 150 is 0.075.

5. Strategy   To find the amount, solve the basic percent equation.
   Percent = n, base = 90, amount = 15

   Solution   Percent · base = amount
   $n \cdot 90 = 15$
   $n = \dfrac{15}{90} = 0.16\dfrac{2}{3}$
   $n = 16\dfrac{2}{3}\%$
   15 is $16\dfrac{2}{3}\%$ of 90.

Section 8.2

7. **Strategy** To find the percent, solve the basic percent equation.
   Percent = n, base = 16, amount = 6

   Solution Percent · base = amount
   $n \cdot 16 = 6$
   $n = \dfrac{6}{16} = 0.375$
   $n = 37.5\%$

9. **Strategy** To find the base, solve the basic percent equation.
   Percent = 10% = 0.10, base = n, amount = 10

   Solution Percent · base = amount
   $0.10 \cdot n = 10$
   $n = \dfrac{10}{0.10}$
   $n = 100$
   10 is 10% of 100.

11. **Strategy** To find the base, solve the basic percent equation.
    Percent = 2.5% = 0.025, base = n, amount = 30

    Solution Percent · base = amount
    $0.025 \cdot n = 30$
    $n = \dfrac{30}{0.025}$
    $n = 1{,}200$
    2.5% of 1,200 is 30.

13. **Strategy** To find the amount, solve the basic percent equation.
    Percent = 10.7% = 0.107, base = 485, amount = n

    Solution Percent · base = amount
    $0.107 \cdot 485 = n$
    $51.895 = n$
    10.7% of 485 is 51.895.

15. **Strategy** To find the amount, solve the basic percent equation.
    Percent = 80% = 0.80, base = 16.25, amount = n

    Solution Percent · base = amount
    $0.80 \cdot 16.25 = n$
    $13 = n$
    80% of 16.25 is 13.

17. **Strategy** To find the percent, solve the basic percent equation.
    Percent = n, base = 2,000, amount = 54

    Solution Percent · base = amount
    $n \cdot 2{,}000 = 54$
    $n = \dfrac{54}{2{,}000} = 0.027$
    $n = 2.7\%$
    54 is 2.7% of 2,000.

19. **Strategy** To find the percent, solve the basic percent equation.
    Percent = n, base = 4.1, amount = 16.4

    Solution Percent · base = amount
    $n \cdot 4.1 = 16.4$
    $n = \dfrac{16.4}{4.1} = 4$
    $n = 400\%$
    16.4 is 400% of 4.1.

21. **Strategy** To find the percent, solve the basic percent equation.
    Percent = 240% = 2.40, base = n, amount = 18

    Solution Percent · base = amount
    $2.40 \cdot n = 18$
    $n = \dfrac{18}{2.40}$
    $n = 7.5$
    18 is 240% of 7.5.

Objective B Exercises

23. Percent = 26, base = 250, amount = n
    $\dfrac{\text{percent}}{100} = \dfrac{\text{amount}}{\text{base}}$
    $\dfrac{26}{100} = \dfrac{n}{250}$
    $26 \cdot 250 = 100 \cdot n$
    $6{,}500 = 100n$
    $65 = n$
    26% of 250 is 65.

© Houghton Mifflin Company. All rights reserved.

25. Percent = n, base = 148, amount = 37
$$\frac{\text{percent}}{100} = \frac{\text{amount}}{\text{base}}$$
$$\frac{n}{100} = \frac{37}{148}$$
$n \cdot 148 = 100 \cdot 37$
$148n = 3{,}700$
$n = \frac{3{,}700}{148}$
$n = 25$
37 is 25% of 148.

27. Percent = 68, base = n, amount = 51
$$\frac{\text{percent}}{100} = \frac{\text{amount}}{\text{base}}$$
$$\frac{68}{100} = \frac{51}{n}$$
$68 \cdot n = 100 \cdot 51$
$68n = 5{,}100$
$n = \frac{5{,}100}{68}$
$n = 75$
68% of 75 is 51.

29. Percent = n, base = 344, amount = 43
$$\frac{\text{percent}}{100} = \frac{\text{amount}}{\text{base}}$$
$$\frac{n}{100} = \frac{43}{344}$$
$n \cdot 344 = 100 \cdot 43$
$344n = 4{,}300$
$n = \frac{4{,}300}{344}$
$n = 12.5$
43 is 12.5% of 344.

31. Percent = 20.5, base = n, amount = 82
$$\frac{\text{percent}}{100} = \frac{\text{amount}}{\text{base}}$$
$$\frac{20.5}{100} = \frac{82}{n}$$
$20.5 \cdot n = 100 \cdot 82$
$20.5n = 8{,}200$
$n = \frac{8{,}200}{20.5}$
$n = 400$
82 is 20.5% of 400.

33. Percent = 6.5, base = 300, amount = n
$$\frac{\text{percent}}{100} = \frac{\text{amount}}{\text{base}}$$
$$\frac{6.5}{100} = \frac{n}{300}$$
$6.5 \cdot 300 = 100 \cdot n$
$1{,}950 = 100n$
$19.5 = n$
19.5 is 6.5% of 300.

35. Percent = n, base = 50, amount = 7.4
$$\frac{\text{percent}}{100} = \frac{\text{amount}}{\text{base}}$$
$$\frac{n}{100} = \frac{7.4}{50}$$
$n \cdot 50 = 100 \cdot 7.4$
$50n = 740$
$n = \frac{740}{50} = 14.8$
7.4 is 14.8% of 50.

37. Percent = 50.5, base = 124, amount = n
$$\frac{\text{percent}}{100} = \frac{\text{amount}}{\text{base}}$$
$$\frac{50.5}{100} = \frac{n}{124}$$
$50.5 \cdot 124 = 100 \cdot n$
$6{,}262 = 100n$
$62.62 = n$
62.62 is 50.5% of 124.

39. Percent = 120, base = n, amount = 6
$$\frac{\text{percent}}{100} = \frac{\text{amount}}{\text{base}}$$
$$\frac{120}{100} = \frac{6}{n}$$
$120 \cdot n = 100 \cdot 6$
$120n = 600$
$n = \frac{600}{120}$
$n = 5$
120% of 5 is 6.

41. Percent = 250, base = 18, amount = n
$$\frac{\text{percent}}{100} = \frac{\text{amount}}{\text{base}}$$
$$\frac{250}{100} = \frac{n}{18}$$
$250 \cdot 18 = 100 \cdot n$
$4{,}500 = 100n$
$45 = n$
250% of 18 is 45.

43. Percent = n, base = 29, amount = 87
$$\frac{\text{percent}}{100} = \frac{\text{amount}}{\text{base}}$$
$$\frac{n}{100} = \frac{87}{29}$$
$n \cdot 29 = 100 \cdot 87$
$29n = 8{,}700$
$n = \frac{8{,}700}{29}$
$n = 300$
87 is 300% of 29.

Objective C Exercises

45. Strategy To find the life of the brakes, use the basic percent equation.
Percent = 12% = 0.12, base = n, amount = 6,000

Solution Percent · base = amount
$0.12 \cdot n = 6,000$
$n = \dfrac{6,000}{0.12}$
$n = 50,000$
The estimated life of the brakes is 50,000 mi.

47. Strategy To find the number of workers in Arkansas, use the basic percent equation.
Percent = 3.3% = 0.033, base = n, amount = 41,000

Solution Percent · base = amount
$0.033 \cdot n = 41,000$
$n = \dfrac{41,000}{0.033}$
$1,242,424 = n$
There are 1,242,424 workers in the labor force in Arkansas.

49. Strategy To find the percent of false alarms, use the basic percent equation.
Percent = n, base = 200, amount = 24

Solution Percent · base = amount
$n \cdot 200 = 24$
$n = \dfrac{24}{200}$
$n = 0.12 = 12\%$
The fire department received 12% false alarms.

51. Strategy To find the percent, use the basic percent equation.
Percent = n, base = 651,700, amount = 948,300

Solution Percent · base = amount
$n \cdot 651,700 = 948,300$
$n = \dfrac{948,300}{651,700}$
$n \approx 1.455 = 145.5\%$
The increase is 145.5% of the 2000 population.

53. Strategy To find the number of pounds of turkey, use the basic percent equation.
Percent = 18.6% = 0.186, base = n, amount = 1,300,000

Solution Percent · base = amount
$0.186n = 1,300,000$
$n = \dfrac{1,300,000}{0.186}$
$n \approx 7,000,000$
The U.S. total turkey production was 7 million pounds.

55. Strategy To find the tax credit, use the basic percent equation.
Percent = 15% = 0.15, base = 85,000, amount = n

Solution Percent · base = amount
$0.15 \cdot 85,000 = n$
$12,750 = n$
The farmer received $12,750 in tax credits.

57. Strategy To find the number of grams, use the basic percent equation.
Percent = 0.05% = 0.0005, base = 30, amount = n

Solution Percent · base = amount
$0.0005(30) = n$
$0.015 = n$
There are 0.015 g of the ingredient in a 30-gram tube.

59. Strategy To find the percent, use the basic percent equation.
Percent = n, base = 8,000, amount = 870,000

Solution Percent · base = amount
$n \cdot 8,000 = 870,000$
$n = \dfrac{870,000}{8,000}$
$n = 108.75 = 10,875\%$
The sun's diameter is 10,875% of Earth's diameter.

61. Strategy → To find the number of boards tested, use the basic percent equation.
Percent = 0.7% = 0.007, base = n, amount = 56
→ To find the boards not defective, subtract 56 from the total number of boards tested.

Solution Percent · base = amount
$0.007 \cdot n = 56$
$n = \dfrac{56}{0.007} = 8{,}000$
$8{,}000 - 56 = 7{,}944$
8,000 computer boards were tested.
7,944 of the boards were not defective.

63. Strategy To find how many more faculty members described their views as liberal than far left,
→ Subtract the percent of Far Left from the percent of Liberal
→ Use the basic percent equation. Percent = difference, base = 32,840, amount = n

Solution Percent · base = amount
$42.3 - 5.3 = 37\% = 0.37$
$0.37 \cdot 32{,}840 = n$
$n = 12{,}150$
There were 12,151 more faculty members who described their views as liberal than far left.

Critical Thinking 8.2

65. Answers will vary.

Find 10% of 100.	$0.10 \cdot 100 = 10$
	$100 - 10 = 90$
Now find 10% of the result	$0.10 \cdot 90 = 9$
	$90 - 9 = 81$
Find 20% of 100.	$0.20 \cdot 100 = 20$
	$100 - 20 = 80$

No, the results of taking two consecutive 10% discounts or one 20% discount is not the same. The 20% discount was on the total of 100. The second 10% discount applied only to 90, not 100, thus the difference in results.

67. For example, consider an initial salary of $20,000 and raises of 5%, 6%, and 7%.
Raise after Year 1: 5%(20,000) = 0.05(20,000) = 1,000
Salary after Year 1: 20,000 + 1,000 = 21,000

Raise after Year 2: 6%(21,000) = 0.06(21,000) = 1,260
Salary after Year 2: 21,000 + 1,260 = 22,260

Raise after Year 3: 7%(22,260) = 0.07(22,260) = 1,558.20
Salary after Year 3: 22,260 + 1,558.20 = 23,818.20

For an initial salary of $20,000 and raises of 6% each year,
Raise after Year 1: 6%(20,000) = 0.06(20,000) = 1,200
Salary after Year 1: 20,000 + 1,200 = 21,200

Raise after Year 2: 6%(21,200) = 0.06(21,200) = 1,272
Salary after Year 2: 21,200 + 1,272 = 22,472

Raise after Year 3: 6%(22,472) = 0.06(22,472) = 1,348.32
Salary after Year 3: 22,472 + 1,348.32 = 23,820.32

23,818.20 < 23,820.32

The raises of 5%, 6%, and 7% result in a lower salary than raises of 6% each year.

69. Student responses will vary. Their answers might include any of the following: a raise in wages; cost of living increases; commissions; exam grades; income tax brackets; sales tax; discounted merchandise; interest rates on car loans, certificates of deposit, or mortgages; a rate of return on an investment. If students are having difficulty finding examples, suggest that they look through a newspaper for ideas.

Section 8.3

Objective A Exercises

1. **Strategy** To find percent increase:
   → Find the increase in the commuting time.
   → Use the basic percent equation. Percent = n, base = 26, amount = amount of increase

   Solution
   $31.2 - 26 = 5.2$
   Percent · base = amount
   $n \cdot 26 = 5.2$
   $n = \dfrac{5.2}{26}$
   $n = 0.20 = 20\%$
   The percent increase in the commuting time is 20%.

3. **Strategy** To find the percent increase:
   → Find the increase in the number of women
   → Use the basic percent equation.
   Percent = n, base = 3.8, amount = amount of increase

   Solution
   $5.2 - 3.8 = 1.4$
   Percent · base = amount
   $n \cdot 3.8 = 1.4$
   $n = \dfrac{1.4}{3.8}$
   $n \approx 0.368 = 36.8\%$
   The percent increase is 36.8%.

5. **Strategy** To find the percent increase:
   → Find the increase in the number of millionaire households.
   → Use the basic percent equation. Percent = n, base = 350,000, amount = amount of increase

   Solution
   $5,600,000 - 350,000 = 5,250,000$
   Percent · base = amount
   $n \cdot 350,000 = 5,250,000$
   $n = \dfrac{5,250,000}{350,000}$
   $n = 15 = 1,500\%$
   The percent increase in the number of millionaire households is 1,500%.

Chapter 8: Percent

7. Strategy
   a. To find the 2-year period when the percent increase is the greatest, find the percent increase for all 2-year periods and compare.
   b. To find the 2-year period when the percent increase is the least, find the percent increase for all 2-year periods and compare.
   c. To determine the type of increase, compare each of the increases.

Solution
   a. $10.4 - 9.6 = 0.8$
   Percent · base = amount
   $n \cdot 9.6 = 0.8$
   $n = \dfrac{0.8}{9.6}$
   $n \approx 0.083 = 8.3\%$ 1998–2000
   $11 - 10.4 = 0.6$
   Percent · base = amount
   $n \cdot 10.4 = 0.6$
   $n = \dfrac{0.6}{10.4}$
   $n \approx 0.058 = 5.8\%$ 2000–2002
   $11.2 - 11 = 0.2$
   Percent · base = amount
   $n \cdot 11 = 0.2$
   $n = \dfrac{0.2}{11}$
   $n \approx 0.0181 = 1.81\%$ 2002–2004
   $11.4 - 11.2 = 0.2$
   Percent · base = amount
   $n \cdot 11.2 = 0.2$
   $n = \dfrac{0.2}{11.2}$
   $n \approx 0.0179 = 1.79\%$ 2004–2006
   $8.3\% > 5.8\% > 1.81\% > 1.79\%$
   The greatest percent increase was in 1998–2000.

   b. $1.79\% < 1.81\% < 5.8\% < 8.3\%$
   The least percent increase was in 2004–2006.

   c. $8.3\% > 5.8\% > 1.81\% > 1.79\%$
   The growth in telecommuting, increases more slowly.

Objective B Exercises

9. Strategy To find the percent decrease in time, use the basic percent equation.
Percent = n, base = 26.5, amount = the decrease in size of the ozone hole

Solution $26.5 - 15.6 = 10.9$
Percent · base = amount
$n \cdot 26.5 = 10.9$
$n = \dfrac{10.9}{26.5}$
$n \approx 0.411 = 41.1\%$
The size of the ozone hole decreased by 41.1%.

11. Strategy To find the decrease in average time waiting:
→Find the decrease in time.
→Use the basic percent equation to find the percent decrease in time waiting.
Percent = n, base = 3.8
amount = decrease in time

Solution $3.8 - 2.5 = 1.3$
Percent · base = amount
$n \cdot 3.8 = 1.3$
$n = \dfrac{1.3}{3.8}$
$n \approx 0.3421$
$n \approx 34.2\%$
The time waiting decreased approximately 34.2%.

13. Strategy To find the value after 1 year:
→Find the decrease in value by using the basic percent equation.
Percent = 30% = 0.30, base = 21,900, amount = n
→Subtract the decrease in value from 21,900.

Solution Percent · base = amount
$0.30 \cdot 21{,}900 = n$
$6{,}570 = n$
$21{,}900 - 6{,}570 = 15{,}330$
The value of the car after 1 year is $15,330.

15. Strategy To find the branch in which the percent decrease was the greatest, find the percent decrease for all military branches and compare:

Solution $751 - 480 = 271$
Percent · base = amount
$n \cdot 751 = 271$
$n = \dfrac{271}{751}$
$n \approx 0.361 = 36.1\%$ Army
$583 - 372 = 211$
Percent · base = amount
$n \cdot 583 = 211$
$n = \dfrac{211}{583}$
$n \approx 0.362 = 36.2\%$ Navy
$539 - 361 = 178$
Percent · base = amount
$n \cdot 539 = 178$
$n = \dfrac{178}{539}$
$n \approx 0.330 = 33.0\%$ Air Force
$197 - 172 = 25$
Percent · base = amount
$n \cdot 197 = 25$
$n = \dfrac{25}{197}$
$n \approx 0.127 = 12.7\%$ Marines
$36.2\% > 36.1\% > 33.0\% > 12.7\%$
The Navy had the greatest decrease of personnel at 36.2%.

Critical Thinking 8.3

17. A 30% discount reduces the price by $900
A further discount of 10% reduces the price by $210. The total discount is $1,110.
The sale price after the two discounts is $1,890.
A single discount of 40% reduces the price to $1,800.
By comparing the sale prices, the 40% discount is $90 greater than the successive discounts of 30% and 10%.
The equivalent discount of the successive discounts is 37%.

$3,000 \cdot 0.30 = 900$
$3,000 - 900 = 2,100$
$2,100 \cdot 0.10 = 210$
$2,100 - 210 = 1,890$

$3,000 \cdot 0.40 = 1,200$
$3,000 - 1,200 = 1,800$

$\dfrac{1,110}{3,000} = 0.37$

Section 8.4

Objective A Exercises

1. **Strategy** To find the markup, solve the formula $M = r \cdot C$ for M.
$r = 55\%$, $C = 110$

    Solution $M = r \cdot C$
    $M = 0.55 \cdot 110$
    $M = 60.50$
    The markup on the cost is $60.50.

3. **Strategy** To find the markup rate:
    →Solve the formula $M = S - C$ for M.
    $S = 156.80$, $C = 98$
    →Solve the formula $M = r \cdot C$ for r.

Chapter 8: Percent

    Solution
$M = S - C$
$M = 156.80 - 98$
$M = 58.80$
$M = r \cdot C$
$58.80 = r \cdot 98$
$\dfrac{58.80}{98} = r$
$0.60 = r$
The markup rate on the cost is 60%.

5. Strategy
To find the markup rate:
→Solve the formula
$M = S - C$ for M.
$S = 520$, $C = 360$
→Solve the formula
$M = r \cdot C$ for r.

    Solution
$M = S - C$
$M = 520 - 360$
$M = 160$
$M = r \cdot C$
$160 = r \cdot 360$
$\dfrac{160}{360} = r$
$0.4444 \approx r$
The markup rate on the cost is approximately 44.4%.

7. Strategy
To find the selling price, solve the formula
$S = (1 + r)C$ for S.
$r = 25\%$, $C = 1{,}750$

    Solution
$S = (1 + r)C$
$S = (1 + 0.25) \cdot 1{,}750$
$s = 1.25 \cdot 1{,}750$
$S = 2{,}187.50$
The selling price of the computer is $2,187.50.

9. Strategy
To find the selling price, solve the formula
$S = (1 + r)$ for S.
$r = 75\%$, $C = 47$

    Solution
$S = (1 + r)C$
$S = (1 + 0.75) \cdot 47$
$S = 1.75 \cdot 47$
$S = 82.25$
The selling price of the PC game is $82.25.

Objective B Exercises

11. Strategy
To find the markdown, solve the formula
$M = R - S$ for M.
$R = 460$, $S = 350$

    Solution
$M = R - S$
$M = 460 - 350$
$M = 110$
The markdown is $110.

13. Strategy
To find the markdown rate:
→Solve the formula
$M = R - S$ for M.
$R = 1{,}295$, $S = 995$
→Solve the formula
$M = r \cdot R$ for r.

    Solution
$M = R - S$
$M = 1{,}295 - 995$
$M = 300$
$M = r \cdot R$
$300 = r \cdot 1{,}295$
$\dfrac{300}{1{,}295} = r$
$0.2317 \approx r$
The discount rate is approximately 23.2%.

15. **Strategy** To find the discount rate:
    → Solve the formula
    $M = R - S$ for M.
    $R = 325$, $S = 201.50$
    → Solve the formula
    $M = r \cdot R$ for r.

    Solution
    $M = R - S$
    $M = 325 - 201.50$
    $M = 123.50$
    $M = r \cdot R$
    $123.50 = r \cdot 325$
    $\dfrac{123.50}{325} = r$
    $0.38 = r$
    The discount rate is 38%.

17. **Strategy** To find the sale price, solve the formula
    $S = (1 - r)R$ for S.
    $r = 30\%$, $R = 1{,}995$

    Solution
    $S = (1 - r)R$
    $S = (1 - 0.30) \cdot 1{,}995$
    $S = 0.70 \cdot 1{,}995$
    $S = 1{,}396.50$
    The sale price is $1,396.50.

19. **Strategy** To find the sale price, solve the formula
    $S = (1 - r)R$ for S.
    $r = 40\%$, $R = 42$

    Solution
    $S = (1 - r)R$
    $S = (1 - 0.40) \cdot 42$
    $S = 0.60 \cdot 42$
    $S = 25.20$
    The sale price is $25.20.

21. **Strategy** To find the regular price, solve the formula
    $S(1 - r)R$ for R.
    $r = 40\%$, $S = 180$

    Solution
    $S = (1 - r)R$
    $180 = (1 - 0.40) \cdot R$
    $180 = 0.60R$
    $\dfrac{180}{0.60} = R$
    $300 = R$
    The regular price is $300.

23. **Strategy** To find the regular price, solve the formula
    $S(1 - r)R$ for R.
    $r = 35\%$, $S = 80$

    Solution
    $S = (1 - r)R$
    $80 = (1 - 0.35) \cdot R$
    $80 = 0.65R$
    $\dfrac{80}{0.65} = R$
    $123.08 \approx R$
    The regular price is approximately $123.08.

Critical Thinking 8.4

25. The sale price after a 20% discount is $4,400.

    $S = (1 - 0.20)5{,}500$
    $= 4{,}400$

    Another 10% discount would give a sale price of $3,960.

    $S = (1 - 0.10)4{,}400$
    $= 3{,}960$

    A single discount of 30% would give a sale price of $3,850.

    $S = (1 - 0.30)5{,}500$
    $= 3{,}850$

    Thus a 30% discount is not equivalent to the successive discounts of 20% and 10%.

    The total amount of the successive discounts is $5{,}500 - 3{,}960 = 1{,}540$. The single discount equivalent is 28%.

    $\dfrac{1{,}540}{5{,}500} = 0.28$

Section 8.5

Objective A Exercises

1. In the simple interest formula $I = Prt$, I is the simple interest earned, P is the principal, r is the annual simple interest rate, and t is the time, in years.

3. **a.** $I = Prt$

    1 month: $I = 5{,}000 \cdot 0.06 \cdot \dfrac{1}{12} = \$25$

    2 month: $I = 5{,}000 \cdot 0.06 \cdot \dfrac{2}{12} = \$50$

    3 month: $I = 5{,}000 \cdot 0.06 \cdot \dfrac{3}{12} = \$75$

    4 month: $I = 5{,}000 \cdot 0.06 \cdot \dfrac{4}{12} = \$100$

    5 month: $I = 5{,}000 \cdot 0.06 \cdot \dfrac{5}{12} = \$125$

    b. $150

    c. $175

    d. $200

    e. $225

5. **Strategy** — To find the simple interest, solve the simple interest formula $I = Prt$ for I.
    $P = 15{,}000$, $t = \dfrac{90}{365}$, $r = 0.074$

    Solution $I = Prt$
    $I = 15{,}000(0.074)\left(\dfrac{90}{365}\right)$
    $I = 273.70$
    The interest on the loan is $273.70.

7. **Strategy** — To find the simple interest, solve the simple interest formula $I = Prt$ for I.
    $P = 100{,}000$, $t = \dfrac{9}{12}$, $r = 0.09$

    Solution $I = Prt$
    $I = (100{,}000)(0.09)\left(\dfrac{9}{12}\right)$
    $I = 6{,}750$
    The interest on the loan is $6,750.

9. **Strategy** — To find the simple interest, solve the simple interest formula $I = Prt$ for I.
    $P = 1{,}250$, $rt = 0.016$ (rate per month)

    Solution $I = Prt$
    $I = (1{,}250)(0.016)$
    $I = 20$
    The interest owed to VISA is $20.

11. **Strategy** — To find the simple interest, solve the simple interest formula $I = Prt$ for I.
    $P = 8{,}000$, $t = 2$, $r = 0.09$

    Solution $I = Prt$
    $I = (8{,}000)(0.09)(2)$
    $I = 1{,}440$
    The interest on the 2-year loan is $1,440.

13. **Strategy** — To find the maturity value:
    →Solve the formula $I = Prt$ for I.
    $P = 150{,}000$, $r = 0.095$, $t = 1$
    →Use the formula for the maturity value of a simple interest loan, $M = P + I$.

    Solution $I = Prt$
    $I = 150{,}000(0.095)(1)$
    $I = 14{,}250$
    $M = P + I$
    $M = 150{,}000 + 14{,}250$
    $M = 164{,}250$
    The maturity value is $164,250.

15. **Strategy** To find the maturity value:
    → Solve the formula $I = Prt$ for I.
    $P = 14{,}000$, $r = 0.1025$, $t = \dfrac{270}{365}$.
    → Use the formula for the maturity value of a simple interest loan, $M = P + I$.

    Solution
    $I = Prt$
    $I = (14{,}000)(0.1025)\left(\dfrac{270}{365}\right)$
    $I = 1{,}061.51$
    $M = P + I$
    $M = 14{,}000 + 1{,}061.51$
    $M = 15{,}061.51$
    The maturity value is $15,061.51.

17. **Strategy** To find the interest rate, solve the formula $I = Prt$ for r.
    $I = 462$, $P = 12{,}000$, $t = \dfrac{6}{12}$

    Solution
    $I = Prt$
    $462 = (12{,}000)(r)\left(\dfrac{6}{12}\right)$
    $462 = 6{,}000r$
    $\dfrac{462}{6{,}000} = r$
    $0.077 = r$
    The simple interest rate is 7.7%.

19. **Strategy** To find the rate, solve the formula $I = Prt$ for r.
    $I = 937.50$, $P = 50{,}000$, $t = \dfrac{75}{365}$

    Solution
    $I = Prt$
    $937.50 = 50{,}000(r)\left(\dfrac{75}{365}\right)$
    $937.50 = \dfrac{750{,}000}{73}r$
    $\dfrac{73 \cdot 937.50}{750{,}000} = r$
    $0.09125 = r$
    The interest rate is 9.125%.

21. Explanations will vary. For example, students may describe interest as the amount of money charged for the privilege of borrowing money so, in this sense, it could be considered a "rental fee".

Chapter Review Exercises

1. $32\% = 32\left(\dfrac{1}{100}\right) = \dfrac{32}{100} = \dfrac{8}{25}$

2. $22\% = 22(0.01) = 0.22$

3. $25\% = 25\left(\dfrac{1}{100}\right) = \dfrac{25}{100} = \dfrac{1}{4}$
   $25\% = 25(0.01) = 0.25$

4. $3\dfrac{2}{5}\% = 3\dfrac{2}{5}\left(\dfrac{1}{100}\right) = \dfrac{17}{5}\left(\dfrac{1}{100}\right) = \dfrac{17}{500}$

5. $\dfrac{7}{40} = \dfrac{7}{40}(100\%) = 17.5\%$

6. $1\dfrac{2}{7} = 1\dfrac{2}{7}(100\%) = \dfrac{9}{7}(100\%) = \dfrac{900}{7}\%$
   $\approx 128.6\%$

7. $2.8 = 2.8(100\%) = 280\%$

8. **Strategy** To find the amount, solve the basic percent equation.
   Percent = $42\% = 0.42$, base = 50, amount = n

   Solution
   Percent · base = amount
   $0.42 \cdot 50 = n$
   $21 = n$
   42% of 50 is 21.

9. **Strategy** To find the percent, solve the basic percent equation.
   Percent = n, base = 3, amount = 15

   Solution
   Percent · base = amount
   $n \cdot 3 = 15$
   $n = \dfrac{15}{3}$
   $n = 5 = 500\%$
   500% of 3 is 15.

10. **Strategy** To find the percent, solve the basic percent equation.
    Percent = n, base = 18, amount = 12

    Solution Percent · base = amount
    $n \cdot 18 = 12$
    $n = \dfrac{12}{18}$
    $n \approx 0.667 = 66.7\%$
    12 is approximately 66.7% of 18.

11. **Strategy** To find the amount, solve the basic percent equation.
    Percent = 150% = 1.50, base = 20, amount = n

    Solution Percent · base = amount
    $1.50 \cdot 20 = n$
    $30 = n$
    150% of 20 is 30.

12. **Strategy** To find the amount, solve the basic percent equation.
    Percent = 18% = 0.18, base = 85, amount = n

    Solution Percent · base = amount
    $0.18 \cdot 85 = n$
    $15.3 = n$
    18% of 85 is 15.3.

13. **Strategy** To find the base, solve the basic percent equation.
    Percent = 32% = 0.32, base = n, amount = 180

    Solution Percent · base = amount
    $0.32 \cdot n = 180$
    $n = \dfrac{180}{0.32}$
    $n = 562.5$
    32% of 562.5 is 180.

14. **Strategy** To find the percent, solve the basic percent equation.
    Percent = n, base = 80, amount = 4.5

    Solution Percent · base = amount
    $n \cdot 80 = 4.5$
    $n = \dfrac{4.5}{80}$
    $n = 0.05625 = 5.625\%$
    4.5 is 5.625% of 80.

15. **Strategy** To find the amount, solve the basic percent equation.
    Percent = 0.58% = 0.0058, base = 2.54, amount = n

    Solution Percent · base = amount
    $0.0058 \cdot 2.54 = n$
    $0.014732 = n$
    0.58% of 2.54 is 0.014732.

16. **Strategy** To find the base, solve the basic percent equation.
    Percent = 0.05% = 0.0005, base = n, amount = 0.0048

    Solution Percent · base = amount
    $0.0005 \cdot n = 0.0048$
    $n = \dfrac{0.0048}{0.0005}$
    $n = 9.6$
    0.0048 is 0.05% of 9.6.

17. **Strategy** To find the percent visiting in China:
    →Add to find the total amount of tourists visiting the four countries.
    →Use the basic percent equation to find the percent.
    Percent = n, base = the total amount of tourists visiting the four countries, amount = 137 million

    Solution $137 + 93 + 71 + 102 = 403$
    Percent · base = amount
    $n \cdot 403 = 137$
    $n = \dfrac{137}{403}$
    $n \approx 0.340 = 34.0\%$
    About 34.0% of the tourists will be visiting China.

18. **Strategy** To find the amount of the budget spent for advertising, use the basic percent equation.
    Percent = 7% = 0.07, base = 120,000, amount = n

    Solution Percent · base = amount
    $0.07 \cdot 120{,}000 = n$
    $8{,}400 = n$
    $8,400 of the budget was spent for advertising.

19. **Strategy** To find the number of phones that were not defective: →Find the number of

defective phones by using the basic percent equation.
Percent = 1.2% = 0.012, base = 4,000, amount = n
→Subtract the number of defective phones from 4,000.

Solution Percent · base = amount
0.012 · 4,000 = n
48 = n
4,000 − 48 = 3,952
3,952 of the telephones were not defective.

20. Strategy To find the percent of the week, use the basic percent equation.
Percent = n, base = 168, amount = 61.35.

Solution Percent · base = amount
n · 168 = 61.35
$$n = \frac{61.35}{168}$$
$n \approx 0.365 = 36.5\%$
The percent is approximately 36.5% of the week.

21. Strategy To find the expected profit, use the basic percent equation.
Percent = 22% = 0.22, base = 750,000, amount = n

Solution Percent · base = amount
0.22 · 750,000 = n
165,000 = n
The expected profit is $165,000.

22. Strategy To find the number of seats added, use the basic percent equation.
Percent = 18% = 0.18, base = 9,000, amount = n

Solution Percent · base = amount
0.18 · 9,000 = n
1,620 = n
1,620 seats were added to the auditorium.

23. Strategy To find the number of tickets sold:
→Use the basic percent equation to find the number of seats overbooked.
Percent = 12% = 0.12, base = 175, amount = n
→Add the number of seats overbooked to 175.

Solution Percent · base = amount
0.12 · 175 = n
21 = n
21 + 175 = 196
The airline would sell 196 tickets.

24. Strategy To find the percent of registered voters that voted, use the basic percent equation.
Percent = n, based = 112,000, amount = 25,400

Solution Percent · base = amount
n · 112,000 = 25,400
$$n = \frac{25,400}{112,000}$$
$n \approx 0.227 = 22.7\%$
Approximately 22.7% of the registered voters voted in the city election.

25. Strategy To find the clerk's new hourly wage:
→Find the increase by using the basic percent equation.
Percent = 8% = 0.08, base = 10.50, amount = n
→Add the increase to 10.50.

Solution Percent · base = amount
0.08 · 10.50 = n
0.84 = n
0.84 + 10.50 = 11.34
The clerk's new wage is $11.34 per hour.

196 Chapter 8: *Percent*

26. Strategy To find the percent decrease in cost:
→Subtract to find the dollar decrease in the cost of the computer.
→Find the percent decrease by using the basic percent equation.
Percent = n, base = 2,400, amount = decrease in price

Solution $2,400 - 1,800 = 600$
Percent · base = amount
$n \cdot 2,400 = 600$
$n = \dfrac{600}{2,400}$
$n = 0.25 = 25\%$
The computer decreased 25% in cost.

27. Strategy To find the selling price of the car:
→Use the basic percent equation to find the markup.
Percent = 6% = 0.06, base = 18,500, amount = n
→Add the markup to 18,500.

Solution Percent · base = amount
$0.06 \cdot 18,500 = n$
$1,110 = n$
$1,110 + 18,500 = 19,610$
The selling price of the car is $19,610.

28. Strategy To find the markup rate:
→Solve the formula $M = S - C$ for M.
$S = 181.50$, $C = 110$
→Solve the formula $M = r \cdot C$ for r.

Solution $M = S - C$
$M = 181.50 - 110$
$M = 71.50$
$M = r \cdot C$
$71.50 = r \cdot 110$
$\dfrac{71.50}{110} = r$
$0.65 = r$
The markup rate of the parka is 65%.

29. Strategy To find the sale price, solve the formula
$S = (1-r)R$ for S.
$r = 0.30$, $R = 80$

Solution $S = (1-r)R$
$S = (1 - 0.30) \cdot 80$
$S = 0.70 \cdot 80$
$S = 56$
The sale price of the tennis racket is $56.

30. Strategy To find the sale price, solve the formula
$S = (1-r)R$ for S.
$r = 0.40$, $R = 650$

Solution $S = (1-r)R$
$S = (1 - 0.40) \cdot 650$
$S = 0.60 \cdot 650$
$S = 390$
The sale price of the ticket is $390.

31. Strategy To find the simple interest, solve the formula
$I = Prt$ for I.
$P = 3,000$, $r = 0.086$, $t = \dfrac{45}{365}$

Solution $I = Prt$
$I = 3,000(0.086)\left(\dfrac{45}{365}\right)$
$I = 31.81$
The interest on the loan is $31.81.

32. Strategy To find the rate, solve the formula $I = Prt$ for r.
$I = 7,397.26$, $P = 500,000$, $t = \dfrac{60}{365}$

Solution $I = Prt$
$7,397.26 = 500,000(r)\left(\dfrac{60}{365}\right)$
$7,397.26 = \dfrac{6,000,000}{73}r$
$\dfrac{73 \cdot 7,397.26}{6,000,000} = r$
$0.09 = r$
The interest rate is 9.00%.

33. Strategy   To find the maturity value:
→Solve the formula $I = Prt$ for I.
$P = 10,000$, $r = 0.084$, $t = \dfrac{9}{12}$.
→Use the formula for the maturity value of a simple interest loan, $M = P + I$.

Solution $I = Prt$
$I = (10,000)(0.084)\left(\dfrac{9}{12}\right)$
$I = 640$
$M = P + I$
$M = 10,000 + 630$
$M = 10,630$
The maturity value is $10,630.

Chapter Test

1. $86.4\% = 86.4(0.01) = 0.864$

2. $0.4 = 0.4(100\%) = 40\%$

3. $\dfrac{5}{4} = \dfrac{5}{4}(100\%) = 125\%$

4. $83\dfrac{1}{3}\% = 83\dfrac{1}{3}\left(\dfrac{1}{100}\right) = \dfrac{250}{3}\left(\dfrac{1}{100}\right) = \dfrac{5}{6}$

5. $32\% = 32\left(\dfrac{1}{100}\right) = \dfrac{32}{100} = \dfrac{8}{25}$

6. $1.18 = 1.18(100\%) = 118\%$

7. Strategy   To find the base, solve the basic percent equation.
Percent = 20% = 0.20, base = n, amount = 18

Solution   Percent · base = amount
$0.20 \cdot n = 18$
$n = \dfrac{18}{0.20} = 90$
20% of 90 is 18.

8. Strategy   To find the amount, solve the basic percent equation.
Percent = 68% = 0.68, base = 73, amount = n

Solution   Percent · base = amount
$0.68 \cdot 73 = n$
$49.64 = n$
68% of 73 is 49.64.

9. Strategy   To find the percent, solve the basic percent equation.
Percent = n, base = 320, amount = 180

Solution   Percent · base = amount
$n \cdot 320 = 180$
$n = \dfrac{180}{320}$
$n = 0.5625 = 56.25\%$
56.25% of 320 is 180.

10. Strategy   To find the base, solve the basic percent equation.
Percent = 14% = 0.14, base = n, amount = 28

Solution   Percent · base = amount
$0.14 \cdot n = 28$
$n = \dfrac{28}{0.14} = 200$
14% of 200 is 28.

11. Strategy   To find the amount of expected accidents, use the basic percent equation.
Percent = 2.2% = 0.022, base = 1,500, amount = n

Solution   Percent · base = amount
$0.022 \cdot 1,500 = n$
$33 = n$
33 accidents are expected.

12. **Strategy** To find the percent answered correctly, use the basic percent equation.
    Percent = n, base = 90, amount = 90 − 16 = 74.

    Solution Percent · base = amount
    $n \cdot 90 = 74$
    $n = \dfrac{74}{90}$
    $n \approx 0.822 = 82.2\%$
    The percent is approximately 82.2% correct.

13. **Strategy** To find the dollar increase:
    →Find the increase by using the basic percent equation.
    Percent = 120% = 1.20, base = n, amount = 480
    →Subtract the increase from 480.

    Solution Percent · base = amount
    $1.20 \cdot n = 480$
    $n = \dfrac{480}{1.20}$
    $n = 400$
    $480 - 400 = 80$
    The dollar increase is $80.

14. **Strategy** To find the percent decrease in cost:
    →Subtract to find the dollar increase in the cost for public tuition.
    →Find the percent increase by using the basic percent equation.
    Percent = n, base = 7,628, amount = increase in price

    Solution 19,143 − 7,628 = 11,515
    Percent · base = amount
    $n \cdot 7{,}628 = 11{,}515$
    $n = \dfrac{11{,}515}{7{,}628}$
    $n \approx 1.510 = 151.0\%$
    The tuition increased 151.0%.

15. **Strategy** To find the percent decrease in cost:
    →Subtract to find the increase in trainees.
    →Find the percent increase by using the basic percent equation.
    Percent = n, base = 36, amount = increase in trainees

    Solution 42 − 36 = 6
    Percent · base = amount
    $n \cdot 36 = 6$
    $n = \dfrac{6}{36}$
    $n \approx 0.1667 = 16\dfrac{2}{3}\%$
    The number of trainees increased by approximately $16\dfrac{2}{3}\%$.

16. **Strategy**
    a. To find the percent decrease in fat:
    →Subtract to find the decrease in fat.
    →Find the percent decrease by using the basic percent equation.
    Percent = n, base = 24, amount = decrease in fat.

    b. To find the percent decrease in cholesterol:
    →Subtract to find the decrease in cholesterol.
    →Find the percent decrease by using the basic percent equation.
    Percent = n, base = 75, amount = decrease in cholesterol.

    c. To find the percent decrease in calories:
    →Subtract to find the decrease in calories.
    →Find the percent decrease by using the basic percent equation.
    Percent = n, base = 280, amount = decrease in calories.

Chapter Test

Solution

a. $24 - 4 = 20$
Percent · base = amount
$n \cdot 24 = 20$
$n = \dfrac{20}{24}$
$n \approx 0.8333 = 83\dfrac{1}{3}\%$

The fat content decreased by $83\dfrac{1}{3}\%$.

b. $75 - 0 = 75$
Percent · base = amount
$n \cdot 75 = 75$
$n = \dfrac{75}{75}$
$n = 1 = 100\%$
The cholesterol percent decreased by 100%.

c. $280 - 140 = 140$
Percent · base = amount
$n \cdot 280 = 140$
$n = \dfrac{140}{280}$
$n = 0.5 = 50\%$
The calorie percent decreased by 50%.

17. Strategy   To find the decrease in personnel:
→Find the decrease in travel expenses.
→Use the basic percent equation to find the percent decrease in expenses. Percent = n, base = 25,000, amount = decrease in expenses.

Solution $25,000 - 23,000 = 2,000$
Percent · base = amount
$n \cdot 25,000 = 2,000$
$n = \dfrac{2,000}{25,000}$
$n = 0.08 = 8\%$
The amount of travel expenses decreased by 8%.

18. Strategy   To find the dollar increase:
→Find the price from last year by using the basic percent equation.
Percent = 125% = 1.25, base = n, amount = 1,500
→Subtract last year's price from 1,500.

Solution   Percent · base = amount
$1.25n = 1,500$
$n = \dfrac{1,500}{1.25} = 1,200$
$1,500 - 1,200 = 300$
The dollar increase is $300.

19. Strategy   To find the markup, solve the formula $M = r \cdot C$ for M.
$r = 60\%$, $C = 21$

Solution $M = r \cdot C$
$M = 0.60 \cdot 21$
$M = 12.60$
The markup on the cost is $12.60.

20. Strategy   To find the markup rate:
→Solve the formula $M = S - C$ for M.
$S = 349$, $C = 225$
→Solve the formula $M = r \cdot C$ for r.

Solution $M = S - C$
$M = 349 - 225$
$M = 124$
$M = r \cdot C$
$124 = r \cdot 225$
$\dfrac{124}{225} = r$
$0.551 \approx r$
The markup rate on the cost is approximately 55.1%.

21. Strategy To find the regular price, solve the formula
$S(1 - r)R$ for R.
$r = 40\%$, $S = 180$

Solution $S = (1 - r)R$
$180 = (1 - 0.40) \cdot R$
$180 = 0.6R$
$\dfrac{180}{0.6} = R$
$300 = R$
The regular price is $300.

22. Strategy To find the discount rate:
→Solve the formula
$M = r \cdot R$ for r, $M = 51.80$, $R = 370$.

Solution $M = r \cdot R$
$51.80 = r \cdot 370$
$\dfrac{51.80}{370} = r$
$0.14 = r$
The discount rate is 14%.

23. Strategy To find the simple interest, solve the interest formula $I = Prt$ for I.
$P = 5,000$, $t = \dfrac{9}{12}$, $r = 0.084$

Solution $I = Prt$
$I = (5,000)(0.084)\left(\dfrac{9}{12}\right)$
$I = 315$
The interest on the loan is $315.

24. Strategy To find the maturity value:
→Solve the formula $I = Prt$ for I.
$P = 40,000$, $r = 0.0925$, $t = \dfrac{150}{365}$.
→Use the formula for the maturity value of a simple interest loan, $M = P + I$.

Solution $I = Prt$
$I = (40,000)(0.0925)\left(\dfrac{150}{365}\right)$
$I = 1,520.55$
$M = P + I$
$M = 40,000 + 1,520.55$
$M = 41,520.55$
The maturity value is $41,520.55.

25. Strategy To find the rate, solve the formula $I = Prt$ for r.
$I = 672$, $P = 12,000$, $t = \dfrac{8}{12}$

Solution $I = Prt$
$672 = 12,000(r)\left(\dfrac{8}{12}\right)$
$672 = 8,000r$
$\dfrac{672}{8,000} = r$
$0.084 = r$
The interest rate is 8.4%.

Cumulative Review Exercises

1. $a - b$
   $102.5 - 77.546 = 24.954$

2. $5^4 = 5 \cdot 5 \cdot 5 \cdot 5$
   $= 625$

3. $(4.67)(3.007) = 14.04269$

4. $(2x - 3)(2x - 5) = 4x^2 - 10x - 6x + 15$
   $= 4x^2 - 16x + 15$

5. $3\frac{5}{8} \div 2\frac{7}{12} = \frac{29}{8} \div \frac{31}{12}$
    $= \frac{29}{8} \cdot \frac{12}{31}$
    $= \frac{29 \cdot 12}{8 \cdot 31}$
    $= \frac{29 \cdot 2 \cdot 2 \cdot 3}{2 \cdot 2 \cdot 2 \cdot 31}$
    $= \frac{87}{62} = 1\frac{25}{62}$

6. $-2a^2b(-3ab^2 + 4a^2b^3 - ab^3)$
    $= 6a^3b^3 - 8a^4b^4 + 2a^3b^4$

7. Strategy   To find the amount, use the basic percent equation.
    Percent = 120% = 1.20, base = 35, amount = n

    Solution   Percent · base = amount
    $1.20 \cdot 35 = n$
    $42 = n$
    120% of 35 is 42.

8. $x - 2 = -5$
    $x - 2 + 2 = -5 + 2$
    $x = -3$
    The solution is -3.

9. $1.005 \times 10^5 = 100{,}500$

10. $-\frac{5}{8} - \left(-\frac{3}{4}\right) + \frac{5}{6} = -\frac{5}{8} + \frac{3}{4} + \frac{5}{6}$
    $= \frac{-15}{24} + \frac{18}{24} + \frac{20}{24}$
    $= \frac{-15 + 18 + 20}{24}$
    $= \frac{23}{24}$

11. $\frac{3 - \frac{7}{8}}{\frac{11}{12} + \frac{1}{4}} = \frac{\frac{24}{8} - \frac{7}{8}}{\frac{11}{12} + \frac{3}{12}} = \frac{\frac{17}{8}}{\frac{14}{12}} = \frac{\frac{17}{8}}{\frac{7}{6}}$
    $= \frac{17}{8} \div \frac{7}{6}$
    $= \frac{17}{8} \cdot \frac{6}{7}$
    $= \frac{17 \cdot 6}{8 \cdot 7} = \frac{51}{28}$
    $= 1\frac{23}{28}$

12. $(-3a^2b)(4a^5b^4) = (-3 \cdot 4)(a^2a^5)(bb^4)$
    $= -12a^7b^5$

13.
x	y
2	1
1	3
0	5

    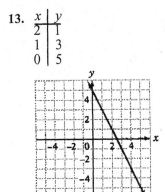

14.
x	y
3	3
0	-2
-3	-7

    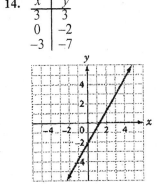

15. $\frac{7}{8} \div \frac{5}{16} = \frac{7}{8} \cdot \frac{16}{5}$
    $= \frac{7 \cdot 16}{8 \cdot 5}$
    $= \frac{7 \cdot 2 \cdot 2 \cdot 2 \cdot 2}{2 \cdot 2 \cdot 2 \cdot 5}$
    $= \frac{14}{5} = 2\frac{4}{5}$

16. $4 - (-3) + 5 - 8 = 4 + 3 + 5 + (-8)$
    $= 7 + 5 + (-8)$
    $= 12 + (-8)$
    $= 4$

17. $\frac{3}{4}x = -9$
    $\frac{4}{3}\left(\frac{3}{4}x\right) = \frac{4}{3}(-9)$
    $x = -12$
    The solution is -12.

18. $6x - 9 = -3x + 36$
    $6x + 3x - 9 = -3x + 3x + 36$
    $9x - 9 = 36$
    $9x - 9 + 9 = 36 + 9$
    $9x = 45$
    $\dfrac{9x}{9} = \dfrac{45}{9}$
    $x = 5$
    The solution is 5.

19. $\dfrac{322.4 \text{ mi}}{5 \text{ h}} = 64.48$ mph

20. $\dfrac{32}{n} = \dfrac{5}{7}$
    $32 \cdot 7 = n \cdot 5$
    $224 = 5n$
    $\dfrac{224}{5} = \dfrac{5n}{5}$
    $44.8 = n$

21. **Strategy**   To find the percent, use the basic percent equation.
    Percent = n, base = 30, amount = 2.5

    Solution   Percent · base = amount
    $n \cdot 30 = 2.5$
    $n = \dfrac{2.5}{30} \approx 0.0833$

    2.5 is approximately 8.3% of 30.

22. **Strategy**   To find the amount, use the basic percent equation.
    Percent = 42% = 0.42, base = 160, amount = n

    Solution   Percent · base = amount
    $0.42 \cdot 160 = n$
    $67.2 = n$
    42% of 160 is 67.2.

23. $44 - (-6)^2 \div (-3) + 2 = 44 - 36 \div (-3) + 2$
    $= 44 - (-12) + 2$
    $= 44 + 12 + 2$
    $= 56 + 2$
    $= 58$

24. $3(x - 2) + 2 = 11$
    $3x - 6 + 2 = 11$
    $3x - 4 = 11$
    $3x - 4 + 4 = 11 + 4$
    $3x = 15$
    $\dfrac{3x}{3} = \dfrac{15}{3}$
    $x = 5$
    The solution is 5.

25. **Strategy**   To find the fraction, convert 10% to a fraction.

    Solution $10\% = 10\left(\dfrac{1}{100}\right) = \dfrac{10}{100} = \dfrac{1}{10}$

    $\dfrac{1}{10}$ of the population aged 75–84 are affected by Alzheimer's Disease.

26. **Strategy**   To find the sale price, solve the formula $S = (1 - r)R$ for S.
    $r = 0.36$, $R = 202.50$

    Solution $S = (1 - r)R$
    $S = (1 - 0.36) \cdot 202.50$
    $S = 0.64 \cdot 202.50$
    $S = 129.60$
    The sale price is $129.60.

27. **Strategy**   To find the cost of the calculator, solve the formula $S = (1 + r)C$ for C.
    $S = 67.20$, $r = 0.60$

    Solution $S = (1 + r)C$
    $67.20 = (1 + 0.60)C$
    $67.20 = 1.60C$
    $\dfrac{67.20}{1.60} = C$
    $42 = C$
    The cost of the calculator is $42.

28. Strategy To find the number of games the team will win, write and solve a proportion using n to represent the number of games won.

Solution $\dfrac{13}{18} = \dfrac{n}{162}$
$13 \cdot 162 = 18 \cdot n$
$2{,}106 = 18n$
$\dfrac{2{,}106}{18} = n$
$117 = n$
The team will win 117 games.

29. Strategy To find the amount of weight to lose:
→Add the amount already lost $\left(3\dfrac{1}{2} + 2\dfrac{1}{4}\right)$.
→Subtract the amount lost from 8.

Solution $3\dfrac{1}{2} + 2\dfrac{1}{4} = 3\dfrac{2}{4} + 2\dfrac{1}{4}$
$= 5\dfrac{3}{4}$
$8 - 5\dfrac{3}{4} = 7\dfrac{4}{4} - 5\dfrac{3}{4}$
$= 2\dfrac{1}{4}$
The wrestler must lose another $2\dfrac{1}{4}$ lb.

30. Strategy To find the speed, substitute 81 for d in the given formula and solve for v.

Solution $v = \sqrt{64d}$
$v = \sqrt{64 \cdot 81}$
$v = \sqrt{64} \cdot \sqrt{81}$
$v = 8 \cdot 9 = 72$
$v = 72$
The speed of the falling object is 72 ft/s.

31. Strategy To convert meters per second to kilometers per hour, use the conversion factors $\dfrac{3{,}600 \text{ s}}{1 \text{ h}}$ and $\dfrac{1 \text{ km}}{1{,}000 \text{ m}}$.

Solution $\dfrac{400 \text{ m}}{43.84 \text{ s}}$
$= \dfrac{400 \text{ m}}{43.84 \text{ s}} \cdot \dfrac{3{,}600 \text{ s}}{1 \text{ h}} \cdot \dfrac{1 \text{ km}}{1{,}000 \text{ m}}$
≈ 32.85 km/h
The speed was approximately 32.85 km/h.

32. Strategy To find the number of hours worked:
→Subtract the cost of materials (192) from the total cost (1,632).
→Divide the cost of labor by 40.

Solution $1{,}632 - 192 = 1{,}440$
$\dfrac{1{,}440}{40} = 36$
The plumber worked for 36 h.

33. Strategy To find the resistance:
→Write the basic inverse variation equations, replace the variables by the given values, and solve for k.
→Write the inverse variation equation, replacing k by its value. Substitute 8 for I and solve for R.

Solution $I = \dfrac{k}{R}$
$2 = \dfrac{k}{20}$
$2 \cdot 20 = k$
$40 = k$
$I = \dfrac{40}{R}$
$8 = \dfrac{40}{R}$
$8R = 40$
$R = \dfrac{40}{8} = 5$
The resistance is 5 ohms.

Chapter 9: Geometry

Prep Test

1. $2(18) + 2(10) = 36 + 20 = 56$

2. abc
   $= (2)(3.14)(9)$
   $= (6.28)(9)$
   $= 56.52$

3. xyz^3
   $= \left(\dfrac{4}{3}\right)(3.14)(3)^3$
   $= 113.04$

4. $x + 47 = 90$
   $x + 47 - 47 = 90 - 47$
   $x = 43$

5. $32 + 97 + x = 180$
   $129 + x = 180$
   $129 - 129 + x = 180 - 129$
   $x = 51$

6. $\dfrac{5}{12} = \dfrac{6}{x}$
   $5x = 12 \times 6$
   $\dfrac{5x}{5} = \dfrac{36}{5}$
   $x = 14.4$

Go Figure

The first figure is a diamond (D) inside a square (S) inside a triangle (T) inside a circle (C), or DSTC.
The second figure is STCD.
The third figure is TCDS.
The next figure would be CDST: a circle inside a diamond inside a square inside a triangle.

Section 9.1

Objective A Exercises

1. The measure of the given angle is approximately 40°. The measure of the angle is between 0° and 90°. The angle is an acute angle.

3. The measure of the given angle is approximately 115°. The measure of the angle is between 90° and 180°. The angle is an obtuse angle.

5. The measure of the given angle is approximately 90°. The angle is a right angle.

7. Strategy Complementary angle are two angles whose sum is 90°. To find the complement, let x represent the complement of a 62° angle. Write an equation and solve for x.

   Solution $x + 62° = 90°$
   $x = 28°$
   The complement of a 62° angle is a 28° angle.

9. Strategy Supplementary angles are two angles whose sum is 180°. To find the supplement, let x represent the supplement of a 162° angle. Write an equation and solve for x.

   Solution $x + 162° = 180°$
   $x = 18°$
   The supplement of a 162° angle is an 18° angle.

11. $AB + BC + CD = AD$
    $12 + BC + 9 = 35$
    $21 + BC = 35$
    $BC = 14$
    $BC = 14$ cm

13. $QR + RS = QS$
    $QR + 3(QR) = QS$
    $7 + 3 \cdot 7 = QS$
    $7 + 21 = QS$
    $28 = QS$
    $QS = 28$ ft

15. $EF + FG = EG$
    $EF + \dfrac{1}{2}(EF) = EG$
    $20 + \dfrac{1}{2}(20) = EG$
    $20 + 10 = EG$
    $30 = EG$
    $EG = 30$ m

17. $\angle LOM + \angle MON = \angle LON$
    $53° + \angle MON = 139°$
    $\angle MON = 139° - 53°$
    $\angle MON = 86°$

© Houghton Mifflin Company. All rights reserved.

Section 9.1   **205**

19. **Strategy** To find the measure of ∠x, write an equation using the fact that the sum of the measure of ∠x and 74° is 145°. Solve for ∠x.

    Solution ∠x + 74° = 145°
    ∠x = 71°
    The measure of ∠x is 71°.

21. **Strategy** To find the measure of ∠x, write an equation using the fact that the sum of the measures of ∠x and ∠2x is 90°. Solve for ∠x.

    Solution x + 2x = 90°
    3x = 90°
    x = 30°
    The measure of x is 30°.

23. **Strategy** To find the measure of ∠x, write an equation using the fact that the sum of x and x + 18° is 90°. Solve for x.

    Solution x + x + 18° = 90°
    2x + 18° = 90°
    2x = 72°
    x = 36°
    The measure of ∠x is 36°.

25. **Strategy** To find the measure of ∠a, write an equation using the fact that the sum of the measure of ∠a and 53° is 180°. Solve for ∠a.

    Solution ∠a + 53° = 180°
    ∠a = 127°
    The measure of ∠a is 127°.

27. **Strategy** The sum of the measures of the three angles shown is 360°. To find ∠a, write an equation and solve for ∠a.

    Solution ∠a + 76° + 168° = 360°
    ∠a + 244° = 360°
    ∠a = 116°
    The measure of ∠a is 116°.

29. **Strategy** The sum of the measures of the three angles shown is 180°. To find x, write an equation and solve for x.

    Solution 3x + 4x + 2x = 180°
    9x = 180°
    x = 20°
    The measure of x is 20°.

31. **Strategy** The sum of the measures of the three angles shown is 180°. To find x, write an equation and solve for x.

    Solution 5x + (x + 20°) + 2x = 180°
    8x + 20° = 180°
    8x = 160°
    x = 20°
    The measure of x is 20°.

33. **Strategy** The sum of the measures of the four angles shown is 360°. To find x, write an equation and solve for x.

    Solution 3x + 4x + 6x + 5x = 360°
    18x = 360°
    x = 20°
    The measure of x is 20°.

35. **Strategy**

    To find the measure of ∠b:
    →Use the fact that ∠a and ∠c are complementary angles.
    →Find ∠b by using the fact that ∠c and ∠b are supplementary angles.

    Solution ∠a + ∠c = 90°
    51° + ∠c = 90°
    ∠c = 39°
    ∠b + ∠c = 180°
    ∠b + 39° = 180°
    ∠b = 141°
    The measure of ∠b is 141°.

Objective B Exercises

37. Strategy The angles labeled are adjacent angles of intersecting lines and are, therefore, supplementary angles. To find x, write an equation and solve for x.

Solution
$x + 74° = 180°$
$x = 106°$
The measure of x is $106°$.

39. Strategy The angles labeled are vertical angles and are, therefore, equal. To find x, write an equation and solve for x.

Solution
$5x = 3x + 22°$
$2x = 22°$
$x = 11°$
The measure of x is $11°$.

41. Strategy
→To find the measure of $\angle a$, use the fact that corresponding angles of parallel lines are equal.
→To find the measure of $\angle b$, use the fact that adjacent angles of intersecting lines are supplementary.

Solution
$\angle a = 38°$
$\angle b + \angle a = 180°$
$\angle b + 38° = 180°$
$\angle b = 142°$
The measure of $\angle a$ is $38°$.
The measure of $\angle b$ is $142°$.

43. Strategy
→To find the measure of $\angle a$, use the fact that alternate interior angles of parallel lines are equal.
→To find the measure of $\angle b$, use the fact that adjacent angles of intersecting lines are supplementary.

Solution
$\angle a = 47°$
$\angle a + \angle b = 180°$
$47° + \angle b = 180°$
$\angle b = 133°$
The measure of $\angle a$ is $47°$.
The measure of $\angle b$ is $133°$.

45. Strategy

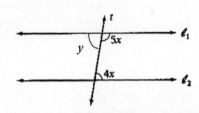

$4x = y$ because alternate interior angles have the same measure. $y + 5x = 180°$ because adjacent angles of intersecting lines are supplementary. Substitute $4x$ for y and solve for x.

Solution
$4x + 5x = 180°$
$9x = 180°$
$x = 20°$
The measure of x is $20°$.

47. Strategy

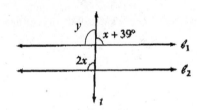

$y = 2x$ because corresponding angles have the same measure. $y + x + 39° = 180°$ because adjacent angles of intersecting lines are supplementary angles. Substitute $2x$ for y and solve for x.

Solution
$2x + x + 39° = 180°$
$3x + 39° = 180°$
$3x = 141°$
$x = 47°$
The measure of x is $47°$.

Objective C Exercises

49. Strategy

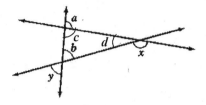

→ To find the measure of ∠y, use the fact that ∠b and ∠y are vertical angles.
→ To find the measure of ∠x:
Find the measure of ∠c by using the fact that the sum of an interior and exterior angle is 180°.
Find the measure of ∠d by using the fact that the sum of the interior angles of a triangle is 180°.
Find the measure of ∠x, by using the fact that the sum of an interior and exterior angle is 180°.

Solution
∠y = ∠b = 70°
∠a + ∠c = 180°
95° + ∠c = 180°
∠c = 85°
∠b + ∠c + ∠d = 180°
70° + 85° + ∠d = 180°
155° + ∠d = 180°
∠d = 25°
∠d + ∠x = 180°
25° + ∠x = 180°
∠x = 155°
The measure of ∠x is 155°.
The measure of ∠y is 70°.

51. Strategy

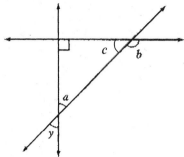

→ To find the measure of ∠a, use the fact that ∠a and ∠y are vertical angles.
→ To find the measure of ∠b:
Find the measure of ∠c by using the fact that the sum of the interior angles of a triangle is 180°.
Find the measure of ∠b by using the fact that the sum of an interior and exterior angle is 180°.

Solution
∠a = ∠y = 45°
∠a + ∠c + 90° = 180°
45° + ∠c + 90° = 180°
∠c + 135° = 180°
∠c = 45°
∠c + ∠b = 180°
45° + ∠b = 180°
∠b = 135°
The measure of ∠a is 45°.
The measure of ∠b is 135°.

208 Chapter 9: Geometry

53. Strategy To find the measure of ∠BOC, use the fact that the sum of the measure of the angles x, ∠AOB, and ∠AOB is 180°. Since $\overline{AO} \perp \overline{OB}$, ∠AOB = 90°

Solution
$x + ∠AOB + ∠BOC = 180°$
$x + 90° + ∠BOC = 180°$
$∠BOC = 90° - x$
The measure of ∠BOC is $90° - x$.

55. Strategy To find the measure of the third angle, use the fact that the sum of the measures of the interior angles of a triangle is 180°. Write an equation using x to represent the measure of the third angle. Solve the equation for x.

Solution
$x + 90° + 30° = 180°$
$x + 120° = 180°$
$x = 60°$
The measure of the third angle is 60°.

57. Strategy To find the measure of the third angle, use the fact that the sum of the measures of the interior angles of a triangle is 180°. Write an equation using x to represent the measure of the third angle. Solve the equation for x.

Solution
$x + 42° + 103° = 180°$
$x + 145° = 180°$
$x = 35°$
The measure of the third angle is 35°.

Critical Thinking 9.1

59. a. The smallest possible whole number of degrees in an angle of a triangle is 1°. For example, the other 2 angles could measure 100° and 79°, and $100° + 79° + 1° = 180°$.

b. The largest possible whole number of degrees in an angle of a triangle is 179°. For example, the other two angles could measure 0.5°, and $0.5° + 0.5° + 179° = 180°$.

61. a.

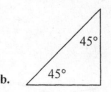

b.

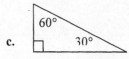

c.

63. Because the sum of the measures of an interior and an exterior angle is 180°:
$∠a + ∠y = 180°$
$∠b + ∠z = 180°$
$∠c + ∠x = 180°$
Adding these 3 equations results in
$∠a+∠y+∠b+∠z+∠c+∠x=180°+180°+180$
$∠a +∠y + ∠b + ∠z + ∠c + ∠x = 540°$
Because the sum of the measures of the interior angles of a triangle is 180°:
$∠a + ∠b + ∠c = 180°$
$(∠a +∠b + ∠c) + (∠y + ∠z + ∠x) = 540°$
$180° + (∠y + ∠z + ∠x) = 540°$
$∠y + ∠z + ∠x = 360°$
$∠x + ∠y + ∠z = 360°$
The sum of the measures of angles x, y, and z is 360°.

65. ∠AOC and ∠BOC are supplementary angles; therefore, $∠AOC + ∠BOC = 180°$. Since $∠AOC = ∠BOC$, by substitution $∠AOC + ∠AOC = 180°$. Therefore $2(∠AOC) = 180°$, and $∠AOC = 90°$. Therefore, $\overline{AB} \perp \overline{CD}$.

Section 9.2

Objective A Exercises

1. The polygon has 6 sides.
The polygon is a hexagon.

3. The polygon has 5 sides.
The polygon is a pentagon.

5. The triangle has no sides equal.
The triangle is a scalene triangle.

7. The triangle has three sides equal.
The triangle is an equilateral triangle.

9. The triangle has one obtuse angle.
The triangle is an obtuse triangle.

11. The triangle has three acute angles.
The triangle is an acute triangle.

13. Strategy   To find the perimeter, use the formula for the perimeter of a triangle. Substitute 12 for a, 20 for b, and 24 for c. Solve for P.

Solution $P = a + b + c$
$P = 12 + 20 + 24$
$P = 56$
The perimeter is 56 in.

15. Strategy   To find the perimeter, use the formula for the perimeter of a square. Substitute 3.5 for s and solve for P.

Solution $P = 4s$
$P = 4 \cdot 3.5$
$P = 14$
The perimeter is 14 ft.

17. Strategy   To find the perimeter, use the formula for the perimeter of a rectangle. Substitute 13 for L and 10.5 for W. Solve for P.

Solution $P = 2L + 2W$
$P = 2 \cdot 13 + 2 \cdot 10.5$
$P = 26 + 21$
$P = 47$
The perimeter is 47 mi.

19. Strategy   To find the circumference, use the circumference formula that involves the radius. For the exact answer, leave the answer in terms of π. For an approximation, use the π key on a calculator. $r = 4$.

Solution $C = 2\pi r$
$C = 2\pi(4)$
$C = 8\pi$
$C \approx 25.13$
The circumference is 8π cm.
The circumference is approximately 25.13 cm.

21. Strategy   To find the circumference, use the circumference formula that involves the radius. For the exact answer, leave the answer in terms of π. For an approximation use the π key on a calculator. $r = 5.5$.

Solution $C = 2\pi r$
$C = 2\pi(5.5)$
$C = 11\pi$
$C \approx 34.56$
The circumference is 11π mi.
The circumference is approximately 34.56 mi.

23. Strategy   To find the circumference, use the circumference formula that involves the diameter. For the exact answer leave the answer in terms of π. For an approximation use the π key on a calculator. $d = 17$.

Solution $C = \pi d$
$C = \pi(17)$
$C = 17\pi$
$C \approx 53.41$
The circumference is 17π ft.
The circumference is approximately 53.41 ft.

25. Strategy   To find the perimeter, use the formula for the perimeter of a triangle. Substitute 3.8 for a, 5.2 for b, and 8.4 for c. Solve for P.

Solution $P = a + b + c$
$P = 3.8 + 5.2 + 8.4$
$P = 17.4$
The perimeter is 17.4 cm.

27. Strategy   To find the perimeter, use the formula for the perimeter of a triangle. Substitute $2\frac{1}{2}$ for a and b, and 3 for c. Solve for P.

Solution $P = a + b + c$
$P = 2\frac{1}{2} + 2\frac{1}{2} + 3$
$P = 8$
The perimeter is 8 cm.

29. Strategy   To find the perimeter, use the formula for the perimeter of a rectangle. Substitute 8.5 for L and 3.5 for W. Solve for P.

Solution $P = 2L + 2W$
$P = 2(8.5) + 2(3.5)$
$P = 17 + 7$
$P = 24$
The perimeter is 24 m.

31. Strategy To find the perimeter, use the formula for the perimeter of a square. Substitute 12.2 for s. Solve for P.

    Solution $P = 4s$
    $P = 4(12.2)$
    $P = 48.8$
    The perimeter is 48.8 cm.

33. Strategy To find the perimeter, multiply the measure of one of the equal sides (3.5) by 5.

    Solution $P = 5(3.5)$
    $P = 17.5$
    The perimeter is 17.5 in.

35. Strategy To find the circumference, use the circumference formula that involves the diameter. Leave the answer in terms of π.
    $d = 1.5$

    Solution $C = \pi d$
    $C = \pi(1.5)$
    $C = 1.5\pi$
    The circumference is 1.5π in.

37. Strategy To find the circumference, use the circumference formula that involves the radius. An approximation is asked for; use the π key on a calculator.
    $r = 36$.

    Solution $C = 2\pi r$
    $C = 2\pi(36)$
    $C = 72\pi$
    $C \approx 226.19$
    The circumference is approximately 226.19 cm.

39. Strategy To find the amount of fencing, use the formula for the perimeter of a rectangle. Substitute 18 for L and 12 for W. Solve for P.

    Solution $P = 2L + 2W$
    $P = 2(18) + 2(12)$
    $P = 36 + 24$
    $P = 60$
    The perimeter of the garden is 60 ft.

41. Strategy To find the amount to be nailed down, use the formula for the perimeter of a rectangle. Substitute 12 for L and 10 for W. Solve for P.

    Solution $P = 2L + 2W$
    $P = 2(12) + 2(10)$
    $P = 24 + 20$
    $P = 44$
    44 ft of carpet must be nailed down.

43. Strategy To find the length, use the formula for the perimeter of a rectangle. Substitute 440 for P and 100 for W. Solve for L.

    Solution $P = 2L + 2W$
    $440 = 2L + 2(100)$
    $440 = 2L + 200$
    $240 = 2L$
    $120 = L$
    The length is 120 ft.

45. Strategy To find the third side of the banner, use the formula for the perimeter of a triangle. Substitute 46 for P, 18 for a, and 18 for b. Solve for c.

    Solution $P = a + b + c$
    $46 = 18 + 18 + c$
    $46 = 36 + c$
    $10 = c$
    The third side of the banner is 10 in.

47. Strategy   To find the length of each side, use the formula for the perimeter of a square. Substitute 48 for P. Solve for s.

    Solution
$$P = 4s$$
$$48 = 4s$$
$$12 = s$$
The length of each side is 12 in.

49. Strategy   To find the length of the diameter, use the circumference formula that involves the diameter. An approximation is asked for; use the π key on a calculator. $C = 8$.

    Solution
$$C = \pi d$$
$$8 = \pi d$$
$$\frac{8}{\pi} = d$$
$$2.55 \approx d$$
The diameter is approximately 2.55 cm.

51. Strategy   To find the length of molding, use the circumference formula that involves the diameter. An approximation is asked for; use the π key on a calculator. $d = 4.2$.

    Solution
$$C = \pi d$$
$$C = \pi(4.2)$$
$$C \approx 13.19$$
The length of molding is approximately 13.19 ft.

53. Strategy   To find the distance:
→Convert the diameter to feet.
→Multiply the circumference by 8. An approximation is asked for; use the π key on a calculator.

    Solution $24 \text{ in.} = 24 \text{ in.} \cdot \frac{1 \text{ ft}}{12 \text{ in.}} = 2 \text{ ft}$
distance $= 8C$
distance $= 8\pi d$
distance $= 8\pi(2)$
distance $= 16\pi$
distance ≈ 50.27
The bicycle travels approximately 50.27 ft.

55. Strategy   To find the circumference of the earth, use the circumference formula that involves the radius. An approximation is asked for; use the π key on a calculator. $r = 6{,}356$.

    Solution
$$C = 2\pi r$$
$$C = 2\pi(6{,}356)$$
$$C = 12{,}712\pi$$
$$C \approx 39{,}935.93$$
The circumference of the earth is approximately 39,935.93 km.

Objective B Exercises

57. Strategy   To find the area, use the formula for the area of a rectangle. Substitute 12 for L and 5 for W. Solve for A.

    Solution
$$A = LW$$
$$A = 12(5)$$
$$A = 60$$
The area is 60 ft^2.

59. Strategy   To find the area, use the formula for the area of the square. Substitute 4.5 for s. Solve for A.

    Solution
$$A = s^2$$
$$A = (4.5)^2$$
$$A = 20.25$$
The area is 20.25 in^2.

61. Strategy   To find the area, use the formula for the area of a triangle. Substitute 42 for b and 26 for h. Solve for A.

    Solution
$$A = \frac{1}{2}bh$$
$$A = \frac{1}{2}(42)(26)$$
$$A = 546$$
The area is 546 ft.^{2}.

63. Strategy To find the area, use the formula for the area of a circle. Substitute 4 for r. Solve for A. For the exact answer, leave the answer in terms of π. For an approximation, use the π key on a calculator.

Solution
$A = \pi r^2$
$A = \pi(4)^2$
$A = 16\pi$
$A \approx 50.27$
The area is 16π cm^2.
The area is approximately 50.27 cm^2.

65. Strategy To find the area, use the formula for the area of a circle. Substitute 5.5 for r. Solve for A. For the exact answer, leave the answer in terms of π. For an approximation, use the π key on a calculator.

Solution
$A = \pi r^2$
$A = \pi(5.5)^2$
$A = 30.25\pi$
$A \approx 95.03$
The area is 30.25π mi^2.
The area is approximately 95.03 mi^2.

67. Strategy To find the area:
→ Find the radius of the circle.
→ Use the formula for the area of a circle. For an exact answer, leave the answer in terms of π. For an approximation, use the π key on a calculator.

Solution
$r = \frac{1}{2}d = \frac{1}{2}(17) = 8.5$
$A = \pi r^2$
$A = \pi(8.5)^2$
$A = 72.25\pi$
$A \approx 226.98$
The area is 72.25π ft^2.
The area is approximately 226.98 ft^2.

69. Strategy To find the area, use the formula for the area of a square. Substitute 12.5 for s. Solve for A.

Solution
$A = s^2$
$A = (12.5)^2$
$A = 156.25$
The area is 156.25 cm^2.

71. Strategy To find the area, use the formula for the area of a rectangle. Substitute 38 for L and 15 for W. Solve for A.

Solution
$A = LW$
$A = 38(15)$
$A = 570$
The area is 570 in^2.

73. Strategy To find the area, use the formula for the area of a parallelogram. Substitute 16 for b and 12 for h. Solve for A.

Solution
$A = bh$
$A = 16(12)$
$A = 192$
The area is 192 in^2.

75. Strategy To find the area, use the formula for the area of a triangle. Substitute 6 for b and 4.5 for h. Solve for A.

Solution
$A = \frac{1}{2}bh$
$A = \frac{1}{2}(6)(4.5)$
$A = 13.5$
The area is 13.5 ft^2.

77. Strategy To find the area, use the formula for the area of a trapezoid. Substitute 35 for b_1, 20 for b_2, and 12 for h. Solve for A.

Solution
$A = \frac{1}{2}h(b_1 + b_2)$
$A = \frac{1}{2} \cdot 12(35 + 20)$
$A = 330$
The area is 330 cm^2.

79. Strategy To find the area, use the formula for the area of a circle. Leave the answer in terms of π. $r = 5$.

Section 9.2

Solution $A = \pi r^2$
$A = \pi(5)^2$
$A = 25\pi$
The area is 25π in^2.

81. Strategy To find the area:
→Find the radius of the telescope. $d = 200$.
→Use the formula for the area of a circle. Leave the answer in terms of π.

Solution $r = \frac{1}{2}d = \frac{1}{2}(200) = 100$
$A = \pi r^2$
$A = \pi(100)^2$
$A = 10,000\pi$
The area is $10,000\pi$ in^2.

83. Strategy To find the area, use the formula for the area of a rectangle. Substitute 14 for L and 9 for W. Solve for A.

Solution $A = LW$
$A = 14(9)$
$A = 126$
The area of the flower garden is 126 ft^2.

85. Strategy To find the amount of turf, use the formula for the area of a rectangle. Substitute 100 for L and 75 for W. Solve for A.

Solution $A = LW$
$A = 100(75)$
$A = 7,500$
7,500 yd^2 of artificial turf must be purchased.

87. Strategy To find the width, use the formula for the area of a rectangle. Substitute 300 for A and 30 for L. Solve for W.

Solution $A = LW$
$300 = 30W$
$10 = W$
The width of the rectangle is 10 in.

89. Strategy To find the length of the base, use the formula for the area of a triangle. Substitute 50 for A and 5 for h. Solve for b.

Solution $A = \frac{1}{2}bh$
$50 = \frac{1}{2}b(5)$
$50 = \frac{5}{2}b$
$20 = b$
The base of the triangle is 20 m.

91. Strategy To find the number of quarts of stain:
→Use the formula for the area of a rectangle to find the area of the deck.
→Divide the area of the deck by the area one quart will cover (50).

Solution $A = LW$
$A = 10(8)$
$A = 80$
$80 \div 15 = 1.6$
Because a portion of a second quart is needed, 2 qt of stain should be purchased.

93. Strategy To find the cost of the wallpaper:
→Use the formula for the area of a rectangle to find the areas of the two walls.
→Add the areas of the two walls.
→Divide the total area by the area in one roll (40) to find the total number of rolls.
→Multiply the number of rolls by 18.50

Solution
$A_1 = LW = 9(8) = 72$
$A_2 = LW = 11(8) = 88$
$A = A_1 + A_2 = 72 + 88 = 160$
$160 \div 40 = 4$
$4 \cdot 18.50 = 74$
The cost to wallpaper the two walls is $74.

95. Strategy To find the increase in area:
→Use the formula for the area of a circle to find the area of a circle with $r = 8$.
→Use the formula for the area of a circle to find the area of a circle with radius $r = 8 + 2 = 10$.
→Subtract the area of the smaller circle from the area of the larger circle. An approximation is asked for; use the π key on a calculator.

Solution
$A_1 = \pi r^2$
$A_1 = \pi(8)^2 = 64\pi$
$A_2 = \pi(10)^2 = 100\pi$
$A_2 - A_1 = 100\pi - 64\pi = 36\pi \approx 113.10$
The area is increased by 113.10 in^2.

97. Strategy To find the cost of the carpet:
→Use the formula of the area of a rectangle to find the area of the carpet.
→Use the conversion factor $\dfrac{1 \text{ yd}^2}{9 \text{ ft}^2}$.
→Multiply the area measured in square yards by 15.95.

Solution
$A = LW = 24(15) = 360$
$360 \text{ ft}^2 = 360 \text{ ft}^2 \cdot \dfrac{1 \text{ yd}^2}{9 \text{ ft}^2}$
$= 40 \text{ yd}^2 = 40(15.95) = 638$
The cost of the carpet is $638.

99. Strategy To find the area of the walkway:
→Use the formula for the area of a rectangle to find the area of the plot of grass. Substitute 30 for L and 20 for W.
→Use the formula for a rectangle to find the area of the total area (walkway + grass). Substitute $30 + 2 + 2 = 34$ for L and $20 + 2 + 2 = 24$ for W.
→Subtract the area of the grass from the total area.

Solution
$A_1 = LW = 30(20) = 600$
$A_2 = LW = 34(24) = 816$
$A_2 - A_1 = 816 - 600 = 216$
The area of the walkway is 216 m^2.

Critical Thinking 9.2

101. If the ratios of the lengths of the sides of the squares 2: 3, then we can let $s_1 = 2$ and $s_2 = 3$.
$A_1 = s_1^2 = 2^2 = 4$, and $A_2 = s_2^2 = 3^2 = 9$. The ratio of the areas of the two squares is 4 : 9.

105. **a.** Consider the three triangles shown below. The perimeter of each triangle is 24 units. The area of the first is 24 square units, the area of the second is 24.5 square units, and the area of the third is 24 square units. The statement is sometimes true.

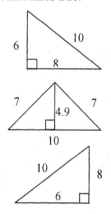

b. Consider the two rectangles shown below. The area of each rectangle is 20 square units. But the perimeter of the first rectangle is 18 units and the perimeter of the second rectangle is 24 units. The statement if sometimes true.

c. Since $A = s^2$, and $\sqrt{s^2} = s$, the length of a side of any square with area s^2 is s. The statement if always true.

d. An isosceles triangle has 2 sides of the same measure and 2 angles of the same measure. An equilateral triangle has 3 equal angles and 3 sides of the same measure. Therefore, 2 angles and 2 sides of an equilateral triangle are of equal measure. The statement is always true.

e. A circle is a plane figure in which all points are the same distance from the center. A radius of a circle is a line segment from the center to a point on the circle. The statement is always true.

f. The distance across a circle through the center of the circle will always be the same distance. The statement is always true.

107. There are a great number of quilt patterns that incorporate regular polygons, for example, Nine Patch Block, Grandmother's Flower Garden, Hour Glasses, Sunshine and Shadow, Field of Diamonds, and Trip Around the World.

Section 9.3

Objective A Exercises

1. **Strategy**   To find the hypotenuse, use the Pythagorean Theorem.
$a = 3, b = 4$

    Solution $c^2 = a^2 + b^2$
$c^2 = 3^2 + 4^2$
$c^2 = 9 + 16$
$c^2 = 25$
$c = \sqrt{25} = 5$
The length of the hypotenuse is 5 in.

3. **Strategy**   To find the hypotenuse, use the Pythagorean Theorem.
$a = 5, b = 7$

    Solution $c^2 = a^2 + b^2$
$c^2 = 5^2 + 7^2$
$c^2 = 25 + 49$
$c^2 = 74$
$c = \sqrt{74}$
$c \approx 8.6$
The length of the hypotenuse if approximately 8.6 cm.

3. Strategy To find the hypotenuse, use the Pythagorean Theorem. $a = 5, b = 7$

Solution
$c^2 = a^2 + b^2$
$c^2 = 5^2 + 7^2$
$c^2 = 25 + 49$
$c^2 = 74$
$c = \sqrt{74}$
$c \approx 8.6$
The length of the hypotenuse if approximately 8.6 cm.

7. Strategy To find the measure of the other leg, use the Pythagorean Theorem. $c = 6, a = 4$

Solution
$a^2 + b^2 = c^2$
$4^2 + b^2 = 6^2$
$16 + b^2 = 36$
$b^2 = 20$
$b = \sqrt{20}$
$b \approx 4.5$
The measure of the other leg is approximately 4.5 cm.

9. Strategy To find the hypotenuse, use the Pythagorean Theorem. $a = 9, b = 9$

Solution
$c^2 = a^2 + b^2$
$c^2 = 9^2 + 9^2$
$c^2 = 81 + 81$
$c^2 = 162$
$c = \sqrt{162}$
$c \approx 12.7$
The length of the hypotenuse is approximately 12.7 yd.

11. Strategy To find the distance, use the Pythagorean Theorem to find the hypotenuse of a right triangle. $a = 3, b = 8$

Solution
$c^2 = a^2 + b^2$
$c^2 = 3^2 + 8^2$
$c^2 = 9 + 64$
$c^2 = 73$
$c = \sqrt{73}$
$c \approx 8.5$
The distance between the holes is approximately 8.5 cm.

13. Strategy To find the perimeter:
→Use the Pythagorean Theorem to find the hypotenuse of the triangle. $a = 5, b = 9$
→Use the formula for the perimeter of a triangle to find the perimeter.

Solution
$c^2 = a^2 + b^2$
$c^2 = 5^2 + 9^2$
$c^2 = 25 + 81$
$c^2 = 106$
$c = \sqrt{106}$
$c \approx 10.3$
$P = a + b + c$
$P = 5 + 9 + 10.3$
$P = 24.3$
The perimeter is approximately 24.3 cm.

Objective B Exercises

15. The ratio is $\frac{5}{10} = \frac{1}{2}$.

17. The ratio is $\frac{6}{8} = \frac{3}{4}$.

19. Strategy To find DE, write a proportion using the fact that in similar triangles, the ratios of corresponding sides are equal. Solve the proportion for DE.

Solution
$\frac{AB}{DE} = \frac{AC}{DF}$
$\frac{4}{DE} = \frac{5}{9}$
$4(9) = (5)DE$
$36 = (5)DE$
$7.2 = DE$
The length of DE is 7.2 cm.

Section 9.3 **217**

21. **Strategy** To find the height of triangle *DEF*, write a proportion using the fact that, in similar triangles, the ratio of corresponding sides equals the ratio of corresponding sides equals the ratio of corresponding heights. Solve the proportion for the height (*h*).

    Solution
    $$\frac{AC}{DF} = \frac{2}{h}$$
    $$\frac{3}{5} = \frac{2}{h}$$
    $$3h = 5(2)$$
    $$3h = 10$$
    $$h \approx 3.3$$
    The height of triangle *DEF* is approximately 3.3 m.

23. **Strategy** To find the perimeter:
    →Find the side *BC* by writing a proportion using the fact that the ratios of corresponding sides of similar triangles are equal.
    →Use the formula for the perimeter of a triangle.

    Solution
    $$\frac{BC}{EF} = \frac{AC}{DF}$$
    $$\frac{BC}{6} = \frac{4}{8}$$
    $$(8)BC = 6(4)$$
    $$(8)BC = 24$$
    $$BC = 3$$
    $$P = a + b + c$$
    $$P = 3 + 4 + 5$$
    $$P = 12$$
    The perimeter of triangle *ABC* is 12 m.

25. **Strategy** To find the perimeter:
    →Find side *BC* by writing a proportion using the fact that the ratios of corresponding sides of similar triangles are equal.
    →Use the formula for the perimeter of a triangle.

    Solution
    $$\frac{BC}{EF} = \frac{AB}{DE}$$
    $$\frac{BC}{15} = \frac{4}{12}$$
    $$(12)BC = 15(4)$$
    $$(12)BC = 60$$
    $$BC = 5$$
    $$P = a + b + c$$
    $$P = 3 + 4 + 5$$
    The perimeter is 12 in.

27. **Strategy** To find the area:
    →Find the height (*h*) of triangle *ABC* by writing a proportion using the fact that, in similar triangles, the ratio of corresponding sides equals the ratio of corresponding heights.
    →Use the formula for the area of a triangle. *b* = 15

    Solution
    $$\frac{AB}{DE} = \frac{h}{20}$$
    $$\frac{15}{40} = \frac{h}{20}$$
    $$15(20) = 40h$$
    $$300 = 40h$$
    $$7.5 = h$$
    $$A = \frac{1}{2}bh$$
    $$A = \frac{1}{2}(15)(7.5) \approx 56.3$$
    The area of triangle *ABC* is approximately 56.3 cm².

29. **Strategy** To find the height, write a proportion using the fact that, in similar triangles, the ratios of corresponding sides are equal.

    Solution
    $$\frac{24}{8} = \frac{\text{height}}{6}$$
    $$24(6) = 8 \cdot \text{height}$$
    $$144 = 8 \cdot \text{height}$$
    $$18 = \text{height}$$
    The height of the flagpole is 18 ft.

© Houghton Mifflin Company. All rights reserved.

31. Strategy To find the height, write a proportion using the fact that, in similar triangles, the ratios of corresponding sides are equal.

Solution $\dfrac{\text{height}}{8} = \dfrac{8}{4}$

height · 4 = 8(8)
4 · height = 64
height = 16
The height of the building is 16 m.

Objective C Exercises

33. Strategy To determine if the triangles are congruent, determine if one of the rules for congruence is satisfied.

Solution $AC = DE$, $AB = EF$, and $\angle A = \angle E$.
Two sides and the included angle of one triangle equal two sides and the included angle of the other triangle.
The triangles are congruent by the SAS rule.

35. Strategy To determine if the triangles are congruent, determine if one of the rules for congruence is satisfied.

Solution $AB = DE$, $AC = EF$, and $BC = DF$.
Three sides of one triangle equal the three sides of the other triangle. The triangles are congruent by the SSS rule.

37. Strategy To determine if the triangles are congruent, determine if one of the rules for congruence is satisfied.

Solution $AC = DF$, $\angle A = \angle D$, and $\angle C = \angle F$.
Two angles and the included side of one triangle equal two angles and the included side of the other triangle.
The triangles are congruent by the ASA rule.

39. Strategy Draw a sketch of the two triangles and determine if one of the rules for congruence is satisfied.

Solution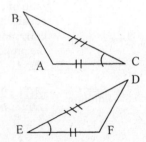
$AC = EF$, $BC = DE$, and $\angle C = \angle E$.
Because two sides and the included angle of one triangle equal two sides and the included angle of the other triangle, the triangles are congruent by the SAS rule.

41. Strategy Draw a sketch of the two triangles and determine if one of the rules of congruence is satisfied.

Solution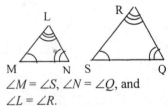
$\angle M = \angle S$, $\angle N = \angle Q$, and $\angle L = \angle R$.
The triangles do not satisfy the SSS rule, the SAS rule, or the ASA rule. The triangles are not necessarily congruent.

43. Strategy   Draw a sketch of the two triangles and determine if one of the rules of congruence is satisfied.

Solution

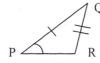

$\angle B = \angle P$, $BC = PQ$, and $AC = QR$.
The triangles do not satisfy the SSS rule, the SAS rule, or the ASA rule. The triangles are not necessarily congruent.

Critical Thinking 9.3

45. To determine if a 25-foot ladder is long enough to reach 24 ft up the side of the home when the bottom of the ladder is 6 ft from the base of the side of the house, use the Pythagorean Theorem to find the hypotenuse of a right triangle with legs that measure 24 ft and 6 ft.
$$c^2 = a^2 + b^2$$
$$c^2 = 24^2 + 6^2$$
$$c^2 = 576 + 36$$
$$c^2 = 612$$
$$c \approx 24.74$$
Compare the leg of the hypotenuse with 25. If the hypotenuse is shorter than 25 ft, the ladder will reach the gutter. If the hypotenuse is longer than 25 feet, the ladder will not reach the gutter.
$$24.74 < 25$$
The hypotenuse is shorter than 25 ft. The ladder will reach the gutters.

Section 9.4

Objective A Exercises

1. Strategy   To find the volume, use the formula for the volume of a rectangular solid. $L = 14$, $W = 10$, $H = 6$.

Solution
$V = LWH$
$V = 14(10)(6)$
$V = 840$
The volume is 840 in^3.

3. Strategy   To find the volume, use the formula for the volume of a pyramid. $s = 3$, $h = 5$.

Solution
$V = \frac{1}{3}s^2 h$
$V = \frac{1}{3}(3^2)(5)$
$V = \frac{1}{3}(9)(5)$
$V = 15$
The volume is 15 ft^3.

5. Strategy   To find the volume:
→Find the radius of the sphere. $d = 3$.
→Use the formula for the volume of sphere.

Solution
$r = \frac{1}{2}d = \frac{1}{2}(3) = 1.5$
$V = \frac{4}{3}\pi r^3$
$V = \frac{4}{3}\pi(1.5)^3$
$V = \frac{4}{3}\pi(3.375)$
$V = 4.5\pi$
$V \approx 14.14$
The volume is 4.5π cm^3.
The volume is approximately 14.14 cm^3.

7. Strategy   To find the volume, use the formula for the volume of a rectangle solid. $L = 6.8$, $W = 2.5$, $H = 2$.

Solution
$V = LWH$
$V = 6.8(2.5)(2)$
$V = 34$
The volume of the rectangular solid is 34 m^3.

9. **Strategy** To find the volume, use the formula for the volume of a cube. $s = 2.5$.

   Solution
   $V = s^3$
   $V = (2.5)^3$
   $V = 15.625$
   The volume of the cube is 15.625 in^3.

11. **Strategy** To find the volume:
    → Find the radius of the sphere. $d = 6$.
    → Use the formula for the volume of a sphere.

    Solution
    $r = \frac{1}{2}d = \frac{1}{2}(6) = 3$
    $V = \frac{4}{3}\pi r^3$
    $V = \frac{4}{3}\pi(3)^3$
    $V = \frac{4}{3}\pi(27)$
    $V = 36\pi$
    The volume of the sphere is $36\pi \text{ ft}^3$.

13. **Strategy** To find the volume:
    → Find the radius of the cylinder. $d = 24$.
    → Use the formula for the volume of a cylinder. $h = 18$.

    Solution
    $r = \frac{1}{2}d = \frac{1}{2}(24) = 12$
    $V = \pi r^2 h$
    $V = \pi(12^2)(18)$
    $V = \pi(144)(18)$
    $V = 2{,}592\pi$
    $V \approx 8{,}143.01$
    The volume of the cylinder is approximately $8{,}143.01 \text{ cm}^3$.

15. **Strategy** To find the volume:
    → Find the radius of the base of the cone. $d = 10$.
    → Use the formula for the volume of a cone. $h = 15$.

    Solution
    $r = \frac{1}{2}d = \frac{1}{2}(10) = 5$
    $V = \frac{1}{3}\pi r^2 h$
    $V = \frac{1}{3}\pi(5)^2(15)$
    $V = \frac{1}{3}\pi(25)(15)$
    $V = 125\pi$
    $V \approx 392.70$
    The volume of the cone is approximately 392.70 cm^3.

17. **Strategy** To find the volume, use the formula for the volume of a pyramid. $s = 9$, $h = 8$.

    Solution
    $V = \frac{1}{3}s^2 h$
    $V = \frac{1}{3}(9^2)(8)$
    $V = \frac{1}{3}(81)(8)$
    $V = 216$
    The volume of the pyramid is 216 m^3.

19. **Strategy** To find the width, use the formula for the volume of a rectangular solid. $V = 52.5$, $L = 7$, $H = 3$.

    Solution
    $V = LWH$
    $52.5 = 7(W)(3)$
    $52.5 = 21W$
    $2.5 = W$
    The width of the freezer is 2.5 ft.

21. Strategy To find the radius, use the formula for the volume of a cylinder. $V = 502.4$, $h = 10$.

Solution
$V = \pi r^2 h$
$502.4 = \pi r^2 (10)$
$50.24 = \pi r^2$
$15.99 \approx r^2$
$\sqrt{15.99} \approx r^2$
→ r is the square root of 15.99.
$4.00 \approx r$
The radius of the cylinder is approximately 4.00 in.

23. Strategy To find the length and the width, use the formula for the volume of a rectangular solid. $V = 125$, $H = 5$, $L = W$.

Solution
$V = LWH$
$125 = LW(5)$
$25 = LW$
$25 = L^2$
Substitute L for W.
$\sqrt{25} = L$
$5 = L$
$5 = W$ because $W = L$.
The length of the rectangular solid is 5 in.
The width of the rectangular solid is 5 in.

25. Strategy To find the amount of oil:
→ Find the radius of the base of the cylinder. $d = 6$.
→ Use the formula for the volume of a cylinder. $h = 4$.
→ Multiply the volume of the cylinder by $\frac{2}{3}$.

Solution
$r = \frac{1}{2}d = \frac{1}{2}(6) = 3$
$V = \pi r^2 h$
$V = \pi (3)^2 (4)$
$V = \pi (9)(4)$
$V = 36\pi$
$\frac{2}{3} V = \frac{2}{3}(36\pi) = 24\pi \approx 75.40$
The storage tank contains approximately 75.40 m³ of oil.

Objective B Exercises

27. Strategy To find the surface area, use the formula for the surface area of a rectangular solid. $L = 5$, $W = 4$, $H = 3$.

Solution
$SA = 2LW + 2LH + 2WH$
$SA = 2(5)(4) + 2(5)(3) + 2(4)(3)$
$SA = 40 + 30 + 24$
$SA = 94$
The surface area of the rectangular solid is 94 m².

29. Strategy To find the surface area, use the formula for the surface area of a pyramid. $s = 4$, $l = 5$.

Solution
$SA = s^2 + 2sl$
$SA = 4^2 + 2(4)(5)$
$SA = 16 + 40$
$SA = 56$
The surface area of the pyramid is 56 m².

31. Strategy To find the surface area, use the formula for the surface area of a cylinder. $r = 6$, $h = 2$.

Solution
$SA = 2\pi r^2 + 2\pi rh$
$SA = 2\pi (6^2) + 2\pi (6)(2)$
$SA = 2\pi (36) + 24\pi$
$SA = 72\pi + 24\pi$
$SA = 96\pi$
$SA \approx 301.59$
The surface area of the cylinder is 96π in². The surface area of the cylinder is approximately 301.59 in².

33. Strategy To find the surface area, use the formula for the surface area of a rectangular solid. $H = 5$, $L = 8$, $W = 4$.

Solution
$SA = 2LW + 2LH + 2WH$
$SA = 2(8)(4) + 2(8)(5) + 2(4)(5)$
$SA = 64 + 80 + 40$
$SA = 184$
The surface area of the rectangular solid is 184 ft².

35. Strategy To find the surface area, use the formula for the surface area of a cube. $s = 3.4$.

Solution
$SA = 6s^2$
$SA = 6(3.4)^2$
$SA = 6(11.56)$
The surface area of the cube is 69.36 m^2.

37. Strategy To find the surface area:
→Find the radius of the sphere. $d = 15$.
→Use the formula for the surface area of a sphere.

Solution
$r = \frac{1}{2}d = \frac{1}{2}(15) = 7.5$
$SA = 4\pi r^2$
$SA = 4\pi(7.5)^2$
$SA = 4\pi(56.25)$
$SA = 225\pi$
The surface area of the sphere is 225π cm^2.

39. Strategy To find the surface area, use the formula for the surface area of a cylinder. $r = 4$, $h = 12$.

Solution
$SA = 2\pi r^2 + 2\pi rh$
$SA = 2\pi(4^2) + 2\pi(4)(12)$
$SA = 2\pi(16) + 96\pi$
$SA = 32\pi + 96\pi$
$SA = 128\pi$
$SA \approx 402.12$
The surface area of the cylinder is approximately 402.12 in^2.

41. Strategy To find the surface area, use the formula for the surface area of a cone. $r = 1.5$, $l = 2.5$.

Solution
$SA = \pi r^2 + \pi rl$
$SA = \pi(1.5^2) + \pi(1.5)(2.5)$
$SA = \pi(2.25) + 3.75\pi$
$SA = 6\pi$
The surface area of the cone is 6π ft^2.

43. Strategy To find the surface area, use the formula for the surface area of a pyramid. $s = 9$, $l = 12$.

Solution
$SA = s^2 + 2sl$
$SA = 9^2 + 2(9)(12)$
$SA = 81 + 216$
$SA = 297$
The surface area of the pyramid is 297 in^2.

45. Strategy To find the width, use the formula for the surface area of a rectangular solid. $SA = 108$, $L = 6$, and $H = 4$.

Solution
$SA = 2LW + 2LH + 2WH$
$108 = 2(6)W + 2(6)(4) + 2W(4)$
$108 = 12W + 48 + 8W$
$108 = 20W + 48$
$60 = 20W$
$3 = W$
The width of the rectangular solid is 3 cm.

47. Strategy To find the number of cans or paint:
→Find the formula for the surface area of a cylinder. $r = 12$, $h = 30$.
→Divide the surface area by 300.

Solution
$SA = 2\pi r^2 + 2\pi rh$
$SA = 2\pi(12^2) + 2\pi(12)(30)$
$SA = 2\pi(144) + 720\pi$
$SA = 288\pi + 720\pi$
$SA = 1{,}008\pi$
$1{,}008\pi \div 300 \approx 10.56$
Because a portion of an eleventh can is needed, 11 cans of paint should be purchased.

49. Strategy   To find the amount of glass, use the formula for the surface area of a rectangular solid. Omit the top of the fish tank. The formula becomes $SA = LW + 2LH + 2WH$. $L = 12$, $W = 8$, $H = 9$.

Solution $SA = LW + 2LH + 2WH$
$SA = 12(8) + 2(12)(9) + 2(8)(9)$
$SA = 96 + 216 + 144$
$SA = 456$
The fish tank requires 456 in² of glass.

51. Strategy   To find the difference in area:
→Use the formula for the surface area of a pyramid. $s = 5$, $l = 8$.
→Use the formula for the surface area of a cone. $r = \frac{1}{2}d = \frac{1}{2}(5) = 2.5$, $l = 8$.
→Subtract the surface area of the cone from the surface area of the pyramid.

Solution   We first calculate the area of the pyramid.
$SA = s^2 + 2sl$
$SA = 5^2 + 2(5)(8)$
$SA = 25 + 80$
$SA = 105$
We now calculate the surface area of the cone.
$SA = \pi r^2 + \pi r l$
$SA = \pi(2.5)^2 + \pi(2.5)(8)$
$SA = \pi(6.25) + 20\pi$
$SA = 26.25\pi$
$SA \approx 82.47$
$105 - 82.47 = 22.53$
The surface area of the pyramid is approximately 22.53 cm² larger than the surface area of the cone.

Critical Thinking 9.4

53. a. The distance from the edge of the base to the vertex of a regular pyramid is longer than the distance, perpendicular to the base, from the base to the vertex. The statement is always true.

b. The distance from the edge of the base of a cone to the vertex is longer than the distance, perpendicular to the base, from the base to the vertex. The statement is never true.

c. The four triangular faces of a regular pyramid could be equilateral triangles, but they could be isosceles triangles that are not equilateral. The statement is sometimes true.

55. a. For example, a cut perpendicular to the top and bottom faces and parallel to two of the sides.

b. For example, beginning at an edge that's perpendicular to the bottom face, cut at an angle through to the bottom face.

c. For example, beginning at the top face, at a distance d from a vertex, cut at an angle to the bottom face, ending at a distance greater than d from the opposite vertex.

d. For example, beginning on the top face, at a distance d from a vertex, cut across the cube to a point just above the opposite vertex.

Section 9.5

Objective A Exercises

1. Strategy   To find the perimeter, add the lengths of the sides.

Solution $P = 19 + 20 + 8 + 5 + 27 + 42$
$P = 121$
The perimeter of the polygon is 121 cm.

224 Chapter 9: Geometry

3. **Strategy** The perimeter is the sum of the two diameters plus two half-circles.

   Solution
   $P = d + d + 2\left(\frac{1}{2}\pi d\right)$
   $P = 2 + 2 + \pi(2)$
   $P = 4 + 2\pi$
   $P \approx 10.28$
   The perimeter is $(4 + 2\pi)$ ft.
   The perimeter is approximately 10.28 ft.

5. **Strategy** The perimeter is the sum of three sides of a rectangle and $\frac{1}{2}$ the circumference of a circle.

   Solution
   $P = 2L + W + \frac{1}{2}(\pi d)$
   $P = 2(15) + 8 + \frac{1}{2}\pi(8)$
   $P = 38 + 4\pi$
   $P \approx 50.57$
   The perimeter is $(38 + 4\pi)$ m.
   The perimeter is approximately 50.57 m.

7. **Strategy** The perimeter is the sum of the measure of the sides of the figure. (Hint: Find the lengths of the two unlabeled sides first.)

   Solution
   $60 - 42 = 18$; $28 - 12 = 16$
   $P = 28 + 60 + 12 + 42 + 16 + 18$
   $P = 176$
   The perimeter is 176 ft.

9. **Strategy** The perimeter is the sum of two sides of a triangle plus one-half the circumference of a circle.

   Solution
   $P = 2 + 2 + \frac{1}{2}(\pi d)$
   $P = 4 + \frac{1}{2}\pi(2)$
   $P = 4 + \pi$
   $P \approx 7.14$
   The perimeter is $(4 + \pi)$ ft.
   The perimeter is approximately 7.14 ft.

11. **Strategy** The perimeter is the sum of three sides of a rectangle and two sides of a triangle. Use the Pythagorean Theorem to find the side of the triangle.

    Solution
    $c^2 = a^2 + b^2$
    $c^2 = 4^2 + 3^2$
    $c^2 = 16 + 9$
    $c^2 = 25$
    $c = \sqrt{25} = 5$
    $P = L + 2W + 2c$
    $P = 8 + 2(3) + 2(5)$
    $P = 8 + 6 + 10$
    $P = 24$
    The perimeter is 24 m.

13. **Strategy** The perimeter is equal to three sides of a rectangle plus one half the circumference of a circle.

    Solution
    $P = 2L + W + \frac{1}{2}\pi d$
    $P = 2(6.5) + 3 + \frac{1}{2}\pi(3)$
    $P = 16 + 1.5\pi$
    $P \approx 20.71$
    Approximately 20.71 ft of weather stripping was used around the door.

15. **Strategy** To find the cost:
    →Find the length of fencing not along the road.
    →Multiply the cost per foot times the length for each type of fence and add.

    Solution
    Length = $2(800) + 1{,}250 = 2{,}850$
    Cost = $5.10(2{,}850) + 6.70(1{,}250)$
    Cost = $22{,}910$
    The cost of the fence is $22,910.

Objective B Exercises

17. Strategy The area is equal to the area of the rectangle minus the area of the triangle. The base of the triangle is $8 - (2 + 2) = 4$.

Solution
$$A = LW - \frac{1}{2}bh$$
$$A = 8(4) - \frac{1}{2}(4)(3)$$
$$A = 32 - 6$$
$$A = 26$$
The area is 26 cm^2.

19. Strategy The area is equal to the area of a square plus $\frac{1}{2}$ the area of a circle. The radius of the circle is one half the length of a side of the square (6).

Solution
$$r = \frac{1}{2}s = \frac{1}{2}(6) = 3$$
$$A = s^2 + \frac{1}{2}\pi r^2$$
$$A = 6^2 + \frac{1}{2}\pi(3)^2$$
$$A = 36 + \frac{1}{2}\pi(9)$$
$$A = 36 + 4.5\pi$$
$$A \approx 50.14$$
The area is $(36 + 4.5\pi)$ in^2.
The area is approximately 50.14 in^2.

21. Strategy The area is equal to $\frac{3}{4}$ of the area of a circle.

Solution
$$A = \frac{3}{4}\pi r^2$$
$$A = \frac{3}{4}\pi(8^2)$$
$$A = \frac{3}{4}\pi(64)$$
$$A = 48\pi$$
$$A \approx 150.80$$
The area is 48π in^2.
The area is approximately 150.80 in^2.

23. Strategy The area is equal to the area of the rectangle plus the area of the triangle.

Solution
$$h = 9 - 6 = 3$$
$$b = W = 4$$
$$A = LW + \frac{1}{2}bh$$
$$A = 6(4) + \frac{1}{2}(4)(3)$$
$$A = 24 + 6 = 30$$
The area is 30 in^2.

25. Strategy The area is equal to the area of a rectangle plus the area of a triangle.

Solution
$$h = 5 - 3 = 2$$
$$b = 4$$
$$A = LW + \frac{1}{2}bh$$
$$A = 4(3) + \frac{1}{2}(4)(2)$$
$$A = 12 + 4$$
$$A = 16$$
The area is 16 in^2.

27. Strategy The area is equal to the area of a triangle plus the area of $\frac{1}{2}$ the area of a circle. The radius of the circle is $\frac{1}{2}$ the diameter.

Solution
$$r = \frac{1}{2}d = \frac{1}{2}(10) = 5$$
$$A = \frac{1}{2}bh + \frac{1}{2}(\pi r^2)$$
$$A = \frac{1}{2}(10)(8) + \frac{1}{2}\pi(5)^2$$
$$A = 40 + \frac{1}{2}\pi(25)$$
$$A = 40 + 12.5\pi$$
$$A \approx 79.27$$
The area is $(40 + 12.5\pi)$ in^2.
The area is approximately 79.27 in^2.

29. Strategy To find the cost:
→Find the area of the room and hallway. The total area is the area of two rectangles.
→Multiply the total area by 28.50.

Solution $A = L_1 W_1 + L_2 W_2$
$A = 6.8(4.5) + (10.8 - 6.8)(1)$
$A = 30.6 + 4(1)$
$A = 34.6$
Cost $= (34.6)(28.50) = 986.10$
The cost of the carpet is $986.10.

31. Strategy To find the amount of hardwood, find the area of the rectangle plus the area of two half circles. The radius of the circle is one half the width of the rectangle.

Solution $r = \frac{1}{2}W = \frac{1}{2}(80) = 40$
$A = LW + 2\left(\frac{1}{2}\pi r^2\right)$
$A = 175(80) + \pi(40^2)$
$A = 14{,}000 + 1{,}600\pi$
$A \approx 19{,}026.55$
19,026.55 ft² of hardwood floor is needed to cover the floor.

Objective C Exercises

33. Strategy The volume is equal to the volume of the rectangular solid minus the volume of the cylinder. The radius of the cylinder is one half the diameter of the circle.

Solution $r = \frac{1}{2}d = \frac{1}{2}(0.4) = 0.2$
$V = LWH - \pi r^2 h$
$V = 2(1.2)(0.8) - \pi(0.2)^2(2)$
$V = 1.92 - \pi(0.04)2$
$V = 1.92 - 0.08\pi$
$V \approx 1.67$
The volume of the solid is $(1.92 - 0.08\pi)$ m³. The volume of the solid is approximately 1.67 m³.

35. Strategy The volume is equal to the volume of a cylinder plus one half the volume of a sphere. The radius of the sphere is one half the diameter of the cylinder.

Solution $r = \frac{1}{2}d = \frac{1}{2}d(6) = 3$
$V = \pi r^2 h + \frac{1}{2}\left(\frac{4}{3}\pi r^3\right)$
$V = \pi(3^2)(12) + \frac{2}{3}\pi(3^3)$
$V = \pi(9)(12) + \frac{2}{3}\pi(27)$
$V = 108\pi + 18\pi = 126\pi$
$V \approx 395.84$
The volume of the solid is 126π ft³. The volume of the solid is approximately 395.84 ft³.

37. Strategy The volume is equal to the volume of the larger rectangular solid minus the volume of the smaller rectangular solid.

Solution $V = L_1 W_1 H_1 - L_2 W_2 H_2$
$V = 8(5)(8) - 8(3)(2)$
$V = 320 - 48$
$V = 272$
The volume of the solid is 272 ft³.

39. Strategy The volume is equal to the volume of the cylinder plus the volume of the cone. The radius of the cone is equal to one half the diameter of the cylinder. $h_1 = 3$, $h_2 = 2$

Solution $r = \frac{1}{2}d = \frac{1}{2}(3) = 1.5$
$V = \pi r^2 h_1 + \frac{1}{3}\pi r^2 h_2$
$V = \pi(1.5)^2(3) + \frac{1}{3}\pi(1.5)^2(2)$
$V = \pi(2.25)(3) + \frac{1}{3}\pi(2.25)(2)$
$V = 6.75\pi + 1.5\pi = 8.25\pi$
$V \approx 25.92$
The volume of the solid is 8.25π in³.
The volume of the solid is approximately 25.92 in³.

41. **Strategy** The volume is equal to the volume of a cone plus one half the volume of a sphere. The radius of the sphere is one half the diameter of the cone.

Solution
$r = \frac{1}{2}d = \frac{1}{2}(6) = 3$
$V = \frac{1}{3}\pi r^2 h + \frac{1}{2}\left(\frac{4}{3}\pi r^3\right)$
$V = \frac{1}{3}\pi(3)^2(6) + \frac{2}{3}\pi(3^3)$
$V = \frac{1}{3}\pi(9)(6) + \frac{2}{3}\pi(27)$
$V = 18\pi + 18\pi = 36\pi$
$V \approx 113.10$
The volume of the solid is 36π in^3.
The volume of the solid is approximately 113.10 in^3.

43. **Strategy** The volume is equal to the large cylinder minus the small cylinder. The radius of each cylinder is one half the diameter of each cylinder.

Solution
$r_1 = \frac{1}{2}d_1 = \frac{1}{2}(18) = 9$
$r_2 = \frac{1}{2}d_2 = \frac{1}{2}(9) = 4.5$
$V = \pi r_1^2 h - \pi r_2^2 h$
$V = \pi(9^2)(24) - \pi(4.5)^2(24)$
$V = \pi(81)(24) - \pi(20.25)(24)$
$V = 1{,}944\pi - 486\pi = 1{,}458\pi$
$V \approx 4{,}580.44$
The volume of the solid is $1{,}458\pi$ cm^3.
The volume of the solid is approximately 4,580.44 cm^3.

45. **Strategy** The volume of the bushing is equal to the volume of the rectangular solid minus one half the volume of the cylinder. The radius of the cylinder is one half the diameter of the cylinder.

Solution
$r = \frac{1}{2}d = \frac{1}{2}(4) = 2$
$V = LWH - \frac{1}{2}\pi r^2 h$
$V = 12(8)(3) - \frac{1}{2}\pi(2^2)(12)$
$V = 288 - \frac{1}{2}\pi(4)(12)$
$V = 288 - 24\pi$
$V \approx 212.60$
The volume of the solid is approximately 212.60 in^3.

47. **Strategy** To find the cost:
→Find the volume. The volume is equal to the volume of a rectangular solid plus one half the volume of a cylinder. The radius of the cylinder is one half the length of the rectangular solid.
→Multiply the volume by 6.15.

Solution
$r = \frac{1}{2}L = \frac{1}{2}(50) = 25$
$V = LWH + \frac{1}{2}\pi r^2 h$
$V = 50(25)(0.5) + \frac{1}{2}\pi(25)^2(0.5)$
$V = 625 + 156.25\pi$
Cost $= 6.15(625 + 156.25\pi)$
$\approx 6{,}862.62$
The cost of the floor is $6,862.62.

Objective D Exercises

49. Strategy The total surface area equals the sum of the surface area of each face of the solid. The front face is split into two squares.

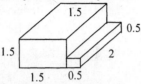

Solution
$SA = 2(1.5)^2 + 2(0.5)^2 + 2(1.5)(2) + 2(2)(0.5) + 1(2) + 2(2)$
$SA = 4.5 + 0.5 + 6 + 2 + 2 + 4 = 19$
The surface area is 19 m².

51. Strategy The total surface area equals the surface area of a cylinder, minus one end of the cylinder, plus the surface area of a cone, minus the bottom of the cone. The radius of the cone is one half the diameter of the cylinder.

Solution
$r = \frac{1}{2}d = \frac{1}{2}(6) = 3$
$SA = (2\pi r^2 + 2\pi rh) - \pi r^2 + (\pi r^2 + \pi rl) - \pi r^2$
$SA = \pi r^2 + 2\pi rh + \pi rl$
$SA = \pi(3)^2 + 2\pi(3)(10) + \pi(3)(8)$
$SA = 9\pi + 60\pi + 24\pi = 93\pi$
$SA \approx 292.17$
The surface area of the solid is 93π cm².
The surface area of the solid is approximately 292.17 cm².

53. Strategy The total surface area equals the surface area of a cylinder, minus one end of the rectangular solid, plus the surface area of the rectangular solid, minus one end of the rectangular solid.

Solution
$SA = (2\pi r^2 + 2\pi rh) - WH + (2LW + 2LH + 2WH) - WH$
$SA = 2\pi r^2 + 2\pi rh + 2LW + 2LH$
$SA = 2\pi(8)^2 + 2\pi(8)(2) + 2(15)(2) + 2(15)(2)$
$SA = 2\pi(64) + 32\pi + 60 + 60$
$SA = 128\pi + 32\pi + 120 = 160\pi + 120$
$SA \approx 622.65$
The surface area of the solid is $(120 + 160\pi)$ m².
The surface area of the solid is approximately 622.65 m².

55. Strategy The total surface area equals the surface area of a cone, minus the bottom of the cone, plus one half the surface area of a sphere. The radius of the sphere is one half the diameter of the base of the cone.

Solution
$r = \frac{1}{2}d = \frac{1}{2}(8) = 4$
$SA = (\pi r^2 + \pi r l) - \pi r^2 + \frac{1}{2}(4\pi r^2)$
$SA = \pi r l + 2\pi r^2$
$SA = \pi(4)(6) + 2\pi(4^2)$
$SA = 24\pi + 2\pi(16)$
$SA = 24\pi + 32\pi = 56\pi$
$SA \approx 175.93$
The surface area of the solid is 56π cm^2.
The surface area of the solid is approximately 175.93 cm^2.

57. Strategy The surface area equals the sum of the surface areas of the ten rectangular faces of the solid.
$V = 2(1.2)(0.8) - \pi(0.2)^2(2)$

Solution
$SA = 2(8)(5) - 2(3)(2) + 2(8)(5) + 2(8)(3) + 2(8)(3) + (8)(2) + 8(8) + 64$
$SA = 80 - 12 + 80 + 48 + 48 + 16 + 64 = 324$
The surface area of the solid is 324 ft^2.

59. Strategy The total surface area equals the surface area of the rectangular solid, minus the cross sectional area of the cylinder (2 · 6), plus one half the surface area of a cylinder. The radius of the cylinder is one half the width of the solid.

Solution
$r = \frac{1}{2}W = \frac{1}{2}(6) = 3$
$SA = (2LW + 2LH + 2WH) - (2 \cdot 6) + \frac{1}{2}(2\pi r^2 + 2\pi r h)$
$SA = 2(9)(6) + 2(9)(1) + 2(6)(1) - 12 + \pi(3^2) + \pi(3)(2)$
$SA = 108 + 18 + 12 - 12 + 9\pi + 6\pi$
$SA = 126 + 15\pi$
$SA \approx 173.12$
The surface area of the solid is $(126 + 15\pi)$ in^2.
The surface area of the solid is approximately 173.12 in^2.

Chapter 9: *Geometry*

61. **Strategy**   To find the number of cans of paint:
   → Find the surface area. The surface area equals the surface area of the sides of the rectangular solid, plus one half the surface area of the cylinder. The radius of the cylinder is one half the width of the rectangular solid.
   → Divide the surface area by 250.

   Solution
   $r = \frac{1}{2}W = \frac{1}{2}(94) = 47$
   $SA = 2LH + 2WH + \frac{1}{2}(2\pi r^2 + 2\pi rh)$
   $SA = 2LH + 2WH + \pi r^2 + \pi rh$
   $SA = 2(125)(32) + 2(94)(32) + \pi(47^2) + \pi(47)(125)$
   $SA = 8{,}000 + 6{,}016 + 2{,}209\pi + 5{,}875\pi$
   $SA = 14{,}016 + 8{,}084\pi$
   $SA \approx 39{,}412.63$
   $39{,}412.63 \div 250 \approx 157.65$
   158 cans should be purchased to paint the auditorium.

63. **Strategy**   To find the cost:
   → Find the surface area. Two walls are 25.5 by 8, two walls are 22 by 8. Subtract the area of the doors and windows.
   → Multiply the surface area by 1.50.

   Solution
   $SA = 2(25.5)(8) + 2(22)(8) - 2(2.5)(7) - 6(2.5)(4)$
   $SA = 408 + 352 - 35 - 60 = 665$
   $665 \cdot 1.50 = 997.50$
   The cost to plaster the room is $997.50.

Critical Thinking 9.5

65. Length = carton width + diameter of bottle + cardboard width + diameter of bottle + cardboard width + diameter of bottle + cardboard width.
   Length = $\frac{1}{8} + 4 + \frac{1}{16} + 4 + \frac{1}{16} + 4 + \frac{1}{8} = 12\frac{3}{8}$ in.
   Width = carton width + diameter of bottle + cardboard width + diameter of bottle + cardboard width
   Width = $\frac{1}{8} + 4 + \frac{1}{16} + 4 + \frac{1}{8} = = 8\frac{5}{16}$ in.
   Height = carton width + height of bottle + carton width
   Height = $\frac{1}{8} + 8 + \frac{1}{8} = 8\frac{1}{4}$ in.

   The dimensions of the shipping carton are $12\frac{3}{8}$ in. × $8\frac{5}{16}$ in. × $8\frac{1}{4}$ in.

67. The vanishing point is a point in a drawing at which parallel lines drawn in perspective converge or seem to converge. Student examples of vanishing point should vary.

69. Many of your students may already be familiar with M.C. Escher's work, since it now appears on everything from postcards to T-shirts. You might have each student select a different work to report on. Some examples include Relativity, House of Stairs, Whirlpools, Stars, Rind, Belvedere, Another World II, Ascending and Descending, and Waterfall.

Chapter Review Exercises

1. Strategy → To find the measure of ∠c, use the fact that the sum of an interior and exterior angle is 180°. $V = LWH - \pi r^2 h$
   → To find the measure of ∠x, use the fact that the sum of the measurements of the interior angles of a triangle is 180°.
   → To find the measure of ∠y, use the fact that the sum of an interior and exterior angle is 180°.

   Solution ∠a + ∠c = 180°
   74° + ∠c = 180°
   ∠c = 106°
   ∠b + ∠c + ∠x = 180°
   52° + 106° + ∠x = 180°
   158° + ∠x = 180°
   ∠x = 22°
   ∠x + ∠y = 180°
   22° + ∠y = 180°
   ∠y = 158°

2. Strategy To find the perimeter:
   → Find AC by writing a proportion, using the fact that in similar triangles, the ratios of corresponding sides are equal.
   → Use the formula for finding the perimeter of a triangle.

   Solution $\dfrac{AC}{DF} = \dfrac{BC}{EF}$
   $\dfrac{AC}{12} = \dfrac{6}{9}$
   9(AC) = 12(6)
   9(AC) = 72
   AC = 8
   P = AB + BC + AC
   P = 10 + 6 + 8
   P = 24
   The perimeter of the triangle is 24 in.

3. Strategy The volume is equal to the volume of the rectangular solid with dimensions 8 by 7 by 6, minus the rectangular solid with dimensions 8 by 4 by 3.

   Solution $V = L_1 V_1 H_1 - L_2 V_2 H_2$
   $V = 8(7)(6) - 8(4)(3)$
   $V = 336 - 96$
   $V = 240$
   The volume of the solid is 240 in^3.

4. Strategy To find the measure of ∠x, use the fact that adjacent angles of intersecting lines are supplementary.

   Solution 112° + ∠x = 180°
   ∠x = 68°
   The measure of ∠x is 68°.

5. Strategy To determine if the triangles are congruent, determine if one of the rules for congruence is satisfied.

   Solution BC = DE, AC = DF,
   ∠C = ∠D
   Two sides and the included angle of one triangle equal two sides and the included angle of the other triangle.
   The triangles are congruent by the SAS rule.

6. **Strategy** The total surface area equals the surface area of a cylinder, minus the surface area of one end of the cylinder, plus one half the surface area of a sphere. The radius of the sphere is one half the diameter of the cylinder.

   Solution
   $r = \frac{1}{2}d = \frac{1}{2}(4) = 2$
   $SA = (2\pi r^2 + 2\pi rh)$
   $\quad\quad - \pi r^2 + \frac{1}{2}(4\pi r^2)$
   $SA = 3\pi r^2 + 2\pi rh$
   $SA = 3\pi(2)^2 + 2\pi(2)(8)$
   $SA = 3\pi(4) + 32\pi$
   $SA = 12\pi + 32\pi$
   $SA = 44\pi$
   $SA \approx 138.23$
   The surface area of the solid is approximately 138.23 m².

7. $AC = AB + BC$
   $AC = 3(BC) + BC$
   $AC = 4(BC)$
   $AC = 4(11)$
   $AC = 44$
   The length of AC is 44 cm.

8. **Strategy** The sum of the measures of the three angles shown is 180°. To find x, write an equation and solve for x.

   Solution
   $4x + 3x + (x + 28°) = 180°$
   $8x + 28° = 180°$
   $8x = 152°$
   $x = 19°$
   The measure of x is 19°.

9. **Strategy** The area is equal to the area of the rectangle plus one half the area of a circle. The radius of the circle is one half the length of the rectangle.

   Solution
   $r = \frac{1}{2}L = \frac{1}{2}(8) = 4$
   $A = LW + \frac{1}{2}\pi r^2$
   $A = 8(4) + \frac{1}{2}\pi(4^2)$
   $A = 32 + \frac{1}{2}\pi(16)$
   $A = 32 + 8\pi$
   $A \approx 57.13$
   The area is approximately 57.13 in².

10. **Strategy** To find the volume, use the formula for the volume of a pyramid. $s = 6$, $h = 8$.

    Solution
    $V = \frac{1}{3}s^2 h$
    $V = \frac{1}{3}(6^2)(8)$
    $V = \frac{1}{3}(36)(8)$
    $V = 96$ cm³.
    The volume of the pyramid is 96 cm³.

11. **Strategy** The perimeter is equal to two sides of a triangle plus one half the circumference of a circle.

    Solution
    $P = a + b + \frac{1}{2}\pi d$
    $P = 16 + 16 + \frac{1}{2}\pi(10)$
    $P = 32 + 5\pi$
    $P \approx 47.71$
    The perimeter is approximately 47.71 in.

12. **Strategy** $\angle a = 138°$ because alternate interior angles of parallel lines are equal.
    $\angle a + \angle b = 180°$ because adjacent angles of intersecting lines are supplementary.

    Solution
    $\angle a = 138°$
    $\angle a + \angle b = 180°$
    $138° + \angle b = 180°$
    $\angle b = 42°$
    The measure of $\angle b$ is 42°.

13. **Strategy** To find the surface area, use the formula for the surface area of a rectangular solid. $L = 10$, $W = 5$, $H = 4$.

    Solution $SA = 2LW + 2LH + 2WH$
    $SA = 2(10)(5) + 2(10)(4) + 2(5)(4)$
    $SA = 100 + 80 + 40$
    $SA = 220$
    The surface area of the solid is 220 ft^2.

14. **Strategy** To find the measure of the other leg, use the Pythagorean Theorem. $c = 12$, $a = 7$.

    Solution $a^2 + b^2 = c^2$
    $7^2 + b^2 = 12^2$
    $49 + b^2 = 144$
    $b^2 = 95$
    $b = \sqrt{95}$
    $b \approx 9.75$
    The other leg is approximately 9.75 ft.

15. **Strategy** To find the volume, use the formula for the volume of a cube. $s = 3.5$.

    Solution $V = s^3$
    $V = (3.5)^3$
    $V = 42.875$
    The volume of the cube is 42.875 in^3.

16. **Strategy** Supplementary angles are two angles whose sum is 180°. To find the supplement, let x represent the supplement of a 32° angle. Write an equation and solve for x.

    Solution $32° + x = 180°$
    $x = 148°$
    The supplement of a 32° angle is a 148° angle.

17. **Strategy** To find the volume, use the formula for the volume of a rectangular solid. $L = 6.5$, $W = 2$, $H = 3$.

    Solution $V = LWH$
    $V = (6.5)(2)(3)$
    $V = 39$
    The volume of the solid is 39 ft^3.

18. **Strategy** To find the third angle, use the fact that the sum of the measures of the interior angles of a triangle is 180°. Let $x =$ the third angle.

    Solution $37° + 48° + x = 180°$
    $85° + x = 180°$
    $x = 95°$
    The third angle is 95°.

19. **Strategy** To find the base, use the formula for the area of a triangle. Substitute 7 for h and 28 for A and solve for b.

    Solution $A = \frac{1}{2}bh$
    $28 = \frac{1}{2}b(7)$
    $56 = 7b$
    $8 = b$
    The base of the triangle is 8 cm.

20. **Strategy** To find the volume, use the formula for the volume of a sphere. The radius of the sphere is one half the diameter.

    Solution $r = \frac{1}{2}d = \frac{1}{2}(12) = 6$
    $V = \frac{4}{3}\pi r^3$
    $V = \frac{4}{3}\pi(6^3)$
    $V = \frac{4}{3}\pi(216)$
    $V = 288\pi$
    The volume of the sphere is 288π mm^3.

21. Strategy To find the length of each side, use the formula for the perimeter of a square. $P = 86$.

Solution
$P = 4s$
$86 = 4s$
$21.5 = s$
A side of the square is 21.5 cm.

22. Strategy To find the number of cans of paint:
→Find the surface area by using the formula for the surface area of a cylinder.
→Divide the surface area by 200.

Solution
$SA = 2\pi r^2 + 2\pi rh$
$SA = 2\pi(6^2) + 2\pi(6)(15)$
$SA = 2\pi(36) + 180\pi$
$SA = 72\pi + 180\pi$
$SA = 252\pi$
$252\pi \div 200 \approx 3.96$
Because a portion of a fourth can is needed, 4 cans of paint should purchased.

23. Strategy To find the amount of fencing, use the formula for the perimeter of a rectangle.

Solution
$P = 2L + 2W$
$P = 2(56) + 2(48)$
$P = 112 + 96$
$P = 208$
208 yd of fencing are needed to fence the park.

24. Strategy To find the area, use the formula for the area of a square. $s = 9.5$.

Solution
$A = s^2$
$A = (9.5)^2$
$A = 90.25$
The area of the patio is 90.25 m².

25. Strategy To find the area of the walkway:
→Find the length and width of the total area.
→Subtract the area of the plot of grass from the total area.

Solution
$L_1 = 40 + 2 + 2 = 44$
$W_1 = 25 + 2 + 2 = 29$
$A = L_1 W_1 - L_2 W_2$
$A = 44(29) - 40(25)$
$A = 1{,}276 - 1{,}000$
$A = 276$
The area of the walkway is 276 m².

Chapter Test

1. Strategy To find the measure of the other leg, use the Pythagorean Theorem. $c = 11$, $a = 8$.

Solution
$a^2 + b^2 = c^2$
$8^2 + b^2 = 11^2$
$64 + b^2 = 121$
$b^2 = 57$
$b = \sqrt{57}$
$b \approx 7.55$
The other leg is approximately 7.55 cm.

2. Strategy To determine if the triangles are congruent, determine if one of the rules for congruence is satisfied.

Solution
$AC = AC$, $AB = AB$,
$\angle A = \angle A$
Two sides and the included angle of one triangle equal two sides and the included angle of the other triangle.
The triangles are congruent by the SAS rule.

3. Strategy To find the area, use the formula for the area of a rectangle. Substitute 15 for L and 7.4 for W. Solve for A.

Solution
$A = LW$
$A = 15(7.4)$
$A = 111$
The area is 111 m^2.

4. Strategy To find the area, use the formula for the area of a triangle. Substitute 7 for b and 12 for h. Solve for A.

Solution
$A = \frac{1}{2}bh$
$A = \frac{1}{2}(7)(12)$
$A = 42$
The area is 42 ft^2.

5. Strategy To find the volume, use the formula for the volume of a cone. $r = 7$, $h = 16$.

Solution
$V = \frac{1}{3}\pi r^2 h$
$V = \frac{1}{3}\pi(7)^2(16)$
$V = \frac{1}{3}\pi(49)(16)$
$V = \frac{784\pi}{3}$
The volume is $\frac{784\pi}{3}$ in^3.

6. Strategy To find the surface area, use the formula for the surface area of a pyramid. $s = 3$, $l = 11$.

Solution
$SA = s^2 + 2sl$
$SA = 3^2 + 2(3)(11)$
$SA = 9 + 66$
$SA = 75$
The surface area of the pyramid is 75 m^2.

7. Strategy The volume is equal to the volume of the cylinder plus the volume of the cone. $h_1 = 30$, $h_2 = 12$, $r = 7$

Solution
$V = \pi r^2 h_1 + \frac{1}{3}\pi r^2 h_2$
$V = \pi(7)^2(30) + \frac{1}{3}\pi(7)^2(12)$
$V = \pi(49)(30) + \frac{1}{3}\pi(49)(12)$
$V = 1{,}470\pi + 196\pi = 1{,}666\pi$
$V \approx 5{,}233.89$
The volume of the solid is $1{,}666\pi$ cm^3.
The volume of the solid is approximately 5,233.89 cm^3.

8. Strategy To find the area, use the formula for the area of a trapezoid. Substitute 20 for b_1, 33 for b_2, and 6 for h. Solve for A.

Solution
$A = \frac{1}{2}h(b_1 + b_2)$
$A = \frac{1}{2} \cdot 6(20 + 33)$
$A = 159$
The area is 159 in^2.

9. Strategy The perimeter is the sum of two sides of a rectangle plus two half circles.

Solution
$P = 2L + 2\left(\frac{1}{2}\pi d\right)$
$P = 2(11) + \pi(6)$
$P = 22 + 6\pi$
$P \approx 40.8$
The perimeter is $(22 + 6\pi)$ ft.
The perimeter is approximately 40.8 ft.

10. **Strategy** The total surface area equals the surface area of the rectangular solid, minus the top of the rectangular solid, plus the surface area of the pyramid, minus the bottom of the pyramid.

    Solution
    $SA = (2LW + 2LH + 2WH)$
    $\quad\quad - LW + (s^2 + 2sl) - s^2$
    $SA = LW + 2LH + 2WH + 2sl$
    $SA = 5(5) + 2(5)(3) + 2(5)(3)$
    $\quad\quad + 2(5)(5)$
    $SA = 25 + 30 + 30 + 50$
    $SA = 135$
    The surface area of the solid is 135 m^2.

11. **Strategy** The angles labeled are adjacent angles of intersecting lines and are, therefore, supplementary angles. To find x, write an equation and solve for x.

    Solution
    $4x + 10° + x = 180°$
    $5x = 170°$
    $x = 34°$
    The measure of x is 34°.

12. The polygon has 8 sides.
    The polygon is an octagon.

13. **Strategy** To determine whether the triangles are congruent, determine whether one of the rules for congruence is satisfied.

    Solution The triangles do not satisfy the SSS Rule, the SAS Rule, or the ASA Rule. The triangles are not necessarily congruent.

14. **Strategy** To find the volume, use the formula for the volume of a rectangular solid. $L = 6$, $W = 7$, $H = 4$.

    Solution
    $V = LWH$
    $V = 6(7)(4) = 168$
    The volume of the rectangular solid is 168 ft^3.

15. **Strategy** To find the hypotenuse, use the Pythagorean Theorem. $a = 4$, $b = 7$

    Solution
    $c^2 = a^2 + b^2$
    $c^2 = 4^2 + 7^2$
    $c^2 = 16 + 49 = 65$
    $c = \sqrt{65}$
    $c \approx 8.06$
    The length of the hypotenuse if approximately 8.06 m.

16. **Strategy**

    $y = a$ because corresponding angles have the same measure. $y + 37° = 180°$ because adjacent angles of intersecting lines are supplementary angles. Substitute a for y and solve for a.

    Solution
    $a + 37° = 180°$
    $a = 143°$
    The measure of a is 143°.

17. **Strategy** To find the surface area, use the formula for the surface area of a cylinder. $r = 10$, $h = 15$.

    Solution
    $SA = 2\pi r^2 + 2\pi rh$
    $SA = 2\pi(10^2) + 2\pi(10)(15)$
    $SA = 2\pi(100) + 300\pi$
    $SA = 200\pi + 300\pi = 500\pi$
    The surface area of the cylinder is 500π cm^2.

18. Strategy

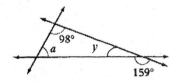

→To find the measure of ∠y, use the fact that ∠y and 159° are supplemental angles.
→To find the measure of ∠a by using the fact that the sum of the interior angles of a triangle is 180°.

Solution ∠y + 159° = 180°
∠y = 21°
21° + 98° + ∠a = 180°
119° + ∠a = 180°
∠a = 61°
The measure of ∠a is 61°.

19. Strategy To find the length of side FG, write a proportion using the fact that, in similar triangles, the ratio of corresponding sides equals the ratio of corresponding heights. Solve the proportion for the height h.

Solution $\dfrac{EF}{BC} = \dfrac{h}{5}$

$\dfrac{12}{9} = \dfrac{h}{5}$

$5(12) = 9h$
$60 = 9h$
$6.67 \approx h$
The length of line segment FG is approximately 6.67 ft.

20. Strategy To find BC, write a proportion using the fact that in similar triangles, the ratios of corresponding sides are equal. Solve the proportion for BC.

Solution $\dfrac{BC}{EF} = \dfrac{AB}{DE}$

$\dfrac{BC}{8} = \dfrac{8}{15}$

$15BC = (8)8$
$15BC = 64$
$BC \approx 4.27$
The length of BC is approximately 4.27 ft.

21. Strategy To find the perimeter, use the formula for the perimeter of a square. Substitute 5 for s. Solve for P.

Solution $P = 4s$
$P = 4(5)$
$P = 20$
The perimeter is 20 m.

22. Strategy To find the perimeter, use the formula for the perimeter of a rectangle. Substitute 8 for L and 5 for W. Solve for P.

Solution $P = 2L + 2W$
$P = 2(8) + 2(5)$
$P = 16 + 10$
$P = 26$
The perimeter is 26 cm.

23. Strategy To find the perimeter:
→Use the Pythagorean Theorem to find the hypotenuse of the triangle. $a = 12, b = 18$
→Use the formula for the perimeter of a triangle to find the perimeter.

Solution $c^2 = a^2 + b^2$
$c^2 = 12^2 + 18^2$
$c^2 = 144 + 324$
$c^2 = 468$
$c = \sqrt{468}$
$c \approx 21.6$
$P = a + b + c$
$P = 12 + 18 + 21.6$
$P = 51.6$
The perimeter is approximately 51.6 ft.

24. Strategy   To find the third angle, use the fact that the sum of the measures of the interior angles of a triangle is 180°. Let x = the third angle.

    Solution $41° + 37° + x = 180°$
                    $78° + x = 180°$
                    $x = 102°$
                  The third angle is 102°.

25. Strategy   The area is equal to the area of a rectangle minus $\frac{1}{2}$ the area of a circle. The width of the rectangle is equal to twice the radius of the circle (3).

    Solution $W = 2r = 2(3) = 6$
                  $A = LW - \frac{1}{2}\pi r^2$
                  $A = 12(6) - \frac{1}{2}\pi(3)^2$
                  $A = 72 - \frac{1}{2}\pi(9)$
                  $A = 72 - 4.5\pi$
                  $A \approx 58.9$
                  The area is $(72 - 4.5\pi)$ ft^2.
                  The area is approximately 57.9 ft^2.

Cumulative Review Exercises

1. Strategy   To find the amount, use the basic percent equation. Percent = 8.5% = 0.085, base = 2,400, amount = n.

    Solution   Percent · base = amount
                  $0.085(2,400) = n$
                  $204 = n$
                  8.5% of 2,400 is 204.

2. $78 \div 1 = 78$
     $78 \div 2 = 39$
     $78 \div 3 = 26$
     $78 \div 6 = 13$
     $78 \div 13 = 6$
     The factors of 78 are 1, 2, 3, 6, 13, 26, 39, and 78.

3. $4\frac{2}{3} \div 5\frac{3}{4} = \frac{14}{3} \div \frac{28}{5}$
                $= \frac{14}{3} \cdot \frac{5}{28}$
                $= \frac{14 \cdot 5}{3 \cdot 28}$
                $= \frac{2 \cdot 7 \cdot 5}{3 \cdot 2 \cdot 2 \cdot 7}$
                $= \frac{5}{6}$

4. $(3x^2 + 5x - 2) + (4x^2 - x + 7)$
     $= (3x^2 + 4x^2) + (5x - x) + (-2 + 7)$
     $= 7x^2 + 4x + 5$

5. $82.93 \div 6.5 \approx 12.8$

6. $0.000029 = 2.9 \times 10^{-5}$

7. Strategy   To find the measure of $\angle x$, use the fact that adjacent angles of intersecting lines are supplementary.

    Solution $\angle x + 49° = 180°$
                  $\angle x = 131°$
                  The measure of $\angle x$ is 131°.

8. Strategy   To find the hypotenuse, use the Pythagorean Theorem. $a = 10$, $b = 24$.

    Solution $c^2 = a^2 + b^2$
                  $c^2 = 10^2 + 24^2$
                  $c^2 = 100 + 576$
                  $c^2 = 676$
                  $c = \sqrt{676}$
                  →C is the square root of 676.
                  $c = 26$
                  The length of the hypotenuse is 26 cm.

9. Strategy   To find the area, use the formula for the area of a trapezoid. $h = 5$, $b_1 = 16$, $b_2 = 4$.

    Solution $A = \frac{1}{2}h(b_1 + b_2)$
                  $A = \frac{1}{2} \cdot 5(16 + 4)$
                  $A = \frac{5}{2}(20) = 50$
                  The area is 50 in^2.

10. Strategy: The volume is equal to the volume of the large cylinder minus the volume of the small cylinder.

Solution
$V = \pi r_1^2 L - \pi r_2^2 L$
$V = \pi(6^2)(14) - \pi(2^2)(14)$
$V = \pi(36)(14) - \pi(4)(14)$
$V = 504\pi - 56\pi = 448\pi$
$V \approx 1,407.43$
The volume is approximately 1,407.43 cm³.

11. $(4x^2y^2)(-3x^3y)$
$= [4(-3)](x^2 \cdot x^3)(y^2 \cdot y)$
$= -12x^5y^3$

12. $3(2x + 5) = 18$
$6x + 15 = 18$
$6x = 3$
$x = \frac{1}{2}$
The solution is $\frac{1}{2}$.

13. Strategy: The perimeter equals three sides of a rectangle plus one half the circumference of a circle.

Solution
$P = 3 + 2 + 3 + \frac{1}{2}(\pi d)$
$P = 8 + \frac{1}{2}\pi(2)$
$P = 8 + \pi$
$P \approx 11.14$
The perimeter is approximately 11.14 cm.

14. The real numbers greater than –3 are to the right of –3 on the number line. Draw a parenthesis at –3. Draw a heavy line to the right of –3. Draw an arrow at the right of the line.

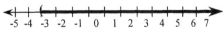

15. $5(2x + 4) - (3x + 2) = 10x + 20 - 3x - 2$
$= 7x + 18$

16. $2x + 3y^2z$
$2(5) + 3(-1)^2(-4) = 2(5) + 3(1)(-4)$
$= 10 + (-12)$
$= -2$

17. $x^2y - 2z$
$\left(\frac{1}{2}\right)^2\left(\frac{4}{5}\right) - 2\left(-\frac{3}{10}\right) = \frac{1}{4}\left(\frac{4}{5}\right) - 2\left(\frac{-3}{10}\right)$
$= \frac{1}{5} - \frac{-3}{5}$
$= \frac{1}{5} + \frac{3}{5}$
$= \frac{4}{5}$

18. $60\text{mph} = \frac{60\text{ mi}}{\text{h}} \cdot \frac{1.61\text{ km}}{1\text{ mi}} = 96.6 \text{km}/\text{h}$

19. $4x + 2 = 6x - 8$
$4x - 6x + 2 = 6x - 6x - 8$
$-2x + 2 = -8$
$-2x + 2 - 2 = -8 - 2$
$-2x = -10$
$\frac{-2x}{-2} = \frac{-10}{-2}$
$x = 5$
The solution is 5.

20.
x	y
2	0
0	3
-2	6

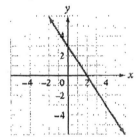

21. 3,482 m = 3.482 km

22. $\frac{3}{8} = \frac{3}{8}(100\%) = 37.5\%$

23. Strategy: To find the simple interest, solve the simple interest formula $I = Prt$ for I.
$P = 20,000$, $t = \frac{270}{365}$, $r = 0.08875$

Solution
$I = Prt$
$I = (20,000)(0.08875)\left(\frac{270}{365}\right)$
$I = 1,313.01$
The interest on the loan is $1,313.01.

24. Strategy To find the amount of coffee:
→ Multiply the number of people (250) by the amount of coffee consumed by each person (12).
→ Use the following conversion factors: $\frac{1c}{8\,oz}$, $\frac{1\,pt}{2\,c}$, $\frac{1\,qt}{2\,pt}$ and $\frac{1\,gal}{4\,qt}$.

Solution $250 \cdot 12 = 3{,}000$ oz
$3{,}000$ oz =
$\frac{3{,}000\,oz}{1} \cdot \frac{1c}{8\,oz} \cdot \frac{1\,pt}{2\,c} \cdot \frac{1\,qt}{2\,pt} \cdot \frac{1\,gal}{4\,qt}$
$= 23.4375$ gal
23 gal of coffee should be prepared for the reception.

25. Strategy To find the charge for phone service, write and solve an equation using n to represent the number of minutes the phone is used.

Solution Cost $= 22 + 0.25n$
$43.75 = 22 + 0.25n$
$21.75 = 0.25n$
$87 = n$
The phone service was used for 87 min.

26. Strategy To find the sales tax, write and solve a proportion using n to represent the sales tax.

Solution $\frac{0.75}{12.50} = \frac{n}{75}$
$0.75(75) = 12.50 \cdot n$
$56.25 = 12.50n$
$\frac{56.25}{12.50} = \frac{12.50n}{12.50}$
$4.50 = n$
The sales tax is $4.50.

27. Strategy To find the percent increase:
→ Subtract 1.48 from 1.56 to find the amount of increase.
→ Use the basic percent equation. Percent $= n$, base $= 1.48$, amount $=$ amount of increase.

Solution $1.56 - 1.48 = 0.08$
Percent $\cdot$ base $=$ amount
$n \cdot 1.48 = 0.08$
$n = \frac{0.08}{1.48}$
$n \approx 0.054 = 5.4\%$
The percent increase is 5.4%.

28. Strategy To find the height of the box, use the formula for the volume of a rectangular solid.
$V = 144$, $L = 12$, $W = 4$.

Solution $V = LWH$
$144 = 12(4)H$
$144 = 48H$
$3 = H$
The height of the box is 3 ft.

29. Strategy To find the pressure, substitute 35 for P in the given equation and solve for D.

Solution $P = 15 + \frac{1}{2}D$
$35 = 15 + \frac{1}{2}D$
$20 = \frac{1}{2}D$
$40 = D$
The depth is 40 ft.

30. Strategy To find the distance, substitute 196 for E and 49 for S and solve for d.

Solution $d = 4{,}000\sqrt{\frac{E}{S}} = 4{,}000$
$d = 4{,}000\sqrt{\frac{196}{49}} - 4{,}000$
$d = 4{,}000\sqrt{4} - 4{,}000$
$d = 4{,}000(2) - 4{,}000$
$d = 8{,}000 - 4{,}000$
$d = 4{,}000$
The explorer is 4,000 mi above the surface.

Chapter 10: Statistics and Probability

Prep Test

1. $\dfrac{3}{2+7} = \dfrac{3}{9} = \dfrac{1}{3}$

2. $\sqrt{13} \approx 3.606$

3. $\dfrac{49}{102} \approx .0480 = 48.0\%$

4. For 2005 – 2006:
   74,418 – 70,206 = 4,212
   For 2006 – 2007:
   78,883 – 74,418 = 4,465
   For 2007 – 2008:
   83,616 – 78,883 = 4,733
   For 2008 – 2009:
   88,633 – 83,616 = 5,017
   For 2009 – 2010:
   93,951 – 88,633 = 5,318
   The greatest increase is between 2009 and 2010 with an increase of $5,318.

5. a. $\dfrac{45}{27} = \dfrac{5}{3}$

   b. 27 : 27 is the same as 1 : 1.

6. a. 3.9, 3.9, 4.2, 4.5, 5.2, 5.5, 7.1

   b. $\dfrac{3.9 + 4.5 + 4.2 + 3.9 + 5.2 + 7.1 + 5.5}{7} = 4.9$
   There is an average of 4.9 million viewers per night.

7. a. $1,400,000 \cdot 0.15 = 210,000$ women

   b. $\dfrac{210,000}{1,400,000} = \dfrac{3}{20}$

Go Figure

Replacing the known values,
```
  271
   51
+ 3HE
─────
  S71
```
In the first column,
1 + 1 + E = 11.
2 + E = 11
E = 9
In the second column,
1 + 7 + 5 + H = 17.
13 + H = 17
H = 4
So then, the third column,
1 + 2 + 3 = S
S = 6

Section 10.1

Objective A Exercises

1. In their descriptions of a frequency table, students should explain that it is a method of organizing data and that the data is organized into classes so that the frequency of each class is readily apparent.

3. Range = 96 – 32 = 64

5. Strategy   To find the class with the greatest frequency, refer to the frequency table in Exercise 4.

   Solution   The class with the greatest frequency is 86–94.

7. Strategy   To find how many universities charge a tuition between $5,000 and $5,800, refer the frequency table in Exercise 4.

   Solution   5 universities charge a tuition between $5,000 and $5,800.

9. **Strategy** To find the percent:
→Use Exercise 4 to find the number of universities that charge a tuition between $9,500 and $10,300.
→Use the basic percent equation.

Solution Number of universities = 4
Percent · base = amount
$p(40) = 4$
$p = \dfrac{4}{40} = 0.10$
10% of the universities charge between $9,500 and $10,300.

11. **Strategy** To find the percent:
→Use Exercise 4 to find the number of universities that charge a tuition greater than or equal to $6800.
→Use the basic percent equation.

Solution Number of universities
$= 4 + 4 + 9 + 4 = 21$
Percent · base = amount
$p(40 = 21$
$p = \dfrac{21}{40} = 0.525$
52.5% of the universities charge tuition greater than or equal to $6,800.

13. **Strategy** To find the frequency distribution table, find the range. Then divide the range by 7, the number of classes to obtain the class width.

Solution range = 127 − 57 = 70
class width = $\dfrac{70}{7} = 10$

Classes	Tally	Frequency
57—67	/////	5
68—78	//////////	10
79—89	difference///////	7
90—100	///////////	11
101—111	//////////	10
112—122	////////	6
123-133	/	1

15. **Strategy** To find the number of hotels with rooms rates between $57 and $67, refer to the frequency table in Exercise 13.

Solution 5 hotels charge a corporate room rate between $57 and $67.

17. **Strategy** To find the number of hotels:
→Refer to the frequency table in Exercise 13 to find the number of hotels with room rates between $47–$67, $68–$78, $79–$89, and $90–$100.
→Add the numbers.

Solution Number of hotels with rates between:
$57–$67: 5
$68–$78: 10
$79–$89: 7
$90–$100: 11
$5 + 10 + 7 + 11 = 33$
33 of the hotels charge less than or equal to $100.

19. **Strategy** To find the percent:
→Refer to the frequency table in Exercise 13 to find the number of hotels with room rates between $90 and $100.
→Use the basic percent equation.

Solution Number of hotels: 11
Percent · base = amount
$p(50) = 11$
$p = \dfrac{11}{50} = 0.22$
22% of the hotels charge a corporate room rate between $90 and $100.

21. Strategy To find the percent:
→Refer to the frequency table in Exercise 13 to find the number of hotels with room rates less than or equal to $78
→Use the basic percent equation.

Solution Number of hotels with rates between:
$57–$67: 5
$68–$78: 10
5 + 10 = 15
Percent · base = amount
$p(50) = 15$
$p = \dfrac{15}{50} = 0.30$
30% of the hotel rooms have rates less than or equal to $78.

Objective B Exercises

23. Strategy Read the histogram to find the number of account balances between $1,500 and $2,000.

Solution There are 13 account balances between $1,500 and $2,000.

25. Strategy To find the percent:
→Read the histogram to find the number of account balances $2000 to $2500.
→Use the basic percent equation.

Solution Number of accounts: 11
Percent · base = amount
$p(50) = 11$
$p = \dfrac{11}{50} = 0.22$
22% of the account balances were between $2000 and $2500.

27. Strategy To find the ratio:
→Read the histogram to find the number of runners with times between 150 min–155 min and 175 min–180 min.
→Write the ratio in simplest form.

Solution Number of runners with times between:
150 min–155 min: 5
175 min–180 min: 10
$\dfrac{\text{Number of runners with times between } 150-155 \text{ min}}{\text{Number of runners with times between } 175 \text{ min} -180 \text{ min}} = \dfrac{5}{10} = \dfrac{1}{2}$
The ratio is $\dfrac{1}{2}$.

29. Strategy To find the percent:
→Read the histogram to find the number of runners with times between 165 min–170 min, 170 min–175 min, and 175 min–180 min.
→Add the numbers.
→Use the basic percent equation.

   Solution Number of runners with times between:
165 min–170 min: 30
170 min–175 min: 20
175 min–180 min: 10
$30 + 20 + 10 = 60$
Percent · base = amount
$p(100) = 60$
$p = \dfrac{60}{100} = 0.60$
60% of the runners had times greater than 165 min.

31. Strategy To find the percent:
→Read the histogram to find number of apartments with rents between $1250 and $1500.
→Use the basic percent equation.

   Solution Number of apartments with rents between $1250 and $1500: 5
Percent · base = amount
$p(40) = 5$
$p = \dfrac{5}{40} = 0.125$
12.5% of the apartments have rents between $1250 and $1500.

33. Strategy To find the percent:
→Read the histogram to find the number of apartments with rents between $1000–$1250, $1250–$1500, and $1500–$1750.
→Add the numbers.
→Use the basic percent equation.

   Solution Number of apartments with rents between:
$1000–$1250: 15
$1250–$1500: 5
$1500–$1750: 2
$15 + 5 + 2 = 22$
Percent · base = amount
$p(40) = 22$
$p = \dfrac{22}{40} = 0.55$
55% of the apartments have rents over $1000.

Objective C Exercises

35. Strategy   To find the number of nurses:
→Read the frequency polygon to find the number of nurses whose score was between 80–90 and between 90–100.
→Add the numbers.

Solution   Number of nurses whose score was between:
80–90: 18
90-100: 4
$18 + 4 = 22$
22 of the nurses had scores greater than 80.

37. Strategy   To find the percent:
→Read the frequency polygon to find the number of nurses whose score was between 70–80 and between 80–90.
→Add the numbers.
→Use the basic percent equation.

Solution   Number of nurses whose score was between:
70–80: 15
80–90: 18
$15 + 18 = 33$
Percent · base = amount
$p(50) = 33$
$p = \frac{33}{50} = 0.66$
66% of the nurses had scores between 70 and 90.

39. Strategy   To find the ratio:
→Read the frequency polygon to find the response times between 6 min and 9 min and between 15 min and 18 min.
→Write the ratio in simplest form.

Solution   Number of response times between:
6 min–9 min: 18
15 min–18 min: 3
$$\frac{\text{response times between } 6-9 \text{ min}}{\text{response times between } 15-18 \text{ min}} = \frac{18}{3} = \frac{6}{1}$$
The ratio is $\frac{6}{1}$.

41. **Strategy** To find the percent:
   → Read the frequency polygon to find the response time between 9 min and 12 min, between 12 min and 15 min, and between 15 min and 18 min.
   → Add the numbers.
   → Use the basic percent equation.

   Solution Number of response times between:
   9 min–12 min: 20
   12 min–15 min: 17
   15 min–18 min: 3
   $20 + 17 + 3 = 40$
   Percent · base = amount
   $p(75) = 40$
   $p = \frac{40}{75}$
   $p \approx 0.533$
   Approximately 53.3% of the response times are greater than 9 min.

Critical Thinking 10.1

43. Answers will vary.

45. Students might explain that data displayed in a histogram can also be displayed in a frequency table. A histogram is bar graph that represents the data in a frequency distribution. Students might note that the comparison of different frequencies is easier to visualize when the data is presented in a histogram.

Section 10.2

Objective A Exercises

1. Students should explain that the mean is determined by adding all the data values and then dividing the sum by the number of data values.
   The median is determined by arranging the numbers in order from smallest to largest (or from largest to smallest). If there are an odd number of data values, the median is the middle number, there are an equal number of data values above and below the median. If the data contain an even number of values, the median is the mean of the two middle numbers.
   To determine the mode, find the data value that occurs most frequently. Students should note that not all sets of data have a mode.

3. **Strategy** To find the mean number of televisions sold per month:
   → Determine the sum of the numbers sold.
   → Divide the sum by 12.
   To find the median number of seats occupied:
   → Arrange the numbers from smallest to largest.
   → Because there is an even number of values, the median is the sum of the two middle numbers divided by 2.

   Solution The sum of the numbers is 228.
   $\bar{x} = \frac{228}{12} = 19$
   The mean number of televisions is 19.
   12 15 15 17 17 19
   20 20 20 22 24 27
   median = $\frac{19+20}{2} = 19.5$
   The median is 19.5 televisions.

5. Strategy To find the mean of the times of the 100-meter dash:
→Determine the sum of the times.
→Divide the sum by 10.
To find the median time for the 100-meter dash:
→Arrange the numbers from smallest to largest.
→Because there is an even number of values, the median is the sum of the two middle numbers divided by 2.

Solution The sum of the numbers is 106.10.
$$\bar{x} = \frac{106.10}{10} = 10.61$$
The mean time for the 100–meter dash is 10.61 s.
10.23 10.26 10.45 10.52 10.57
10.64 10.74 10.78 10.90 11.01
$$\text{median} = \frac{10.57 + 10.64}{2} = 10.605$$
The median time for the 100–meter dash is 10.605 s.

7. Strategy To find the mean score:
→Determine the sum of the numbers.
→Divide the sum by 6.
To find the median score:
→Arrange the numbers from smallest to largest.
→The median is the middle term.

Solution The sum of the numbers is 524.
$$\bar{x} = \frac{524}{6} \approx 87.33$$
The mean is approximately 87.33.
77 78 88
92 94 95
$$\text{median} = \frac{88 + 92}{2} = 90$$
The mean score is 87.33 and the median score is 90. The instructor using the mean score would give the student a B. Using the median, the instructor would give the student an A.

9. Strategy To find the number of yards to be gained:
→Let n represent the number of yards to be gained.
→Use the formula for the mean to solve for n.

Solution $$100 = \frac{98 + 105 + 120 + 90 + 111 + 104 + n}{7}$$
$700 = 628 + n$
$72 = n$
The running back must gain 72 yd.

11. Strategy To find the score:
→Let n represent the score on the sixth round.
→Use the formula for the mean to solve for n.

Solution $$78 = \frac{78 + 82 + 75 + 77 + 79 + n}{6}$$
$468 = 391 + n$
$77 = n$
The score must be 77 on the sixth round.

248 Chapter 10: *Statistics and Probability*

13. Strategy To find the modal response, write down the category that received the most responses.

Solution Because a response of brown was recorded most frequently, the modal response was brown.

15. Strategy To find the modal response, write down the category that received the most responses.

Solution Because a response of very good was recorded most frequently, the modal response was very good.

Objective B Exercises

17. Strategy To draw the box-and-whiskers plot:
→Arrange the data from smallest to largest, and then find the median.
→Find Q_1, the median of the lower half of the data.
→Find Q_3, the median of the upper half of the data.
→Determine the smallest and largest data values.
→Draw the box-and-whiskers plot.

Solution
7.35   7.45   7.46   7.70   8.09   8.12   8.64   8.94
9.02   9.03   9.31   9.85   10.05   10.35   11.40   11.50

median $= \dfrac{8.94 + 9.02}{2} = 8.98$

$Q_1 = \dfrac{7.70 + 8.09}{2} = 7.895$, $Q_3 = \dfrac{9.85 + 10.05}{2} = 9.95$

Smallest value: 7.35
Largest value: 11.50

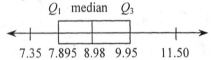

7.35 7.895 8.98 9.95   11.50

19. Strategy To draw the box-whiskers plot:
→Arrange the data from smallest to largest, and then find the median.
→Find Q_1, the median of the lower half of the data.
→Find Q_3, the median of the upper half of the data.
→Determine the smallest and largest data values.
→Draw the box-and-whiskers plot.

Solution
16   17   19   20   20   21   21   22   24   25
26   26   28   30   30   31   31   32   33

median = 25
$Q_1 = 20$
$Q_3 = 30$
Smallest value: 16
Largest value: 33

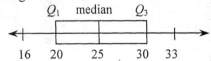

16   20   25   30   33

Section 10.2

21. Strategy   To draw the box-and-whiskers plot:
   → Arrange the data from smallest to largest, and then find the median.
   → Find Q_1, the median of the lower half of the data.
   → Find Q_3, the median of the upper half of the data.
   → Determine the smallest and largest data values.
   → Draw the box-and-whiskers plot.

   Solution   2.6 3.1 3.5 4.3 4.3 4.8 4.9 5.1
                     5.3 5.3 5.4 6.0 6.2 6.7 6.8 8.0

   median $\frac{5.1+5.3}{2} = 5.2$

   $Q_1 = \frac{4.3+4.3}{2} = 4.3$

   $Q_3 = \frac{6.0+6.2}{2} = 6.1$

   Smallest value: 2.6
   Largest value: 8.0

   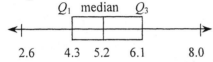

   2.6   4.3 5.2 6.1   8.0

Objective C Exercises

23. Strategy   To calculate the standard deviation:
   → Find the mean of the times.
   → Use the procedure for calculating standard deviation.

   Solution $\bar{x} = \frac{12+18+20+14+16}{5} = 16$

x	$(x-\bar{x})^2$	
12	$(12-16)^2$	16
18	$(18-16)^2$	4
20	$(20-16)^2$	16
14	$(14-16)^2$	4
16	$(16-16)^2$	0
	Total	= 40

   $\frac{40}{5} = 8$

   $\sigma = \sqrt{8} \approx 2.828$

   The standard deviation of the times is approximately 2.828 min.

25. Strategy
To calculate the standard deviation:
→Find the mean number of rooms.
→Use the procedure for calculating standard deviation.

Solution
$$\bar{x} = \frac{234+321+222+246+312+396}{6} = \frac{1,731}{6} = 288.5$$

x	$(x-\bar{x})^2$	
234	$(234-288.5)^2$	2,970.25
321	$(321-2.885)^2$	1,056.25
222	$(222-288.5)^2$	4,422.25
246	$(246-288.5)^2$	1,806.25
312	$(312-288.5)^2$	552.25
396	$(396-288.5)^2$	11,556.25
	Total	= 22,363.5

$$\frac{22,363.5}{6} \approx 3,727.25$$

$$\alpha = \sqrt{3,727.25} \approx 61.051$$

The standard deviation of the number of rooms is approximately 61.051 rooms.

27. Strategy
To determine the place of the greater standard deviation of temperatures:
→Find the amount of the temperatures from the two places.
→Use the procedure for calculating standard deviation.
→Compare the answers and write the larger standard deviation.

Solution
For the desert resort:
$$\bar{x} = \frac{95+98+98+104+97+100+96+97+108+93+104}{11}$$

$$\bar{x} = \frac{1090}{11} \approx 99.091$$

x	$(x-\bar{x})^2$	
95	$(95-99.091)^2$	16.7363
98	$(98-99.091)^2$	1.1903
98	$(98-99.091)^2$	1.1903
104	$(104-99.091)^2$	24.0983
97	$(7-99.091)^2$	4.3723
100	$(100-99.091)^2$	0.8263
96	$(96-99.091)^2$	9.5543
97	$(97-99.091)^2$	4.3723
108	$(108-99.091)^2$	79,3703
98	$(93-99.091)^2$	37.1003
104	$(104-99.091)^2$	24,0983
	Total	= 202.9093

$$\frac{202.9093}{11} \approx 18.4463$$
$$\sigma = \sqrt{18.4463} \approx 4.295$$

For the Antarctic:
$$\bar{x} = \frac{27+28+28+30+28+27+30+25+24+26+21}{11} = \frac{294}{11} \approx 26.727$$

x	$(x-\bar{x})^2$	
27	$(27-26.727)^2$	0.0745
28	$(28-26.727)^2$	1.6205
28	$(28-26.727)^2$	1.6205
30	$(30-26.727)^2$	10.7125
28	$(28-26.727)^2$	1.6205
27	$(27-26.727)^2$	0.0745
30	$(30-26.727)^2$	10.7125
25	$(25-26.727)^2$	2.9825
24	$(24-26.727)^2$	7.4365
26	$(26-26.727)^2$	0.5285
21	$(21-26.727)^2$	32.7985
	Total	= 70.1815

$$\frac{70.1815}{11} \approx 6.3801$$
$$\sigma = \sqrt{6.3801} \approx 2.526$$

The standard deviation of the temperature is greater at the desert resort.

Critical Thinking 10.2

29. $\bar{x}_1 = \frac{85+92+86+89}{4} = \frac{352}{4} = 88$

    $\bar{x}_2 = \frac{90+97+91+94}{4} = \frac{372}{4} = 93$

    The mean scores of the two students are not the same.
    The mean score of the second student is 5 points higher.

Student 1:

x	$(x-\bar{x})^2$	
85	$(85-88)^2$	9
92	$(92-88)^2$	16
86	$(86-88)^2$	4
89	$(89-88)^2$	1
	Total	= 30

Student 2:

x	$(x-\bar{x})^2$	
90	$(90-93)^2$	9
97	$(97-93)^2$	16
91	$(91-93)^2$	4
94	$(94-93)^2$	1
	Total	= 30

$\frac{30}{4} = 7.5$

$\sigma_1 = \sqrt{7.5} \approx 2.74$

$\frac{30}{4} = 7.5$

$\sigma_2 = \sqrt{7.5} = 2.74$

Yes. The standard deviations of the two sets of scores are the same.

31. Answers will vary; for example, 10, 14, 14, 14, 16, 18, 19, 23.

Section 10.3

Objective A Exercises

1. Answers will vary. For example, lotteries and weather forecasts.

3. Possible outcomes of tossing 4 coins:

Q_1	Q_2	Q_3	Q_4
H	H	H	H
H	H	H	T
H	H	T	H
H	T	H	H
T	H	H	H
H	H	T	T
H	T	T	H
T	T	H	H
H	T	H	T
T	H	T	H
T	H	H	T
H	T	T	T
T	T	T	H
T	T	H	T
T	H	T	T
T	T	T	T

5. Possible outcomes from tossing two tetrahedral die:

(1, 1)   (2, 1)   (3, 1)   (4, 1)
(1, 2)   (2, 2)   (3, 2)   (4, 2)
(1, 3)   (2, 3)   (3, 3)   (4, 3)
(1, 4)   (2, 4)   (3, 4)   (4, 4)

7. No, because the dice are weighted so that some numbers occur more often than other numbers.

9. **Strategy**   To find the probability:
   → Refer to Exercise 3 to count the number of possible outcomes of the experiment.
   → Count the outcomes of the experiment that are favorable to the event HHTT.
   → Use the probability formula.

   Solution   There are 16 possible outcomes.
   There is 1 outcome favorable to E.
   $P(E) = \dfrac{\text{number of favorable outcomes}}{\text{number of possible outcomes}}$ $P(E) = \dfrac{1}{16}$
   The probability of HHTT is $\dfrac{1}{16}$.

11. Strategy   To find the probability:
→Refer to Exercise 3 to count the number of possible outcomes of the experiment.
→Count the outcomes of the experiment that are favorable to the event.
→Use the probability formula.

Solution   There are 16 possible outcomes.
There are 6 outcomes favorable to E: HHTT, HTTH, TTHH, HTHT, THTH, THHT

$$P(E) = \frac{\text{number of favorable outcomes}}{\text{number of possible outcomes}}$$

$$P(E) = \frac{1}{16} = \frac{3}{8}$$

The probability of the event is $\frac{3}{8}$.

13. Strategy   To find the probability:
→Refer to the table on page 606 to count the number of possible outcomes of the experiment.
→Count the outcomes of the experiment that are favorable to the event the sum is 5.
→Use the probability formula.

Solution   There are 36 possible outcomes.
There are 4 outcomes favorable to E: (1, 4), (2, 3), (3, 2), (4, 1)

$$P(E) = \frac{\text{number of favorable outcomes}}{\text{number of possible outcomes}}$$

$$P(E) = \frac{4}{36} = \frac{1}{9}$$

The probability that the sum is 5 is $\frac{1}{9}$.

15. Strategy   To find the probability:
→Refer to the table on page 606 to count the number of possible outcomes of the experiment.
→Count the outcomes of the experiment that are favorable to the event the sum is 15.
→Use the probability formula.

Solution   There are 36 possible outcomes.

$$P(E) = \frac{\text{number of favorable outcomes}}{\text{number of possible outcomes}}$$

$$P(E) = \frac{0}{36} = 0$$

The probability that the sum is 15 is 0.

17. Strategy   To find the probability:
→Refer to the table on page 606 to count the number of possible outcomes of the experiment.
→Count the outcomes of the experiment that are favorable to the event the sum is 2.
→Use the probability formula.

Solution   There are 36 possible outcomes.
There is 1 outcome favorable to E: (1, 1)

$$P(E) = \frac{\text{number of favorable outcomes}}{\text{number of possible outcomes}}$$

$$P(E) = \frac{1}{36}$$

The probability that the sum is 2 is $\frac{1}{36}$.

© Houghton Mifflin Company. All rights reserved.

19. Strategy   To find the probability:
→Count the number of possible outcomes of the experiment.
→Count the outcomes of the experiment that produce an 11.
→Use the probability formula.

Solution   A dodecahedral die has 12 sides.
There is 1 outcome favorable to E: 11

$$P(E) = \frac{\text{number of favorable outcomes}}{\text{number of possible outcomes}}$$

$$P(E) = \frac{1}{12}$$

The probability that the number is 11 is $\frac{1}{12}$.

21. Strategy   To find the probability:
→Refer to Exercise 5 to count the number of possible outcomes of the experiment.
→Count the outcomes of the experiment that are favorable to the event the sum is 4.
→Use the probability formula.

Solution   There are 16 possible outcomes.
There are 3 outcomes favorable to E: (1, 3), (2, 2), (3, 1)

$$P(E) = \frac{\text{number of favorable outcomes}}{\text{number of possible outcomes}}$$

$$P(E) = \frac{3}{16}$$

The probability that the sum is 4 is $\frac{3}{16}$.

23. Strategy   To find the probability:
→Count the number of possible outcomes of the experiment.
→Count the outcomes of the experiment that produce a number divisible by 4.
→Use the probability formula.

Solution   A dodecahedral die has 12 sides.
There are 3 outcomes favorable to E: 4, 8, 12

$$P(E) = \frac{\text{number of favorable outcomes}}{\text{number of possible outcomes}}$$

$$P(E) = \frac{3}{12} = \frac{1}{4}$$

The probability that the side is divisible by 4 is $\frac{1}{4}$.

25. Strategy   To find the empirical probability, use the probability formula and divide the number of observations of E (37) by the total number of observations (95).

Solution

$$P(E) = \frac{\text{number of observations of } E}{\text{total number of observations}}$$

$$P(E) = \frac{37}{95}$$

The probability is $\frac{37}{95}$ that a person prefers a cash discount.

Section 10.3   **255**

27. Strategy   To find the probability:
→Count the number of possible outcomes of the experiment.
→Count the number of outcomes of the experiment favorable to the event E, the light is green.
→Use the probability formula.

Solution   There are $5\frac{1}{4}\left(3+\frac{1}{4}+2\right)$ min for the lights to proceed through a complete cycle.

The green light lasts for a duration of 3 min.
$$P(E) = \frac{3}{5\frac{1}{4}} = \frac{3}{\frac{21}{4}} = 3 \div \frac{21}{4} = \frac{3}{1} \cdot \frac{4}{21} = \frac{4}{7}$$

The probability of having a green light is $\frac{4}{7}$.

29. Strategy   To find the probability:
→Count the number of possible outcomes of the experiment.
→Count the number of outcomes of the experiment favorable to the event E, the service is satisfactory or excellent.
→Use the probability formula.

Solution   There are 377(98 + 87 + 129 + 42 + 21) outcomes of the survey.
There are 185(98 + 87) outcomes favorable to E.
$$P(E) = \frac{185}{377}$$

The probability is $\frac{185}{377}$ that the cable service is rated satisfactory or excellent.

31. Strategy   To find the odds:
→Count the favorable outcomes.
→Count the unfavorable outcomes.
→Use the odds in favor of an event formula.

Solution   Number of favorable outcomes: 1
Number of unfavorable outcomes: 1
$$\text{Odds in favor} = \frac{\text{number of favorable outcomes}}{\text{number of unfavorable outcomes}}$$
$$\text{Odds in favor} = \frac{1}{1}$$

The odds of showing heads is 1 TO 1.

33. Strategy   To calculate the probability of winning, use the odds in favor fraction. The probability of winning is the ratio of the numerator to the sum of the numerator and denominator.

Solution   Probability of winning $= \frac{3}{3+2} = \frac{3}{5}$

The probability of winning the election is $\frac{3}{5}$.

35. **Strategy** To find the odds:
→ Count the favorable outcomes of the experiment.
→ Count the unfavorable outcomes of the experiment.
→ Use the formula for the odds in favor of an event.

Solution Use the table on page 630 to:
Count the favorable outcomes: 6
Count the unfavorable outcomes: 30
Odds in favor $= \frac{6}{30} = \frac{1}{5}$
The odds in favor of rolling a 7 are $\frac{1}{5}$.

37. **Strategy** To find the odds:
→ Determine the unfavorable outcomes.
→ Determine the favorable outcomes.
→ Use the formula for the odds against an event.

Solution There are 48 ways of not picking an ace.
There are 4 way of picking an ace.
Odds against $= \frac{48}{4} = \frac{12}{1}$
The odds against picking an ace are $\frac{12}{1}$.

39. **Strategy** To calculate the probability of winning:
→ Restate the odds against as odds in favor.
→ Using the odds in favor fraction, the probability of winning is the ratio of the numerator to the sum of the numerator and denominator.

Solution The odds against winning are 40 TO 1. Therefore, the odds in favor of winning are 1 TO 40.
Probability of winning:
$\frac{1}{1+40} = \frac{1}{41}$
The probability of winning the Super Bowl is $\frac{1}{41}$.

41. **Strategy** To calculate the probability of the stock going down:
→ Restate the odds of the stock going up as the odds of the stock going down.
→ Using the odds in favor of the stock going down, the probability of going down is the ratio of the numerator to the sum of the numerator and denominator.

Solution The odds in favor of the stock going up: 2 TO 1
The odds in favor of the stock going down: 1 TO 2
Probability of going down:
$= \frac{1}{1+2} = \frac{1}{3}$
The probability of the stock going down is $\frac{1}{3}$.

Critical Thinking 10.3

43. The sum of the probabilities that a ball is chosen is 1.
$\frac{1}{2} + \frac{1}{3} + \frac{1}{9} = \frac{18}{36} + \frac{12}{36} + \frac{4}{36} = \frac{34}{36}$. The sum of the probabilities is not 1.

45. The probability of tossing a fair coin and having it land heads is $\frac{1}{2}$. This does not mean that if the coin is tossed 100 times, it will land heads 50 times. Students should explain the difference between theoretical probabilities, which are obtained from logical reasoning, and empirical probabilities, which are obtained from experimental data.

Chapter Review Exercises

1. Strategy To prepare the frequency distribution table, use 7 as the class width and 12 as the lower class boundary.

Solution Number of Students in Math Classes

Classes	Tally	Frequency
12–19	/////	5
20–27	/////////	9
28–35	///////////	11
36–43	/////////	9
44–51	////	4
52–59	//	2

2. Strategy To find the class with the greatest frequency, refer to the frequency table in Exercise 1.

Solution The class with the greatest frequency is 28–35.

3. Strategy To find the number of classes:
→Use Exercise 1 to find the number of classes with 12–19 students, 20–27 students, and 28–35 students.
→Add the numbers.

Solution Number of classes with 12–19 students: 5
Number of classes with 20–27 students: 9
Number of classes with 28–35 students: 11
$5 + 9 + 11 = 25$
25 of the math classes have 35 or fewer students.

4. Strategy To find the percent:
→Use Exercise 1 to find the number of classes with 44–51 students and with 52–59 students.
→Add the numbers.
→Use the basic percent equation.

Solution Number of classes with 44–51 students: 4
Number of classes with 52–59 students: 2
$4 + 2 = 6$
$pB = A$
$P(40) = 6$
$p = \dfrac{6}{40} = 0.15$
15% of the math classes have 44 or more students.

5. Strategy To find the percent:
→Use Exercise 1 to find the number of classes with 12–19 students and with 20–27 students.
→Add the numbers.
→Use the basic percent equation.

Solution Number of classes with 12–19 students: 5
Number of classes with 20–27 students: 9
$5 + 9 = 14$
$pB = A$
$p(40) = 14$
$p = \dfrac{14}{40} = 0.35$
35% of the math classes have 27 or fewer students.

6. Strategy To find the number of days the temperature was 45° or above:
→Read the histogram to find the number of days the temperature was 45°–50° and the number of days the temperature was 50°–55°.
→Add the numbers.

Solution Number of days the temperature was 45°–50°: 15
Number of days the temperature was 50°–55°: 10
$15 + 10 = 25$
The temperature was 45° or above on 25 days.

7. **Strategy** To find the number of days the temperature was 25° or below:
→Read the histogram to find the number of days the temperature was 10°–15°, the number of days the temperature was 15°–20°, and the number of days the temperature was 20°–25°.
→Add the numbers.

   Solution Number of days the temperature was 10°–15°: 2
Number of days the temperature was 15°–20°: 15
Number of days the temperature was 20°–25°: 12
2 + 15 + 12 = 29
The temperature was 25° or below on 29 days.

8. **Strategy** To calculate the mean:
→Calculate the sum of the cholesterol levels.
→Divide by 11.
To calculate the median:
→Arrange the numbers from smallest to largest. The median is the middle number.

   Solution The sum of the numbers is 2,360.
$\bar{x} = \frac{2,360}{11} = 214.\overline{54}$
The mean of the cholesterol levels is $215.\overline{54}$.
160 180 190 200 210 210
220 230 230 250 280
The median cholesterol level is 210.

9. **Strategy** To calculate the mean:
→Calculate the sum of the weights.
→Divide by 10.
To calculate the median:
→Arrange the numbers from smallest to largest.
→Because there is an even number of values, the median is the sum of the two middle numbers divided by 2.

   Solution The sum of the numbers is 71.7.
$\bar{x} = \frac{71.7}{10} = 7.17$.
The mean weight of the babies is 7.17 lb.
5.6   5.9   6.3   6.5   6.9
7.2   7.2   8.1   8.9   9.1
median = $\frac{6.9 + 7.2}{2} = 7.05$
The median weight of the babies is 7.05 lb.

10. **Strategy** To find the modal response:
→Find the response that was recorded most frequently.

    Solution The response "good" was mentioned most frequently and thus is the modal response.

11. **Strategy** To find the number of shares:
→Read the frequency polygon to find the number of shares sold between 7 A.M.–8 A.M., 8 A.M.–9 A.M., and 9 A.M.–10 A.M.
→Add the numbers.

    Solution Number of shares sold 7 A.M.–8 A.M.: 25
Number of shares sold 8 A.M.–9 A.M.: 13
Number of shares sold 9 A.M.–10 A.M.: 17
25 + 13 + 17 = 55
55 million shares of stock were sold between 7 A.M. and 10 A.M.

12. Strategy   To find the number of shares:
   → Read the frequency polygon to determine when less than 15 million shares sold.

   Solution   Less than 15 million shares sold between 8 A.M. and 9 A.M.

13. Strategy   To find the ratio:
   → Read the frequency polygon to find the number of shares sold between 10 A.M.–11 A.M. and between 11 A.M.–12 P.M.
   → Write the ratio in lowest terms.

   Solution   Number of shares sold 10 A.M.–11 A.M.: 15
   Number of shares sold 11 A.M.–12 P.M.: 25

   $$\frac{15 \text{ million}}{25 \text{ million}} = \frac{15}{25} = \frac{3}{5}$$

   The ratio is $\frac{3}{5}$.

14. Strategy   To prepare the box-and-whiskers plot:
   → Arrange the data from smallest to largest. Then find the median.
   → Find Q_1, the median of the lower half of the data.
   → Find Q_3, the median of the upper half of the data.
   → Determine the smallest and largest data values.
   → Draw the box-and-whiskers plot.

   Solution   89   99   102   105   109   110   110   111
   116   120   121   124   124   131   134
   median = 111
   $Q_1 = 105$
   $Q_3 = 124$
   Smallest value: 89
   Largest value: 134

   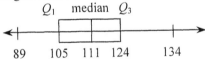

15. **Strategy** To calculate the standard deviation:
   → Find the mean of the average miles per gallon.
   → Use the procedure for calculating the standard deviation.

   Solution $\bar{x} = \dfrac{177}{6} = 29.5$

x	$(x-\bar{x})^2$	
24	$(24-29.5)^2$	30.25
28	$(28-29.5)^2$	2.25
22	$(22-29.5)^2$	56.25
35	$(35-29.5)^2$	30.25
41	$(41-29.5)^2$	132.25
27	$(27-29.5)^2$	6.25
	Total	= 257.5

   $\dfrac{257.5}{6} \approx 42.9167$

   $\sigma = \sqrt{42.9167} \approx 6.55$
   The standard deviation is 6.551 mpg.

16. **Strategy** Use the probability formula.

    Solution There are 2,500 possible outcomes.
    There are 5 outcomes favorable to E.
    $P(E) = \dfrac{\text{number of favorable outcomes}}{\text{total number of outcomes}}$
    $P(E) = \dfrac{5}{2{,}500} = \dfrac{1}{500}$
    The probability of winning the television is $\dfrac{1}{500}$.

17. **Strategy** Use the formula for finding the odds in favor.

    Solution Number of favorable outcomes: 15
    Number of unfavorable outcomes: 50 − 15 = 35
    Odds in Favor $= \dfrac{\text{number of favorable outcomes}}{\text{number of unfavorable outcomes}}$
    $= \dfrac{15}{35} = \dfrac{3}{7}$
    The odds of the ball being red are $\dfrac{3}{7}$.

18. **Strategy** To calculate the probability:
    → Restate the odds against as odds in favor.
    → Using the odds in favor fraction, the probability of winning is the ratio of the numerator to the sum of the numerator and denominator.

    Solution The odds against winning are 5 TO 2, therefore the odds in favor of winning are 2 TO 5.
    Probability of winning $= \dfrac{2}{2+5} = \dfrac{2}{7}$
    The probability of winning is $\dfrac{2}{7}$.

19. **Strategy** To calculate the probability:
   → Count the number of possible outcomes.
   → Count the number of favorable outcomes of the experiment E.
   → Use the probability formula.

   Solution There are 12 possible outcomes of the experiment.
   There are 2 outcomes favorable to E: 6, 12
   $$P(E) = \frac{2}{12} = \frac{1}{6}$$
   The probability is $\frac{1}{6}$ that the number will be divisible by 6.

20. **Strategy** To calculate the probability:
   → Find the number of possible outcomes.
   → Find the number of favorable outcomes.
   → Use the probability formula.

   Solution There are 14(3 + 4 + 5 + 2) possible outcomes.
   There are 5 favorable outcomes.
   $$P(E) = \frac{5}{14}$$
   The probability is $\frac{5}{14}$ that the student is a junior.

Chapter Test

1. **Strategy** To find the number of residences where the cost was $60 or above:
   → Read the histogram to find the number of customers who had monthly cost of over $60 was $60–$80 and the number of customers when the cost was $80–$100 and the number of customers when the cost was $100–$120.
   → Add the numbers.

   Solution Number of customers when the cost was $60–$80: 40
   Number of customers when the cost was $80–$100: 15
   Number of customers when the cost was $100–$120: 10
   40 + 15 + 10 = 65
   The number of residences having costs for telephone service of $60 or more is 65.

2. **Strategy** To find the total gross sales:
   → Read the frequency polygon to find the total gross sales in January and February.
   → Add the numbers.

   Solution Total gross sales in January: 25,000
   Total gross sales in February: 30,000
   25,000 + 30,000 = 55,000
   The total gross sales for Jan. and Feb. was $55,000.

3. **Strategy** To find the percent:
   → Use the frequency table to find the number of restaurants that had annual sales between $750,000 and $1,000,000.
   → Use the basic percent equation.

   Solution Number of restaurants = 8
   Percent · base = amount
   $p(50) = 8$
   $$p = \frac{8}{50} = 0.16$$
   16% of the restaurants had annual sales between $750,000 and $1,000,000.

4. **Strategy** To find the mean bowling score:
→Determine the sum of the bowling scores.
→Divide the sum by 8.

   Solution The sum of the scores is 1,223.
   $\bar{x} = \dfrac{1,223}{8} = 152.875$
   The mean bowling score is 152.875.

5. **Strategy** To find the median response time:
→Arrange the numbers from smallest to largest.
→The median is the middle term.

   Solution 8 8 11 11 14 15 17 21 22
   median = 14
   The median response time is 14 minutes.

6. **Strategy** To find the modal response, write down the category that received the most responses.

   Solution Because a response of very good was recorded most frequently, the modal response is very good.

7. **Strategy** To find the number to be sold:
→Let n represent the number of to be sold.
→Use the formula for the mean to solve for n.

   Solution $35 = \dfrac{34 + 28 + 31 + 36 + 38 + n}{6}$
   $210 = 167 + n$
   $43 = n$
   In the sixth month, 43 digital assistants must be sold.

8. **Strategy** To find the first quartile:
→Arrange the data from smallest to largest, and then find the median.
→Find Q_1, the median of the lower half of the data.

   Solution 6.5 8.6 9.3 9.8 9.8 10.5
   10.5 11.2 11.9 17.3 18.5
   19.6 20.3 2.10
   median $= \dfrac{10.5 + 11.2}{2} = 10.85$
   $Q_1 = 9.8$
   The first quartile is 9.8.

9. **Strategy** To find the range, subtract the smallest value from the largest value.
   To find the median, look at the median of the box-and-whisker plot.

   Solution $26 - 4 = 22$
   The range is 22.
   The median is 14 vacation days.

10. **Strategy** To draw the box-and-whiskers plot:
→Arrange the data from smallest to largest, and then find the median.
→Find Q_1, the median of the lower half of the data.
→Find Q_3, the median of the upper half of the data.
→Determine the smallest and largest data values.
→Draw the box-and-whiskers plot.

Solution
68 69 70 70 70 71 72
73 73 74 74 75 76 80

$$\text{median} = \frac{72+73}{2} = 72.5$$

$Q_1 = 70$
$Q_3 = 74$
Smallest value: 68
Largest value: 80

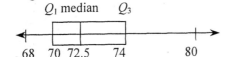

11. **Strategy** To calculate the standard deviation:
→Find the mean number of incorrect answers.
→Use the procedure for calculating standard deviation.

Solution
$$\bar{x} = \frac{2+0+3+1+0+4+5+1+3+1}{10} = \frac{20}{10} = 2$$

x	$(x-\bar{x})^2$	
2	$(2-2)^2$	0
0	$(0-2)^2$	4
3	$(3-2)^2$	1
1	$(1-2)^2$	1
0	$(0-2)^2$	4
4	$(4-2)^2$	4
5	$(5-2)^2$	9
1	$(1-2)^2$	1
3	$(3-2)^2$	1
1	$(1-2)^2$	1
	Total	= 26

$$\frac{26}{10} = 2.6$$

$\sigma = \sqrt{2.6} \approx 1.612$
The standard deviation of the number of incorrect answers is approximately 1.612.

12. The possible outcomes of tossing a coin and then a regular die: (H, 1), (H, 2), (H, 3), (H, 4), (H, 5), (H, 6), (T, 1), (T, 2), (T, 3), (T, 4), (T, 5), (T, 6). There are 12 elements in the sample space:

13. The possible outcomes of stacking a nickel, dime and quarter: (N, D, Q), (N, Q, D), (D, N, Q), (D, Q, N), (Q, N, D), (Q, D, N)

14. Strategy   To calculate the probability:
→Find the number of possible outcomes.
→Find the number of favorable outcomes.
→Use the probability formula.

Solution   There are 248(14 + 32 + 202) possible outcomes.
There are 32 favorable outcomes.
$$P(E) = \frac{32}{248} = \frac{4}{31}$$
The probability is $\frac{4}{31}$ that the person is in business class.

15. Strategy   To calculate the probability:
→Find the number of possible outcomes.
→Find the number of favorable outcomes.
→Use the probability formula.

Solution   There are 6 possible outcomes: (A, K, Q), (A, Q, K), (K, A, Q), (K, Q, A), (Q, A, K), (Q, K, A).
There are 2 favorable outcomes.
$$P(E) = \frac{2}{6} = \frac{1}{3}$$
The probability is $\frac{1}{3}$ that the ace is on top of the stack.

16. Strategy   To calculate the probability:
→Find the number of possible outcomes.
→Find the number of favorable outcomes.
→Use the probability formula.

Solution   There are 8 possible outcomes.
There is 1 favorable outcomes.

$$P(E) = \frac{1}{8}$$
The probability is $\frac{1}{8}$ that the student will answer all three questions correctly.

17. Strategy   To calculate the probability:
→Find the number of possible outcomes.
→Find the number of favorable outcomes.
→Use the probability formula.

Solution   There are 45(15 + 20 + 10) possible outcomes.
There are 30(20 + 10) favorable outcomes.
$$P(E) = \frac{30}{45} = \frac{2}{3}$$
The probability is $\frac{2}{3}$ that the seed is not for a red flower.

18. Strategy   To calculate the probability of winning by using the odds in favor fraction, the probability of winning is the ratio of the numerator to the sum of the numerator and denominator.

Solution   Probability of winning:
$$\frac{1}{1+12} = \frac{1}{13}$$
The probability of winning the lottery is $\frac{1}{13}$.

19. Strategy   To find the odds:
→Find the unfavorable outcomes.
→Use the odds in favor of an event formula.

Solution   There is 1 favorable outcome.
There are 8(9 −1) unfavorable outcomes:
Odds in favor = $\frac{1}{8}$
The odds of in favor of rolling a nine is 1 TO 8.

20. Strategy   To calculate the probability:
    → Find the number of possible outcomes.
    → Find the number of favorable outcomes.
    → Use the probability formula.

    Solution   There are 12 possible outcomes
                There are 5 favorable outcomes
                $P(E) = \dfrac{5}{12}$

                The probability is $\dfrac{5}{12}$ that the number on the upward face is less than six.

Cumulative Review Exercises

1. $\sqrt{200} = \sqrt{100 \cdot 2} = \sqrt{100} \cdot \sqrt{2} = 10\sqrt{2}$

2. $7p - 2(3p - 1) = 5p + 6$
   $7p - 6p + 2 = 5p + 6$
   $p + 2 = 5p + 6$
   $p - 5p + 2 = 5p - 5p + 6$
   $-4p + 2 = 6$
   $-4p + 2 - 2 = 6 - 2$
   $-4p = 4$
   $\dfrac{-4p}{-4} = \dfrac{4}{-4}$
   $p = -1$
   The solution is -1.

3. $3a^2b - 4ab^2$
   $3(-1)^2(2) - 4(-1)(2^2)$
   $= 3(1)(2) - 4(-1)(4)$
   $= 3(2) - 4(-1)(4)$
   $= 6 - 4(-1)(4)$
   $= 6 - (-4)(4)$
   $= 6 - (-16)$
   $= 6 + 16$
   $= 22$

4. $-2[2 - 4(3x - 1) + 2(3x - 1)]$
   $= -2[2 - 12x + 4 + 6x - 2]$
   $= -2[-6x + 4]$
   $= 12x - 8$

5. $-\dfrac{2}{3}y - 5 = 7$
   $-\dfrac{2}{3}y - 5 + 5 = 7 + 5$
   $-\dfrac{2}{3}y = 12$
   $-\dfrac{3}{2}\left(-\dfrac{2}{3}\right)y = -\dfrac{3}{2}(12)$
   $y = -18$
   The solution is -18.

6. $-\dfrac{4}{5}\left(\dfrac{3}{4} - \dfrac{7}{8} - \left(\dfrac{2}{3}\right)^2\right) = -\dfrac{4}{5}\left(\dfrac{3}{4} - \dfrac{7}{8} - \dfrac{4}{9}\right)$
   $= -\dfrac{4}{5}\left(\dfrac{54}{72} - \dfrac{63}{72} + \dfrac{-32}{72}\right)$
   $= -\dfrac{4}{5}\left(\dfrac{54 - 63 - 32}{72}\right)$
   $= -\dfrac{4}{5}\left(\dfrac{-41}{72}\right)$
   $= \dfrac{41}{90}$

7.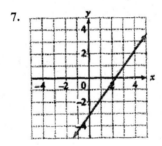

8.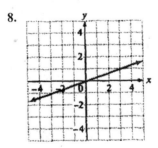

9. $(7y^2 + 5y - 8) - (4y^2 - 3y + 1)$
   $= (7y^2 + 5y - 8) + (-4y^2 + 3y - 1)$
   $= 3y^2 + 8y - 9$

10. $(4a^2b)^3 = 4^{1 \cdot 3} a^{2 \cdot 3} b^{1 \cdot 3} = 4^3 a^6 b^3$
    $= 64a^6b^3$

11. Strategy To find the base, solve the basic percent equation.
   Percent = $16\frac{2}{3}\% = \frac{50}{300} = \frac{1}{6}$, base = n, amount = 24.

   Solution Percent · base = amount
   $\frac{1}{6}n = 24$
   $\left(\frac{6}{1}\right)\left(\frac{1}{6}\right)n = \left(\frac{6}{1}\right)(24)$
   $n = 144$
   $16\frac{2}{3}\%$ of 144 is 24.

12. $\frac{9}{8} = \frac{3}{n}$
   $9n = 8(3)$
   $9n = 24$
   $n = \frac{24}{9}$
   $n = \frac{8}{3}$
   The solution is $\frac{8}{3}$.

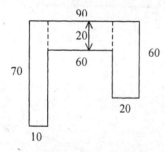

13. $87,600,000,000 = 8.76 \times 10^{10}$

14. Strategy The area of the composite figure is equal to the area of three rectangles, as shown in the figure.

   Solution $A = LW$
   $A_1 = 70(10) = 700$
   $A_2 = 20(60) = 1,200$
   $A_3 = 20(60) = 1,200$
   $A = 700 + 1,200 + 1,200$
   $A = 3,100$
   The area of the patio is 3,100 ft².

15. $(5c^2d^4)(-3cd^6)$
   $= [5(-3)](c^{2+1})(d^{4+6})$
   $= -15c^3d^{10}$

16. 40 km = 40,000 m

17. Strategy

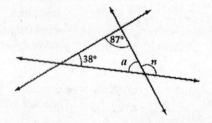

   →To find the measure of $\angle a$, use the fact that the sum of the interior angles of a triangle is 180°.
   →To find the measure of $\angle n$, use the fact that the sum of an interior and an exterior angle is 180°.

   Solution $38° + 87° + \angle a = 180°$
   $125° + \angle a = 180°$
   $\angle a = 55°$
   $\angle a + \angle n = 180°$
   $55° + \angle n = 180°$
   $\angle n = 125°$
   The measure of $\angle n$ is 125°.

18. Strategy To find the area, use the formula for the area of a parallelogram. Substitute 8 for b and 4 for h. Solve for A.

   Solution $A = bh$
   $A = 8(4)$
   $A = 32$
   The area of the parallelogram is 32 m².

19. Strategy To find the simple interest, solve the formula $I = Prt$ for I.
   $P = 25,000$, $r = 0.075$, $t = \frac{3}{12}$

   Solution $I = Prt$
   $I = 25,000(0.075)\left(\frac{3}{12}\right)$
   $I = 468.75$
   The simple interest on the loan is $468.75.

20. Strategy To calculate the probability:
→Count the number of possible outcomes of the experiment.
→Count the number of outcomes that are favorable to the event of the ball is not white.
→Use the probability formula.

Solution There are 36(12 + 15 + 9) possible outcomes.
There are 24(15 + 9) outcomes favorable to E.
$$P(E) = \frac{24}{36} = \frac{2}{3}$$
The probability is $\frac{2}{3}$ that the ball is not white.

21. Strategy To calculate the mean score:
→Calculate the sum of the scores.
→Divide by the number of scores.
To find the median score:
→Arrange the scores from smallest to largest.
→Because there is an even number of values, the median is the sum of the two middle numbers divided by 2.

Solution The sum of the numbers is 186.
$$\bar{x} = \frac{186}{6} = 31$$
The mean score on the six tests is 31.
22 24 31
34 37 38
median $= \frac{31 + 34}{2} = 32.5$
The median score on the six tests is 32.5.

22. Strategy To find the percent:
→Find the number of voters registered who did not vote.
→Use the basic percent equation.
Percent = n, base = 230,000, amount = number not voting.

Solution 230,000 − 55,000 = 175,000
percent · base = amount
$n \cdot 230{,}000 = 175{,}000$
$n = \frac{175{,}000}{230{,}000} \approx 0.761$
76.1% of the registered voters did not vote.

23. Strategy To find the circumference of the earth, write and solve a proportion using n to represent the circumference.

Solution $\frac{7.5}{360} = \frac{1{,}600}{n}$
$7.5n = 360(1{,}600)$
$7.5n = 576{,}000$
$n = \frac{576{,}000}{7.5} = 76{,}800$
The circumference of the earth is approximately 76,800 km.

24. Strategy To find the standard deviation:
→Find the mean of the rainfall totals.
→Use the procedure for calculating the standard deviation.

Solution $\bar{x} = \frac{12 + 16 + 20 + 18 + 14}{5} = 16$

x	$(x - \bar{x})^2$	
12	$(12 - 16)^2$	16
16	$(16 - 16)^2$	0
20	$(20 - 16)^2$	16
18	$(18 - 16)^2$	4
14	$(14 - 16)^2$	4
	Total	= 40

$\frac{40}{5} = 8$

$\sigma = \sqrt{8} = 2.828$
The standard deviation of the rainfall totals is 2.828 in.

25. Strategy   To find the wage before the increase, write and solve an equation, using n to represent the wage before the increase.

   Solution $0.10n + n = 19.80$
   $1.10n = 19.80$
   $n = \dfrac{19.80}{1.10}$
   $n \approx 18.00$
   The hourly wage before the increase was $18.00 per hour.

FINAL EXAMINATION

1. $\begin{array}{rcr} 672 & \to & 700 \\ 843 & \to & 800 \\ 509 & \to & 500 \\ 417 & \to & +400 \\ \hline & & 2{,}400 \end{array}$

2. $18 + 3(6-4)^2 \div 2 = 18 + 3(2)^2 \div 2$
   $= 18 + 3(4) \div 2$
   $= 18 + 12 \div 2$
   $= 18 + 6$
   $= 24$

3. $-8 - (-13) - 10 + 7 = -8 + 13 + (-10) + 7$
   $= 5 + (-10) + 7$
   $= -5 + 7$
   $= 2$

4. $|a - b| - 3bc^3$
   $|-2 - 4| - 3(4)(-1)^3 = |-6| - 3(4)(-1)^3$
   $= 6 - 3(4)(-1)^3$
   $= 6 - 3(4)(-1)$
   $= 6 - (12)(-1)$
   $= 6 - (-12)$
   $= 6 + 12$
   $= 18$

5. $5\frac{3}{8} - 2\frac{11}{16} = 5\frac{6}{16} - 2\frac{11}{16}$
   $= 4\frac{22}{16} - 2\frac{11}{16}$
   $= 2\frac{11}{16}$

6. $\frac{7}{9} \div \frac{5}{6} = \frac{7}{9} \cdot \frac{6}{5}$
   $= \frac{7 \cdot 6}{9 \cdot 5}$
   $= \frac{7 \cdot 2 \cdot 3}{3 \cdot 3 \cdot 5}$
   $= \frac{14}{15}$

7. $\dfrac{\frac{3}{4} - \frac{1}{2}}{\frac{5}{8} + \frac{1}{2}} = \dfrac{\frac{3}{4} - \frac{2}{4}}{\frac{5}{8} + \frac{4}{8}}$
   $= \dfrac{\frac{1}{4}}{\frac{9}{8}} = \frac{1}{4} \div \frac{9}{8}$
   $= \frac{1}{4} \cdot \frac{8}{9}$
   $= \frac{1 \cdot 8}{4 \cdot 9}$
   $= \frac{1 \cdot 2 \cdot 2 \cdot 2}{2 \cdot 2 \cdot 3 \cdot 3}$
   $= \frac{2}{9}$

8. $\frac{5}{16} = 0.3125$
   $0.3125 < 0.313$
   $\frac{5}{6} < 0.313$

9. $-10qr$
   $-10(-8.1)(-9.5) = 81(-9.5)$
   $= -769.5$

10. $-15.32 \div 4.67 \approx -3.28$

11. $-90y = 45$
    $\dfrac{-90(-0.5)\ |\ 45}{45 = 45}$
    Yes, -0.5 is a solution of the equation.

12. $\sqrt{162} = \sqrt{81 \cdot 2}$
    $= \sqrt{81} \cdot \sqrt{2}$
    $= 9\sqrt{2}$

13. Draw a bracket at -4.
    Draw a heavy line to the right of -4.
    Draw an arrow at the right of the line.

14. $-\frac{5}{6}(-12t) = 10t$

15. $2(x - 3y) - 4(x + 2y) = 2x - 6y - 4x - 8y$
    $= (2x - 4x) + (-6y - 8y)$
    $= -2x - 14y$

16. $(5z^3 + 2z^2 - 1) - (4z^3 + 6z - 8)$
    $= (5z^3 + 2z^2 - 1) + (-4z^3 - 6z + 8)$
    $= z^3 + 2z^2 - 6z + 7$

17. $(4x^2)(2x^5 y) = (4 \cdot 2)(x^{2+5})(y) = 8x^7 y$

18. $2a^2b^2(5a^2 - 3ab + 4b^2)$
$= 2a^2b^2(5a^2) - (2a^2b^2)(3ab)$
$+ (2a^2b^2)(4b^2)$
$= 10a^4b^2 - 6a^3b^3 + 8a^2b^4$

19. $(3x - 2)(5x + 3)$
$= 3x(5x) + 3x(3) - 2(5x) - 2(3)$
$= 15x^2 + 9x - 10x - 6$
$= 15x^2 - x - 6$

20. $(3x^2y)^4 = 3^{1 \cdot 4} x^{2 \cdot 4} y^{1 \cdot 4} = 3^4 x^8 y^4$
$= 81x^8y^4$

21. $4^{-3} = \dfrac{1}{4^3} = \dfrac{1}{64}$

22. $\dfrac{m^5 n^8}{m^3 n^4} = m^{5-3} n^{8-4} = m^2 n^4$

23. $2 - \dfrac{4}{3}y = 10$
$2 - 2 - \dfrac{4}{3}y = 10 - 2$
$-\dfrac{4}{3}y = 8$
$\left(-\dfrac{3}{4}\right)\left(-\dfrac{4}{3}y\right) = -\dfrac{3}{4}(8)$
$y = -6$
The solution is -6.

24. $6z + 8 = 5 - 3z$
$6z + 3z + 8 = 5 - 3z + 3z$
$9z + 8 = 5$
$9z + 8 - 8 = 5 - 8$
$9z = -3$
$\dfrac{9z}{9} = \dfrac{-3}{9}$
$z = -\dfrac{1}{3}$
The solution is $-\dfrac{1}{3}$.

25. $8 + 2(6c - 7) = 4$
$8 + 12c - 14 = 4$
$12c - 6 = 4$
$12c - 6 + 6 = 4 + 6$
$12c = 10$
$\dfrac{12c}{12} = \dfrac{10}{12}$
$c = \dfrac{5}{6}$
The solution is $\dfrac{5}{6}$.

26. $2.48 \text{ m} = 248$ cm

27. $2.6 \text{ mi} = \dfrac{2.6 \text{ mi}}{1} \cdot \dfrac{5,280 \text{ ft}}{1 \text{ mi}} = 13,728$ ft

28. $\dfrac{n + 2}{8} = \dfrac{5}{12}$
$(n + 2) \cdot 12 = 8 \cdot 5$
$12n + 24 = 40$
$12n = 16$
$n = \dfrac{16}{12}$
$n = \dfrac{4}{3}$
The solution is $\dfrac{4}{3}$.

29. Strategy

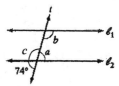

→ To find the measure of ∠a, use the fact that ∠a and the 74° angle are vertical angles.
→ Find the measure of ∠c by using the fact that adjacent angles of intersecting lines are supplementary.
→ To find the measure of ∠b, use the fact that ∠b and ∠c are alternate interior angles of parallel lines.

Solution ∠a = 74°
∠a + ∠c = 180°
74° + ∠c + 180°
∠c = 106°
∠b = 106°
The measure of ∠a is 74°.
The measure of ∠b is 106°.

270 Chapter 10: *Statistics and Probability*

30. Strategy To find the hypotenuse, use the Pythagorean Theorem. $a = 7, b = 8$.

Solution
$c^2 = a^2 + b^2$
$c^2 = 7^2 + 8^2$
$c^2 = 49 + 64$
$c^2 = 113$
$c = \sqrt{113}$
$c \approx 10.6$
The length of the hypotenuse is approximately 10.6 ft.

31. Strategy The perimeter is equal to three sides of a rectangle plus one half the circumference of a circle.

Solution
$P = 2L + W + \frac{1}{2}\pi d$
$P = 2(7) + 6 + \frac{1}{2}\pi(6)$
$P = 14 + 6 + 3\pi$
$P = 20 + 3\pi$
$P \approx 29.42$
The perimeter is approximately 29.42 cm.

32. Strategy The volume is equal to the volume of the rectangular solid minus one half the volume of the cylinder. The radius of the cylinder is one half the diameter of the cylinder.

Solution
$r = \frac{1}{2}d = \frac{1}{2}(1) = \frac{1}{2}$
$V = LWH - \frac{1}{2}\pi r^2 h$
$V = 8(4)(3) - \frac{1}{2}\pi\left(\frac{1}{2}\right)^2(8)$
$V = 96 - \pi$
$V \approx 92.86$
The volume of the solid is approximately 92.86 in^3.

33.

x	y
-1	5
1	1
3	-3

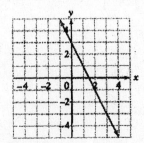

34.

x	y
0	-4
5	-1
2	$-2\frac{4}{5}$

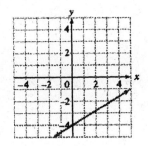

35. Strategy To find the ground speed, substitute 22 for h and 386 for a in the given formula and solve for g.

Solution
$g = a - h$
$g = 386 - 22$
$g = 364$
The ground speed of the airplane is 364 mph.

Final Examination 271

36. **Strategy** To find the number products:
    → Convert 8 h to minutes using the conversion factor $\frac{60 \text{ min}}{1 \text{ h}}$.
    → Divide the total number of minutes by $1\frac{1}{2}$.

    Solution $8 \text{ h} = \frac{8 \text{ h}}{1} \cdot \frac{60 \text{ min}}{1 \text{ h}} = 480 \text{ min}$
    $480 \div 1\frac{1}{2} = 480 \div \frac{3}{2}$
    $= 480 \cdot \frac{2}{3}$
    $= 320$
    The worker can inspect 320 products in one day.

37. **Strategy** To find the difference, subtract the melting point of bromine (−7.2°) from the boiling point of bromine (58.78°).

    Solution $58.78° - (-7.2°)$
    $= 58.78° + 7.2°$
    $= 65.98°$
    The difference between the melting point and the boiling point is 65.98°C.

38. $5,880,000,000,000 = 5.88 \times 10^{12}$

39. **Strategy** The distance of the fulcrum from the 50-pound child: x
    The distance of the fulcrum from the 75-pound child: $10 - x$
    To find the placement of the fulcrum, replace the variables F_1, F_2, and d by the given variables and solve for x.

    Solution $F_1 x = F_2(d - x)$
    $50 \cdot x = 75(10 - x)$
    $50x = 750 - 75x$
    $125x = 750$
    $x = \frac{750}{125}$
    $x = 6$
    The fulcrum is 6 ft from the 50-pound child.

40. **Strategy** To find the number of tickets, write and solve an equation, using x to represent the number of tickets.

    Solution $10.50 + 52.50x = 325.50$
    $52.50x = 315$
    $x = \frac{315}{52.50}$
    $x = 6$
    You purchased 6 tickets.

41. **Strategy** To find the property tax, write and solve a proportion using n to represent the amount of tax.

    Solution $\frac{3,750}{250,000} = \frac{n}{314,000}$
    $3,750(314,000) = 250,000(n)$
    $117,750,000 = 250,000n$
    $\frac{117,750,000}{250,000} = n$
    $4,710 = n$
    The property tax is $4,710.

42. **Strategy** To find the percent of the states that have a land area of 75,000 mi² or more:
    → Add the number of states that have a land area of 75,000–100,000 mi² (8) and the number that have a land area of 100,000 mi² (8).
    → Use the basic percent equation.
    Percent = n, base = 50, amount = the sum found in Step 1

    Solution $8 + 8 = 16$
    Percent · base = amount
    $n \cdot 50 = 16$
    $n = \frac{16}{50}$
    $n = 0.32 = 32\%$
    32% of the states have a land area of 75,000 mi² or more.

© Houghton Mifflin Company. All rights reserved.

43. Strategy To find the revolutions per minute:
→Write the basic inverse variation equation, replace the variables by the given values, and solve for k.
→Write the inverse variation equation, replacing k by its value. Substitute 24 for the number of teeth and solve for the number of revolutions per minute.

Solution $s = \dfrac{k}{t}$
$12 = \dfrac{k}{32}$
$12(32) = k$
$384 = k$
$s = \dfrac{384}{t}$
$s = \dfrac{384}{24}$
$s = 16$
The gear will make 16 revolutions per minute.

44. Strategy To find the total cost of the car:
→Find the sales tax by using the basic percent equation. Percent = 5.5% = 0.055, base = 32,500, amount = n
→Add the tax to 32,500.

Solution Percent · base = amount
$0.055(32,500) = n$
$1787.50 = n$
$1787.50 + 32,500 = 34,287.50$

45. Strategy To find the percent decrease:
→Subtract 96 from 124 to find the decrease.
→Use the basic percent equation. Percent = n, base = 124, amount = the decrease.

Solution $124 - 96 = 28$
Percent · base = amount
$n \cdot 124 = 28$
$n = \dfrac{28}{124}$
$n \approx 0.226$
The housing starts decreased approximately 22.6%.

46. Strategy To find the sale price, solve the formula $S = (1 - r)R$ for S.
$r = 35\%$, $R = 245$.

Solution $S = (1 - r)R$
$S = (1 - 0.35)(245)$
$S = 0.65(245)$
$S = 159.25$
The sale price of the necklace is $159.25.

47. Strategy To find the simple interest, solve the formula $I = Prt$ for I.
$P = 25,000$ $r = 0.086$, $t = \dfrac{9}{12}$.

Solution $I = Prt$
$I = 25,000(0.086)\left(\dfrac{9}{12}\right)$
$I = 1,612.50$
The interest on the loan is $1,612.50.

48. Strategy To find the percent:
→Use the line graph to find the number of students that work more than 15 hours per week.
→Solve the basic percent equation for percent.

Solution Number of students who work 15–20 h: 15
Number of students who work 20–25 h: 10
Number of students who work 25–30 h: 5
Total $15 + 10 + 5 = 30$
Percent · base = amount
$p(80) = 30$
$p = \dfrac{30}{80}$
$p = 0.375$
37.5% of the students work more than 15 h per week.

49. Strategy To calculate the mean rate:
→Calculate the sum of the rates for the insurance.
→Divide by the number of quotes.
To calculate the median rate:
→Arrange the numbers from smallest to largest.
→The median is the middle number.

Solution The sum of the rates is 1,674.
$$\bar{x} = \frac{1,674}{5} = 334.8$$
The mean rate for the insurance is $334.80.
281 297 309
362 425
median = 309
The median rate for the insurance is $309.

50. Strategy To calculate the probability:
→Refer to the table on page 606 to count the number of possible outcomes.
→Count the outcomes of the experiment that are favorable to the event that the sum is divisible by 3.
→Use the probability formula

Solution There are 36 possible outcomes of the experiment.
There are 12 outcomes favorable to E:
(1, 2), (2, 1), (1, 5), (2, 4), (3, 3), (4, 2), (5, 1), (3, 6), (4, 5), (5, 4), (6, 3), (6, 6)
$$P(E) = \frac{12}{36} = \frac{1}{3}$$
The probability is $\frac{1}{3}$ that the sum will be divisible by 3.